Lecture Notes in Mathematics

Edited by A. Dold and B. Eckmann

1270

C. Carasso P.-A. Raviart D. Serre (Eds.)

Nonlinear Hyperbolic Problems

Proceedings of an Advanced Research Workshop
held in St. Etienne, France
January 13–17, 1986

Springer-Verlag
Berlin Heidelberg New York London Paris Tokyo

Editors

Claude Carasso
Denis Serre
Analyse Numérique,
23 rue du Dr. Paul Michelon
42023 Saint Etienne Cedex 2, France

Pierre-Arnaud Raviart
Université Pierre et Marie Curie
Analyse Numérique
4, place Jussieu
75230 Paris Cedex 05, France

Mathematics Subject Classification (1980): 35L, 65M, 76N

ISBN 3-540-18200-4 Springer-Verlag Berlin Heidelberg New York
ISBN 0-387-18200-4 Springer-Verlag New York Berlin Heidelberg

Printed in Germany

Printing and binding: Druckhaus Beltz, Hemsbach/Bergstr.
2146/3140-543210

CONTENTS

PREFACE

The International Conference on Non-linear Hyperbolic Problems was held in Saint-Etienne on January 13-17, 1986. Our aim was to get an overview of a branch of applied mathematics which has been expanding very fast over the last few years. As we all know applications are possible in many fields (aerodynamics, multifluid flows, combustion, detonics...), but this involves difficulties of a theoretical as well as a numerical nature. It is therefore not surprising that, in the end, papers dealing with theoretical problems raised by the resolution of non-linear hyperbolic systems outnumbered those which concentrated on numerical methods.

The number of participants, the diversity of the teams, researchers and countries represented do not leave any doubt as to the opportuneness of such a conference. It was particularly useful in as much as it produced the original papers contained in this volume and also because it helped establish conditions of scientific cooperation between research teams. We are confident that the wish expressed by many participants to see the conference become a regular fixture will be fulfilled.

The success of such an undertaking requires the help of many people and organizations. We are particularly grateful to :

le Centre National de la Recherche Scientifique (CNRS),
the National Science Foundation (NSF),
la Direction des Recherches et Etudes Techniques (DRET, contrat n°85/1392),
the European Research Office of the United States Army,
l'Académie des Sciences de Chine,
le Commissariat à l'Energie Atomique (CEA),
le Centre National d'Etudes Spatiales (CNES),
la Société de Mathématiques Appliquées et Industrielles (SMAI),
la Société Mathématique de France (SMF),
les collectivités territoriales (Mairie and Conseil général),

for their precious support.

We also wish to thank Nohra RAHMANI and Hélène CHABREUIL who greatly contributed to the success of the conference by attending to its practical organization.

Claude CARASSO, Pierre-Arnaud RAVIART, Denis SERRE

PREFACE

Le Congrès International sur les Problèmes Hyperboliques Non Linéaires s'est déroulé à Saint-Etienne du 13 au 17 janvier 1986. Nous souhaitions, par cette rencontre, faire le point sur un secteur des Mathématiques Appliquées qui s'est considérablement développé ces dernières années. Les domaines d'application sont en effet multiples (aérodynamique, écoulements multiphases, combustion, détonique...) mais également difficiles, tant du point de vue théorique que numérique. Il n'est d'ailleurs pas étonnant que les communications sur les problèmes théoriques soulevés par la résolution des systèmes hyperboliques non linéaires aient, en définitive, été plus nombreuses que celles portant sur les méthodes numériques.

Le nombre de participants, la variété des équipes, des chercheurs et des pays représentés montrent à l'évidence qu' une telle rencontre était nécéssaire. Elle a été utile d'une part par les contributions originales contenues dans ce volume, et d'autre part par les liens de coopération scientifique qui ont pu s'établir entre les équipes de chercheurs. Nous sommes certains que le souhait des participants d'une rencontre périodique sur le même thème pourra être exaucé.

La réalisation d'une telle rencontre nécessite des aides multiples, qu'elles soient matérielles ou humaines. Nous remercions particulièrement:

le Centre National de la Recherche Scientifique (CNRS),
the National Science Foundation (NSF),
la Direction des Recherches et Etudes Techniques (DRET, contrat n°85/1392),
the European Research Office of the United States Army,
l'Académie des Sciences de Chine,
le Commissariat à l'Energie Atomique (CEA),
le Centre National d'Etudes Spatiales (CNES),
la Société de Mathématiques Appliquées et Industrielles (SMAI),
la Société Mathématique de France (SMF),
les collectivités territoriales (Mairie et Conseil général),

pour leur aide, ainsi que Nohra RAHMANI et Hélène CHABREUIL qui ont assuré le succès matériel du Congrès.

Claude CARASSO, Pierre-Arnaud RAVIART, Denis SERRE

P.S. Ce Workshop était classé U.S.-French conference, ainsi que Journées SMF. Le présent document a été établi en exécution du contrat n°85/1392 passé par la direction des recherches et études techniques, direction scientifique, section soutien à la recherche (Délégation générale pour l'armement).

Liste des participants

AFIF	Mohamed	Université Kadiayad, faculté des sciences, Marrakech, Maroc
ALVES	A.S	Lab. de mécanique théorique, Univ. P. et M. Curie, 4 place Jussieu, 75230 Paris Cedex
AMIRAT	Youcef	INRIA, Domaine de Rocquencourt, BP 105, 78135 Le Chesnay
AROSIO	Alberto	Dipart. Matematica , V. Buonaroti 2, 56100 Pisa, Italie
ARRIGO	Jean-Luc	E. Polytechnique Fédérale de Lausanne,DMA Cité Universitaire, CH-1015 Lausanne
AUBRY	D.	E.Centrale des art et manufactures, 92295 Chatenay Malabry Cédex
AUDOUNET	Jacques	Université P.Sabatier, 118 route de Narbonne, 31062 Toulouse Cédex
AYADI	Abdelhamid	Inst. de Math., Univ. de Constantine, Algérie
BALABANE	Mikhaël	CMA, Ecole Normale Supérieure, 45 rue d'Ulm, 75230 Paris Cedex
BALLMAN	Joseph	RWTH Aachen, Templergraben 64 D-5100 Aachen
BARAS	Pierre	TIM 3, BP 68, 38402 St-Martin d'Hères
BARDOS	Claude	CMA, Ecole Normale Supérieure, 45 rue d'Ulm 75005 Paris
BATE-AYE	Hans	Université de Yaoundé, Dépt. de Maths., Fac. des Sciences, BP 812, Yaoundé Cameroun
BENABDALLAH	Assia	CMA, Ecole Normale Sup., 45 rue d'Ulm, 75230 Paris Cedex 05
BENBOUCHTA	Adellatif	Univ. Scientifique et Médicale de Grenoble, Lab. de Physique des Plasmas, 38402 St-Martin d'Hères
BERNARD	Michèle	CEA/Cadarache DRELSTRE 13108 St-Palu lez Durane
BERNARDI	Christine	Lab. d'Analyse Numérique, Tour 55-65, 5è étage, 4 place Jussieu, 75252 Paris
BESNAULT		DRET, 26 Bd Victor 75996, Paris Armées
BIGOT	Jean-Yves	Lab. de Spectroscopie et d'optique du corps solide, 5 r. de l'Université, 67084 Strasbourg Cédex
BONNEFILLE	Max	Analyse Numérique, 23 rue Dr Paul Michelon, 42023 St-Etienne Cedex 2
BOUCHON	Hélène	CEA, BP 7, 77181 Courtry

BOURDEL		ONERA-CERT, 2 Av. Edouard Belin Toulouse
BOURGEAT	Alain	Analyse Numérique, 23 rue du Dr Paul Michelon, 42023 St-Etienne Cedex
BOUZIT	Mohamed	Univ. de Constantine, Inst. de Maths Algérie
BRAUNER	Cl-Michel	Département M.I.S., Ecole Centrale de Lyon, BP 163, 69131 ECULLY Cédex
BRENIER	Yann	INRIA, Domaine de Rocquencourt, BP 105, 78135 Le Chesnay
BRUNEAU	Ch-Henri	Université Paris Sud, Bât.425, 91405 Orsay Cedex
CABANNES	Henri	Université de Paris 6, Mécanique Théorique, 4 place Jussieu 75005 Paris
CALLEJA	Pierre	C.E.A, BP 7, 77181 Courtry
CARASSO	Claude	Analyse Numérique, 23 rue Dr Paul Michelon 42023 St-Etienne Cedex 2
CHALABI	Abdallah	TIM 3, BP 68, 38402 St-Martin d'Hères
CHAMBAT	Michèle	UER Math de Lyon 1, 69622 Villeurbanne Cédex
CHARNAY	Michel	INSA Lyon, Centre de Maths., 20 Av. A. Einstein, 69621 Villeurbanne Cedex
CHARRIER	Pierre	UER Maths. Info., Univ. Bordeaux 1, Talence 33405
CHATTOT	J-Jacques	MATRA BP 1 78146 Velizy-Villacoublay Cédex
DARU	Virginie	ENSAM, 151 Bd de l'hôpital, 75013 Paris
DEGOND	Pierre	CMAP, Ecole Polytechnique, 91128 Palaiseau Cédex
DELANNOY	Guy	SNPE/Centre de Recherches du Bouchet, BP 2, 91710 Vert le Petit
DELORME	Philippe	Direction de l'Energétique (DE) ONERA BP 72 92322 Chatillon Cédex
DOMINGUEZ DE LA BASILLA	Jesus	HISPANOIL C/PEZ Volador, 28007 Madrid Espagne
DOUGLAS Jr	Jim	Univ. of Chicago Mathematics 5734 University Av., Chicago IL 60637 USA
DUBOIS	François	CMAP, Ecole Polytechnique, 91118 Palaiseau Cédex
DUBROCA	B.	CEA-CESTA, BP 2 Le Barp, 33830 Belin-Beliet
DURAND	Gilbert	DRET/SDR/G61-26 Bd Victor 75996 Paris Armées

EINFELT	Bernd	RWTH Aachen, Templergraben 55,D-5100 Aachen
EL DABAGHI	Fadi	INRIA, Bât 16, 78153 Le Chesnay
ENDRE	Suli	Oxford Univ. Computing Lab. 8-11 Keble Road Oxford OX1 3QD Angleterre
ERIKSSON	KENNETH	Department of Maths. Chalmers Univ. of Tech. S-41296 Goteborg Suède
FORESTIER	Alain	CEA, Saclay 911191 Gif sur Yvette Cedex
FREISTHULER	H.	RWTH Aachen, Templergraben 55, D-5100 Aachen
FRESNEL	Christophe	Univ. de Bordeaux 1, UER de Maths. 351 cours de La Libération 33405 Talence
GALEONE	Luciano	Universita, Dipartimento de Matematica Via Nicolai 2 70100 Bari Italie
GALLOUET	Thierry	Maths. Univ., 73000 Chambéry
GAUDY	Catherine	CEA, Saclay 91191 Gif sur Yvette
GILQUIN	Hervé	Analyse Numérique, 23 rue du Dr Paul Michelon, 42023 St Etienne Cedex 2
GODOUNOV	S.K.	Inst. of Math., Siberian Div. ofthe Academy of Sciences, 630090 Novossibirsk 90 USSR
GOUTAL	Nicole	6 quai Watier 78400 Chatou
GREENBERG	James M.	University of Maryland, Baltimore Country Catonsville, Maryland 21228 USA
GRUNPAHN	Klaus	RWTH Aachen, Templergraben 64, D-5100 Aachen
GUSTAFSSON	Bertil	Univ. of Uppsala, Comp. Dept. , Sturegatan 4B, Uppsala, Suède
HALPERN	Laurence	CMAP, Ecole Polytechnique, 91128 Palaiseau Cédex
HAMDACHE	Kamel	CNRS et ENSTA-GHN, Centre de l'Yvette, Chemin de la Hunière, 91120 Palaiseau
HANOUZET	Bernard	Univ. de Bordeaux 1, UER de Math. et Info.,351 cours de la Libération, 33405 Talence Cedex
HARTEN	Ami	University ot California, Dpt of Mathematics, Los Angeles 90024 USA
HENRY	Jean-Joseph	DOWELL-SCHLUMBERGER Z.I. Molina la Chazotte 42000 Saint Etienne
HERVOUET	Jean-Michel	EDF/AEE/LNH, 6 quai Watier, 78400 Chatou
HIRSCH	C.	Wrije Univ. Bruxelles, Depart. Fluid Mechanics, Pleinland 2, Bruxelles, Belgique

HOFF	David	Indiana Univ. Dept of Math., Bloomington,Indiana 47405 USA
JETSCH		RWTH Aachen, Templergraben 55, D-5100
JOLY	Jean-Luc	Univ. de Bordeaux 1, UER de Math. et Info., 351 Cours de la Libération, 33405 Talence Cedex
JOHNSON	Claes	Maths. Dept., Chalmers Univ. of Technology, 41296 Göteborg, Suède
JOSTEIN	Alvestad	STATOIL, Varlikroken 25N-4 000 Stavanger Norvège
KEYFITZ	Barbara	Maths. ,Univ. of Houston, Univ. Park, Houston, TX 77004 USA
KLEIN	Rupert	Inst. für Allgemeine Mechanik, Templergraben 55, D-5100 Aachen
KLINGENBERG	Christian	Dept. of Applied Maths., Univ. of Heidelberg, Im Neuenheimer feld 294, Heidelberg, FRG
LACAS	François	Labo. EM2C, Ecole Centrale de Paris, Gde voie des Vignes, 92295 Chatenay-Malabry Cédex
LAMARQUE	Cl-Henri	DRET, Section Formation, Bureau du Personnel 26 Bd Victor 75996 Paris Armées
LARGILLIER	Alain	Analyse Numérique, 23 rue du Dr Paul Michelon, 42023 Saint Etienne Cédex 2
LATROBE	André	Info. Intern, 57 rue pierre Sémard, 38000 Grenoble
LAYEC	Yves	Dept. G. M. P. / I. U. T. de Brest, rue de la Grandière, 29287 Brest Cédex
LE COQ	Gérard	EDF/DER/PHR, 1 avenue du Général De Gaulle, 91141 Clamart Cédex
LE FLOCH	Philippe	Ecole Polytechnique, CMAP, 91128 Palaiseau
LEQUESNE	Paul	6 Quai Watier, 78400 Chatou
LENNART	Egnesund	Uppsala Universitat, Formansgatan 17, S-72466 Vasteras Suède
LERAT	Alain	Ecole Nationale Sup. des Arts et Métiers, 151 Bd. de l'hopital, 75640 Paris Cédex
LEROUX	Alain-Yves	Univ. de Bordeaux 1, Maths et Infor, 351 Cours de la Libération 33400 Talence
LESAINT	Pierre	Lab. d'Analyse Numérique, Faculté des Sciences, Route de Gray, 25030 Besançon
LEVEQUE	Randal	Dpt of Maths., Univ. of Washington, Seattle, WHA 98195 USA
LINDBERG	Hans	Dept. of Computer Science, Sturegatan 4B, S-75234 Uppsala Suède
LIU	Tai Ping	Univ. of Maryland, Dept. of Maths., College Park, MD 20742 USA

LOCHAK	Pierre	CMA, Ecole Normale Sup., 45 rue d'Ulm, 75230 Paris Cédex 05
LOESENER	Charles	BRGM, BP 6009 , 45060 Orléans Cédex 2
MADAY	Yvon	Lab. d'Analyse Numérique, Tour 55-65, 5è étage, 4 pl. Jussieu, 75252 Paris Cédex 05
MAGGI-VIDELAINE	Pascale	USMG, Lab. de physique des Plasmas, Domaine Universitaire, BP 68, 38402 St-Martin d'Hères
MALET	Sylvie	MATRA, BP 1, 78146 Velizy-Villacoublay Cédex
MARX	Yves	E.N.S.M. Nantes, 1 rue de la Noé, 44072 Nantes
MAS-GALLIC	Sylvie	CNRS Lab. d'Analyse Numérique, Tour 55-65, 4 place Jussieu, 75252 Paris Cédex 2
MASLOV	V.P.	Moscow Institute of Electronic, Machine building B. Vusovsky 3/12 Moscou USSR
MAZET	Pierre-Alain	ONERA-CERT, 2 avenue Edouard Belin, 32000 Toulouse
MILANI	Albert	Dipartimento di Matematica, Universita di Torino, V. Principe Amadeo, 8 10123 Torino
MORICE	Phlippe	Off. Nat. d'Etudes et de Recherches Aérospatiales, BP 72, 92322 Chatillon Cédex
MUNZ	Cl-Dieter	Mathematisches Institut II, Universität Karlsruhe, Englerstr. 2, D-7500 Karlsruhe 1
NATALINI	Roberto	Math. et Info. , 351 Cours de la Libération, Univ. de Bordeaux 1 , 33405 Talence Cédex
NAPOLITANO	M.	Inst. Macchine, Univ. di Bari, Via re David Ado, 0125 Bari, Italie
NEDELEC	Jean-Claude	CMAP, Ecole Polytechnique, 91128 Palaiseau Cédex
NEGRO	Angelo	Dipar. di Mat. Univ. di Torino, V. Principe Amadeo 8, 10123 Torino Italie
ORAN	Elaine	Labo. for Computer Physics Naval Research Labo code 4040 D.C. 20375 USA
OSHER	Stanley	Maths., UCLA Los Angeles, CA 90024 USA
PART	Eva	Dept. of Computer Sci., Uppsala Univ., Sturegatan 4B, 2 TR S-752 23 Uppsala Suède
PAUMIER	Jean-Claude	TIM 3, BP 68, 38402 Saint Martin d'Heres
PERIAUX	Jacques	Avions M. Dassault/BA/DEA, 78 Quai Carnot, 92214 Saint Cloud
PERTHAME	Benoît	CMA, Ecole Normale Supérieure, 45 rue d'Ulm, 75230 Paris Cédex 05

PFERTZEL	Agnès	40 rue de la Sablière, 75014 Paris
PHAM NGOC-DINH		19 rue Erard, 75012 Paris
PIRONNEAU	Olivier	Univ. Paris 6, Analyse Numérique, 4 place Jussieu, 75252 Paris
PIROZZI	Antonietta	Instituto di Maths., Fac Di Scienze Nautiche, Via Acton 38, Napoli Italie
POGU	Marc	I.N.S.A. de Rennes, 20 avenue des Buttes de Coesmes, 35043 Rennes Cédex
POINSARD	Thierry	CNET, 38-40 rue de Général Leclerc, 92131 Issy les Moulineaux
POINSOT	Thierry	CNRS Labo de thermique, Ecole Centrale de Paris 92290 Chatenay Malabry
RASCLE	Michel	Analyse Numérique, 23 rue du Dr Paul Michelon, 42023 Saint Etienne
RAUCH	Jeffrey	Mathematics, Univ. of Michigan, Ann Arbor, MI 48109 USA
RAVIART	P-Arnaud	Univ. P. et M. Curie, Analyse Numérique, 4 place Jussieu, 75230 Paris Cédex 05
REBOURCET	Bernard	CEA-V, BP 27, 94190 Villeneuve St-Georges
REGGIO	Marcelo	Ecole Polytechnique de Montréal, Maths. Appl., Case Postale 6079, Succursale A, Montréal Canada H3C 3A7
RIGAL	Alain	Analyse Numérique, Univ. P. Sabatier, 118 route de Narbonne, 31062 Toulouse Cédex
RIOUX	Françoise	CNRS Lab. d'Elec. , Bât 214, 91405 Orsay
ROCHE	Jean	Univ. Nancy 1, BP 239, 54506 Vandoeuvre les Nancy
ROE	P.L.	College of Aeronautics, Cranfield Inst. of Technology, Cranfield, Bedford, MK 43 OAL
ROMSTEDT	Peter	Gesellschaft für Reaktorsicherheit D-8046 Garching
ROSS	Catherine	Ecole Normale Sup., C.M.A. , 45 rue d'Ulm 75005 Paris
SAIAC	Hervé	C.N.A.M., 117 Bd Jourdan, 75014 Paris
SAMADEN	Guy	C.E.A., BP 12, 91680 Bruyères le Chatel
SANTOS	Raphael	Paris XIII, Dept. Math, Av. J.B. Clément, 93430 Villetaneuse
SAUVETERRE	Maryline	Analyse Numérique, 23 rue du Dr Paul Michelon, 42023 Saint Etienne Cédex 2
SCHATZMAN	Michelle	Univ. Lyon 1, 43 Bd du 11 Novembre 1918 69622 Villeurbanne Cédex

SCHMIDT-LAINE	Claudine	Ecole Centrale de Lyon BP 163 69131 Ecully Cédex
SCHOENAUER	M	CMAP Ecole Polytechnique 91128 Palaiseau
SHERWOOD	J.D.	Etudes et fabrication Dowell-Schlumberger BP 90, 42003 Saint Etienne Cédex 1
SERRE	Denis	Analyse Numérique , 23 rue du Dr Paul Michelon, 42023 Saint Etienne Cédex 2
SHOKIN	Youri	Computing Center. USSR Academy of Science, 660036 Krasnoyarsk 36
SIBE	Evelyne	C.E.A. , BP 27, 94190 Villeneuve St-Georges
SIDERIS	Thomas	Courant Institute, 251 Mercer Steet, New-York N.Y. 10012 USA
SLEMROD	Marshall	Depar. of Math. Science, Rensselear Poly. Institute Troy. New York 12181 USA
SOLER	Juan	Lab. d'Analyse Numérique,Tour 55-65,5ème étage, Univ. P. et M. Curie,75005 Paris
SJOGREEN	Bjorn	Dept. of Computer Science, Sturegatan 4B, S-752 23 Uppsala Suède
STEPHAN	Yann	Informatique Internationale , BP 24, 91190 Gif sur Yvette
SURRY	Claude	Ecole Nationale d'Ingénieurs de Saint Etienne, rue Jean Parot, 42100 Saint Etienne
SWEBY	P.K.	Department of Mathematics Univ. of Reading P.O. Box 220 Reading , RG 6 2AX , U.K.
TADMOR	Eitan	Tel-Aviv Univ. Tel Aviv Israel
TAPIERO	Roland	Analyse Numérique, Univ. Lyon 1 , 43 Bd du 11 novembre 1918, 69622 Villeurbanne Cédex
TARTAR	Luc	CEA, Centre de Limeil-Valenton, BP 27, 94190 Villeneuve Saint Georges
TEIPEL	Ing. I	Univ. of Hannover , Appelstr. 11, D-3000 Hannover 12
TEMAM	Roger	Analyse Numérique, Bât 425, Univ. Paris Sud 91405 Orsay
TEMPLE	Blake	Math. Res. Center, 610 Walnut Street, Madison Wis. 53705 USA
TOURE	HAMIDOU	Inst. de Maths. BP 7021 Ouagadoudou Burkina Faso
TRUC	Joël	CNIN, BP 161, 83507 La Seyne sur Mer
VAN HARTEN	Aart	Maths. Inst. , Rijksuniversiteit Utrecht Postbus 80.010 3508 Ta Utrecht Nederland
VILA	Jean-Paul	CEMAGREF Div. Nivologie, Centre Universitaire, BP 76, Saint Martin d'Heres

VINCENT	Alain	CNES, 18 avenue E. Belin 31055 Toulouse Cédex
WAGNER	Alain	CNES, Rond - Point des Poètes, 91000 Evry
WAGNER	David	Dept. of Mathematics, University of Houston, University Park, Houston, Texas 77004
WANG JHING HUA	Inst. of Systeme	Science, Academia SINICA Beijing 0 100080 China
WAZNER	Alain	TIM3 BP 68 38402 Saint Martin d'Heres Cédex
WERNER	Klaus	RWTH Aachen, Templergraben 55,D-5100 Aachen
WITOMSKI	Patrick	Univ. Grenoble 1 Labo. IMAG-TIM3 BP 68 38402 Saint Martin d'Hères Cédex
WOODWARD	Paul	Dep. of Astronomy , Univ. of Minnesota, 116 Church Street SO, Minneapolis, Minnesota 55455
ZITI	Cherif	Analyse Numérique , 23 rue du Dr Paul Michelon , 42023 Saint Etienne Cédex 2

APPROXIMATE SOLUTION OF THE GENERALIZED RIEMANN PROBLEM AND APPLICATIONS.

A. Bourgeade(*) , P. Le Floch(**) and P.A. Raviart(***)

(*) C.E.A. Centre d'Etudes de Limeil
B.P. 27, 94190 Villeneuve Saint Georges, France.

(**) Centre de Mathématiques Appliquées, Ecole Polytechnique,
91128 Palaiseau Cedex, France.

(***) Analyse Numérique, Université Pierre et Marie Curie,
4 Place Jussieu, 75252 Paris Cedex 05, France.

1. INTRODUCTION.

Let us consider the nonlinear hyperbolic system of conservation laws

$$\frac{\partial U}{\partial t} + \frac{\partial}{\partial x} f(x,t,U) + g(x,t,U) = 0 \quad , \quad x \in \mathbb{R} \ , \ t > 0 \tag{1.1}$$

where U is a vector with p componants and f,g are smooth functions from $\mathbb{R} \times (0,\infty) \times \mathbb{R}^p$ into $\mathbb{R}^p$. We are looking for the "entropy" solution of the Cauchy problem for the system (1.1) corresponding to the initial condition

$$U(x,0) = \begin{cases} U_L(x) & , \quad x < 0 \\ U_R(x) & , \quad x > 0 \ . \end{cases} \tag{1.2}$$

In (1.2), U_L (resp. U_R) is a smooth function from $(-\infty,0]$ into $\mathbb{R}^p$ (resp. from $[0,\infty)$ into $\mathbb{R}^p$) such that $U_L(0) \neq U_R(0)$. We shall refer to Problem (1.1), (1.2) as the generalized Riemann problem for (1.1). In fact, when the functions U_L and U_R are constant, we obtain the classical Riemann problem which can be solved explicitely at least in the case $f = f(U)$, $g = 0$ (See [6] for instance). The purpose of this lecture is to derive an explicit approximation of the "entropy" solution U of Problem (1.1), (1.2) in a neighborhood of the origin.

This question arises naturally in connexion with the now classical Van Leer's scheme [7] of numerical approximation of the Cauchy problem for (1.1). Let us briefly recall the main features of Van Leer's method. Given a uniform space increment Δx and a time step Δt , we set :

$$\lambda = \frac{\Delta t}{\Delta x} \quad , \quad x_j = j\,\Delta x \quad , \quad t_n = n\,\Delta t \ .$$

Then, starting from an approximation V^n of the exact solution $U(\cdot,t_n)$ at the time t_n of the form

$$V^n(x) = V_j^n + \frac{1}{\Delta x}(x - x_j)\, S_j^n \quad , \quad x_{j-1/2} < x < x_{j+1/2} \tag{1.3}$$

(i.e., V^n is a piecewise affine function which is discontinuous in general at the points $x_{j+1/2} = (j+\frac{1}{2})\Delta x$) , we define V^{n+1} as follows :

(i) we solve the Cauchy problem

$$\frac{\partial W}{\partial t} + \frac{\partial}{\partial x} f(x, t_n + t, W) + g(x, t_n + t, W) = 0 \ , \ x \in \mathbb{R} \ , \ t > 0 \tag{1.4}$$

$$W(x,0) = V^n(x) \quad ; \tag{1.5}$$

(ii) we "project" $W(\cdot,\Delta t)$ (which is indeed an approximation of $U(\cdot,t_{n+1})$) onto the space of piecewise affine functions, which gives V^{n+1} .

In fact, the two steps (i) and (ii) of the method cannot be performed exactly and need to be approximated. However, in any case, we define V_j^{n+1} by

$$V_j^{n+1} = \frac{1}{\Delta x} \int_{x_{j-1/2}}^{x_{j+1/2}} W(x,\Delta t)dx \ . \tag{1.6}$$

Hence, integrating (1.5) in the rectangle $[x_{j-1/2}, x_{j+1/2}] \times [0,\Delta t]$ in the (x,t)-plane gives

$$\begin{aligned} V_j^{n+1} = V_j^n - \lambda \frac{1}{\Delta t} \int_0^{\Delta t} \{ & f(x_{j+1/2}, t_n + t, W(x_{j+1/2}, t)) - \\ & - f(x_{j-1/2}, t_n + t, W(x_{j-1/2}, t)\}dt \\ & - \frac{1}{\Delta x} \int_0^{\Delta t} \int_{x_{j-1/2}}^{x_{j+1/2}} g(x, t_n + t, W(x,t))dx\, dt \ . \end{aligned}$$

Next, if we approximate the time and space integrals by means of the mid-point rule and the trapezoidal rule respectively, we obtain the finite-difference scheme

$$\begin{aligned} V_j^{n+1} = V_j^n - \lambda\{ & f(x_{j+1/2}, t_{n+1/2}, W(x_{j+1/2}, \tfrac{\Delta t}{2})) \\ & - f(x_{j-1/2}, t_{n+1/2}, W(x_{j-1/2}, \tfrac{\Delta t}{2}))\} \\ & - \tfrac{\Delta t}{2}\{g(x_{j+1/2}, t_{n+1/2}, W(x_{j+1/2}, \tfrac{\Delta t}{2})) + g(x_{j-1/2}, t_{n+1/2}, W(x_{j-1/2}, \tfrac{\Delta t}{2}))\} . \end{aligned} \tag{1.7}$$

Therefore, we have only to determine $W(x_{j+1/2}, \frac{\Delta t}{2})$ for $j \in \mathbb{Z}$. Now, if we assume that Δt is chosen small enough, the various waves issued from the points $x_{j+1/2}$ do not interact during the time interval $(0,\Delta t)$. Hence, we need only to solve locally generalized Riemann problems of the form (1.1), (1.2) where U_L and U_R are affine functions. Thus, a critical stage in Van Leer's scheme consists in deriving an approximation of $U(0,t)$ for t small enough where U is the "entropy" solution of

(1.1), (1.2). It has been first reckognized by Ben-Artzi and Falcovitz [1] that, in the case of the gas dynamics equations in plane symmetry, one could explicitely compute $\frac{\partial U}{\partial t}(0,0)$ and therefore obtain analytically an $O(t^2)$ approximation of $U(0,t)$. Moreover, the associated Van Leer's scheme appears to be very robust and to provide a fairly precise numerical treatment of very strong shock waves. The purpose of this paper is to show that the method of Ben-Artzi and Falcovitz is indeed completely general and to stress the underlying mathematical features.

2. AN ASYMPTOTIC EXPANSION METHOD.

Let us assume that the nonlinear system of conservation laws (1.1) is strictly hyperbolic in the sense that the Jacobian matrix $\frac{\partial f}{\partial U}(x,t,U)$ has, for every triple $(x,t,U) \in \mathbb{R} \times \mathbb{R}_+ \times \mathbb{R}^p$, p real distinct eigenvalues

$$\lambda_1(x,t,U) < \lambda_2(x,t,U) < \ldots < \lambda_p(x,t,U) .$$

We assume as usual that the k-th characteristic field is either genuinely nonlinear or linearly degenerate (cf. again [6]). Then, it is well known that, if

$$f(x,t,U) = f(U) \quad , \quad g = 0 \quad ,$$

the entropy solution of the classical Riemann problem is a self-similar solution which consists of at most $(p+1)$ constant states separated by k-waves, $1 \leqslant k \leqslant p$ (a k-shock or a k-rarefaction wave if the k-th characteristic field is genuinely nonlinear, a k-contact discontinuity if the k-th characteristic field is linearly degenerate).

In the general case of Problem (1.1), (1.2), the entropy solution U is no longer self-similar but has still the structure of the solution of the classical Riemann problem for t small enough : U consists of at most $(p+1)$ smooth states separated by k-waves which may be again shock waves or rarefaction waves or contact discontinuities (see Figure 1). The transition curves are now smooth curves passing through the origin which are characteristic in the two latter cases. For a detailed description of these results, see Li Ta-tsien and Yu Wen-ci [5] and Harabetian [4] who considers the case of several space dimensions.

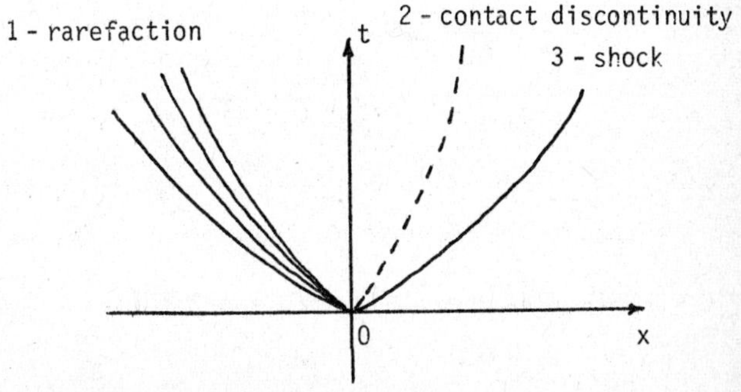

Figure 1.

Now, we want to find the entropy solution U of Problem (1.1), (1.2) in the form of an asymptotic expansion

$$U(\tfrac{x}{t}) = U_o(\tfrac{x}{t}) + t\, U_1(\tfrac{x}{t}) + t^2\, U_2(\tfrac{x}{t}) + \dots , \tag{2.1}$$

this expansion being valid in a neighborhood of the origin in the (x,t)-plane and in each domain of smoothness of U. In fact, we shall show in the sequel how to determine the first two terms U_o and U_1 in the expansion (2.1). More precisely, we shall check that U_o is indeed the entropy solution of the classical Riemann problem

$$\begin{cases} \dfrac{\partial U_o}{\partial t} + \dfrac{\partial}{\partial x} f(0,0,U_o) = 0 \\ U_o(x,0) = \begin{cases} U_L(0) , & x < 0 \\ U_R(0) , & x > 0 \end{cases} \end{cases} \tag{2.2}$$

and U_1 can be determined as the solution of a <u>linear</u> problem.

First, assuming that the asymptotic expansion (2.1) holds and setting $\xi = \frac{x}{t}$, we may write in each domain of smoothness of U :

$$\frac{\partial U}{\partial t} = -\frac{1}{t}\,\xi\, U_o' + (U_1 - \xi\, U_1') + O(t) \quad ,$$

$$\frac{\partial}{\partial x} f(x,t,U) = \frac{1}{t}\frac{d}{d\xi} f(0,0,U_o) +$$

$$+ \frac{d}{d\xi}\Big(\xi\, \frac{\partial f}{\partial x}(0,0,U_o) + \frac{\partial f}{\partial t}(0,0,U_o) + \frac{\partial f}{\partial U}(0,0,U_o)U_1\Big) + O(t)$$

and

$$g(x,t,U) = g(0,0,U_o) + O(t) .$$

By identifying the coefficients of t^{-1} and t^0 respectively in the expansion of (1.1), we obtain

$$-\xi\, U_o' + \frac{d}{d\xi} f(0,0,U_o) = 0 \tag{2.3}$$

and

$$\begin{cases} -\xi\, U_1' + \dfrac{d}{d\xi}\Big(\dfrac{\partial f}{\partial U}(0,0,U_o)U_1\Big) + U_1 + \\ \quad + \dfrac{d}{d\xi}\Big(\xi\, \dfrac{\partial f}{\partial x}(0,0,U_o) + \dfrac{\partial f}{\partial t}(0,0,U_o)\Big) + g(0,0,U_o) = 0 . \end{cases} \tag{2.4}$$

Next, let $x = \phi(t)$ be a transition curve Γ which separates two smoothness regions Ω_i , $i = 1,2$, of the solution U . Since $\phi(0) = 0$, we can write

$$\phi(t) = \sigma_o\, t + \sigma_1\, t^2 + \dots \tag{2.5}$$

Assume first that U is continuous across Γ. In each region Ω_i, we have

$$U(\phi(t),t) = U_0(\sigma_0 + \sigma_1 t + \ldots) + t\, U_1(\sigma_0 + \sigma_1 t + \ldots) + \ldots =$$
$$= U_0(\sigma_0) + t(U_1(\sigma_0) + \sigma_1 U_0'(\sigma_0)) + \ldots$$

so that the <u>continuity of</u> U <u>across</u> Γ gives in $\xi = \sigma_0$:

$$[\, U_0 \,] = 0 \quad , \tag{2.6}$$

and

$$[\, U_1 \,] + \sigma_1 [\, U_0' \,] = 0 \quad , \tag{2.7}$$

where $[\, V \,]$ denotes the jump of V. Observe that U_1 is generally discontinuous in $\xi = \sigma_0$.

Assume now that U <u>is discontinuous across</u> Γ (i.e., Γ is a shock or contact discontinuity). Starting from the Rankine-Hugoniot conditions

$$\phi'(t)\, [\, U \,] = [\, f(x,t,U) \,] \quad , \quad x = \phi(t)$$

and replacing U and ϕ by their asymptotic expansions (2.1) and (2.5) gives in $\xi = \sigma_0$:

$$\sigma_0 [\, U_0 \,] = [\, f(0,0,U_0) \,] \tag{2.8}$$

and

$$\left\{ \begin{aligned} & [\, (\frac{\partial f}{\partial U}(0,0,U_0) - \sigma_0) U_1 \,] = \sigma_1 [\, U_0 \,] - \\ & - [\, \sigma_0 \frac{\partial f}{\partial x}(0,0,U_0) + \frac{\partial f}{\partial t}(0,0,U_0) \,] \end{aligned} \right. \tag{2.9}$$

It remains to derive the conditions satisfied by the functions U_0 and U_1 as $\xi \to \pm \infty$. Consider for instance the case $\xi \to +\infty$. We know that the function U is sufficiently smooth for $\frac{x}{t} \geq \xi_0$ large enough and t small enough. By using a Taylor expansion at the origin, we obtain

$$U(x,t) = U(0,0) + x \frac{\partial U}{\partial x}(0,0) + t \frac{\partial U}{\partial t}(0,0) + \frac{x^2}{2} \frac{\partial^2 U}{\partial x^2}(0,0) + \ldots =$$
$$= U(0,0) + t(\frac{x}{t} \frac{\partial U}{\partial x}(0,0) + \frac{\partial U}{\partial t}(0,0)) + t^2(\frac{1}{2} \frac{x^2}{t^2} \frac{\partial^2 U}{\partial x^2}(0,0) + \ldots) + \ldots$$

where $V(0,0)$ stands for

$$V(0,0) = \lim_{\substack{x,t \to 0 \\ \frac{x}{t} \geq \xi_0}} V(x,t) .$$

By comparing with the expansion (2.1), we obtain that, for $\xi \geqslant \xi_o$, U_i is a polynomial of degree i in the variable ξ and

$$U_i^{(i)} = U_R^{(i)}(0) \quad , \quad i = 0,1,\ldots \tag{2.10}$$

Similarly, we obtain for $\xi \leqslant -\xi_o$, ξ_o large enough

$$U_i^{(i)} = U_L^{(i)}(0) \quad , \quad i = 0,1,\ldots \tag{2.11}$$

As a first and expected consequence of the above formal derivation, it follows from (2.3), (2.6), (2.8), (2.10) and (2.11) that U_o is indeed the entropy solution of the classical Riemann problem (2.2) which is known to exist and to be unique provided that $|U_R(0) - U_L(0)|$ is small enough.

3. DETERMINATION OF THE FUNCTION U_1 .

Concerning the function U_1 , we can state the following result.

THEOREM. Assume that $|U_R(0) - U_L(0)|$ is small enough. Then, there exists a unique function U_1 which satisfies the equation (2.4) in the domains of smoothness of the function U_o together with the jump conditions (2.7) and (2.9) at a discontinuity of U_o' and U_o respectively and the conditions at infinity

$$\lim_{\xi\to+\infty} U_1'(\xi) = U_R'(0) \quad , \quad \lim_{\xi\to-\infty} U_1'(\xi) = U_L'(0) \ . \tag{3.1}$$

Proof. For the sake of brevity, we only sketch the proof. Recall again that the function U_o consists of at most $(p+1)$ constant states separated by shock waves, rarefaction waves or contact discontinuities.

Consider first an interval where the function U_o is constant. In that case, (2.4) becomes

$$(A(U_o)-\xi)U_1' + U_1 + \frac{\partial f}{\partial x}(0,0,U_o) + g(0,0,U_o) = 0 \tag{3.2}$$

where

$$A(U) = \frac{\partial f}{\partial U}(0,0,U) \ .$$

Denote by $\lambda_1(U) < \lambda_2(U) < \ldots < \lambda_p(U)$ the eigenvalues of the matrix $A(U)$ and by $r_1(U), r_2(U), \ldots, r_p(U)$ a set of corresponding eigenvectors

$$A(U)\ r_k(U) = \lambda_k(U)\ r_k(U) \ .$$

If we set

$$U_1 = \sum_{k=1}^{p} \alpha_k\ r_k(U_o) \tag{3.3}$$

and

$$\frac{\partial f}{\partial x}(0,0,U_0) + g(0,0,U_0) = \sum_{k=1}^{p} \gamma_k \, r_k(U_0) \ , \tag{3.4}$$

the equation (3.2) gives

$$(\lambda_k(U_0)-\xi)\alpha_k' + \alpha_k + \gamma_k = 0 \quad , \quad 1 \leq k \leq p$$

so that

$$\alpha_k = a_k(\lambda_k(U_0) - \xi) - \gamma_k \quad , \quad 1 \leq k \leq p \ , \tag{3.5}$$

where a_k , $1 \leq k \leq p$, are arbitrary constants. Note that U_1 is an affine function of the variable ξ .

Consider next an interval where U_0 is a i-rarefaction wave, i.e.,

$$U_0'(\xi) = r_i(U_0(\xi)) \quad , \quad \xi = \lambda_i(U_0(\xi)) \tag{3.6}$$

(Here, we have assumed the usual normalization condition $\nabla\lambda_i(U)\cdot r_i(U) \equiv 1$). Then, (2.4) becomes

$$(A(U_0)-\xi)U_1' + (\frac{d}{d\xi} A(U_0) + 1)U_1 = \text{known terms.} \tag{3.7}$$

Setting again (3.3), we obtain a linear differential system of the first order in the α_k's , $k \neq i$, which can be solved explicitely in a number of practical cases. The solution of (3.7) is then uniquely defined up to (p-1) arbitrary constants.

It remains to specify the jump conditions at a discontinuity σ_0 of U_0 or U_0' . Assume first that U_0 is continuous but U_0' is discontinuous in σ_0 . This means that σ_0 separates a constant state and a i-rarefaction wave. Using (2.7) and (3.6), we obtain

$$U_1(\sigma_0 + 0) = U_1(\sigma_0 - 0) \pm \sigma_1 \, r_i(U_0(\sigma_0)) \ , \tag{3.8}$$

where the sign + (resp. the sign -) corresponds to the case of a constant state located on the right-hand side (resp. on the left-hand side) of σ_0 . In any case, it follows from (3.8) that α_k is continuous across σ_0 for $k \neq i$ while α_i has a jump discontinuity in σ_0 . Moreover, the jump of α_i gives the curvature σ_1 of the transition curve $x = \phi(t)$ at the origin.

Assume next that U_0 is continuous in σ_0 . This means that σ_0 separates two constant states and corresponds either to a shock wave or a contact discontinuity. Using (2.9) gives

$$[\,(A(U_0)-\sigma_0)U_1\,] - \sigma_1\,[\,U_0\,] = \text{known terms.} \tag{3.9}$$

This leads to a system of p equations in the (2p+1) unknowns $\alpha_k(\sigma_0 + 0)$, $\alpha_k(\sigma_0 - 0)$, $1 \leq k \leq p$, and σ_1 .

Finally, we deal with the conditions (3.1). Consider for instance the case $\xi \to +\infty$. Since, for ξ large enough, $U_o(\xi) = U_R(0)$ is constant, we have by (3.5)

$$\alpha_k' = - a_k .$$

Hence, setting

$$U_R'(0) = - \sum_{k=1}^{p} a_{Rk} \, r_k(U_R(0)) ,$$

we obtain for ξ large enough

$$\alpha_k = a_{Rk}(\lambda_k(U_R(0)) - \xi) - \gamma_k \quad , \quad 1 \leqslant k \leqslant p , \tag{3.10}$$

which yields

$$U_1(\xi) = (\xi - A(U_R(0))U_R'(0) + (\frac{\partial f}{\partial x} + g)(0,0,U_R(0)) .$$

Similarly, we get for $-\xi$ large enough

$$\alpha_k = a_{Lk}(\lambda_k(U_L(0)) - \xi) - \gamma_k \quad , \quad 1 \leqslant k \leqslant p , \tag{3.11}$$

where

$$U_L'(0) = - \sum_{k=1}^{p} a_{Lk} \, r_k(U_L(0))$$

so that

$$U_1(\xi) = (\xi - A(U_L(0)))U_L'(0) + (\frac{\partial f}{\partial x} + g)(0,0,U_L(0)) .$$

Now, it is a purely technical matter to check that the linear equations (3.5), (3.7), (3.8), (3.9), (3.10) and (3.11) enable us to determine the vector-valued function U_1 or equivalently the p scalar functions α_k , $1 \leqslant k \leqslant p$, in a unique way at least when $|U_R(0) - U_L(0)|$ is small enough. ∎

One can generalize easily the above analysis in order to determine the functions U_k , $k \geqslant 2$, which appear in the expansion (2.1). In fact, in a similar way, one can derive formally the equations and jump conditions satisfied by U_k and then prove that, for $|U_R(0) - U_L(0)|$ small enough, such a function U_k is uniquely defined. It remains to prove that the asymptotic expansion is indeed convergent. For a detailed analysis, we refer to [3].

Let us very briefly consider the application of the above theory to the gas dynamics equations

$$\begin{cases} \frac{\partial}{\partial t}(x^\alpha \rho) + \frac{\partial}{\partial x}(x^\alpha \rho u) = 0 , \\ \frac{\partial}{\partial t}(x^\alpha \rho u) + \frac{\partial}{\partial x}(x^\alpha (\rho u^2 + p)) - \alpha x^{\alpha-1} p = 0 , \\ \frac{\partial}{\partial t}(x^\alpha \rho e) + \frac{\partial}{\partial x}(x^\alpha (\rho e + p) u) = 0 , \end{cases}$$

where $\alpha = 0$ for slab symmetry, $\alpha = 1$ for cylindrical symmetry and $\alpha = 2$ for spherical symmetry. If we assume for instance that the equation of state is the ideal gas γ-law, one can compute explicitely the first two terms U_0 and U_1 of the expansion (2.1) of the solution of the generalized Riemann problem in a neighborhood of the point $(x_0,0)$ $(x_0 > 0$ if $\alpha = 1,2)$. In particular, using asymptotic expansions for the velocity u and the pressure p

$$u = u_0 + t\, u_1 + \dots \quad , \quad p = p_0 + t\, p_1 + \dots \quad ,$$

we obtain again the explicit expressions of u_1 and p_1 at the contact discontinuity that have been derived in [1] and [2] by a different method and which are the key tools for implementing Van Leer's method [7]. We refer again to [3] for a complete derivation of (ρ_1, u_1, p_1) in a neighborhood of $(x_0, 0)$.

REFERENCES.

[1] M. Ben-Artzi and J. Falcowitz, A second order Godunov-type scheme for compressible fluid dynamics, J. Comp. Phys. 55 (1984), 1-32.

[2] M. Ben-Artzi and J. Falcowitz, An upwind second-order scheme for compressible duct flows, SIAM J. Sci. Stat. Comp. (to appear).

[3] A. Bourgeade, P. Le Floch and P.A. Raviart (to appear).

[4] E. Harabetian, A Cauchy-Kowalevsky theorem for strictly hyperbolic systems of conservation laws with piecewise analytic initial data, Ph.D. dissertation, University of California Los Angeles, 1984.

[5] Li Ta-Tsien and Yu Wen-ci, Boundary value problems for quasilinear hyperbolic systems, Duke University Mathematics series, 1985.

[6] J. Smoller, Shock waves and reaction-diffusion equations, Springer-Verlag, New York, 1983.

[7] B. Van Leer, Towards the ultimate conservative difference scheme V, J. Comp. Phys. 32 (1979), 101-136.

COMPUTATION OF OSCILLATORY SOLUTIONS TO PARTIAL DIFFERENTIAL EQUATIONS

Björn Engquist
Department of Mathematics, UCLA

Abstract

Numerical approximations of hyperbolic partial differential equations with oscillatory solutions are studied. Convergence is analyzed in the practical case for which the continous solution is not well resolved on the computational grid. Averaged difference approximations of linear problems and particle method approximations of semilinear problems are presented. Highly oscillatory solutions to the Carleman and Broadwell models are considered. The continous and the corresponding numerical models converge to the same homogenized limit as the frequency in the oscillation increases.

1. Introduction

There are many practical computational problems with highly oscillatory solutions, e.g. computation of high frequency acoustic and electro-magnetic fields, properties of composite materials and turbulent flow. Sometimes the original equations describing these problems can be replaced by effective equations modelling some average quantity without the oscillations. Geometrical optics, homogenization of composite materials and turbulence models are such examples. See e.g. [1].

Whenever effective equations are applicable they are very useful also for computational purposes. There are however many situations for which we do not have well-posed effective equations or for which the solution contains different frequences such that effective equations are not practical. In these cases we would like to approximate the original equations directly.

We shall show that it is sometimes possible to numerically approximate the original equations directly without resolving all scales in the solutions. The computed solution may still contain useful information. This is not always obvious since a good classical numerical approximation requires a substantial number of grid points on elements per wave length in the oscillation [5], [7]. The requirement of this good resolution is often unrealistic in practice.

It is natural to compare our problem with the numerical approximation of discontinous solutions of nonlinear conservation laws. Shockcapturing methods do not produce the correct shock profiles but the overall solution may still be good. For this the schemes must satisfy certain conditions such as e.g. conservation form. We are here interested in analogous conditions on algorithms for oscillatory solutions. These conditions should ideally guarantee that the numerical approximation in some sense is close to the solution of the corresponding effective equation when the wave length in the oscillation tends to zero.

When we talk about convergence of an approximation to an oscillatory solution we need a new definition. In traditional convergence the error decreases for fixed data as the grid step size decreases. This concept is too weak in practice and does not discriminate between solutions which are highly oscillatory and those which are not. We need the error to be small when the computational grid step size is small more or less independent of the wave length in the oscillation. See section 1 for a precise formulation.

On the other hand we can not expect the approximation to be pointwise well behaved. It is enough if the continous solution and its discrete approximation have similar local or moving averages. See figure 1 and section 1.

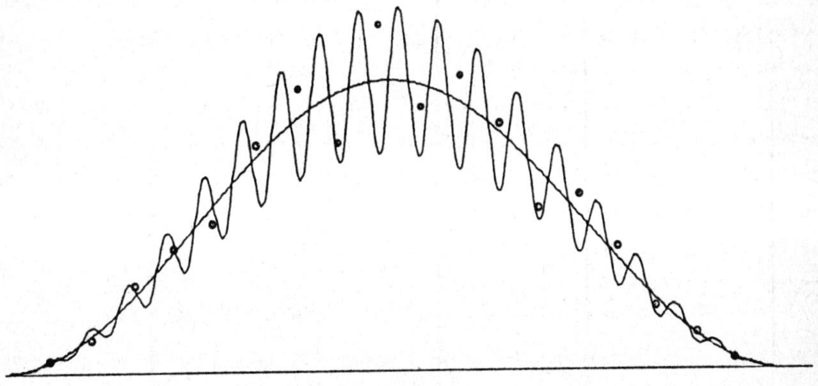

Figure 1. Oscillatory solution and numerical approximation with similar moving average.

There are three central sources of problems for discrete approximations of highly oscillatory solutions.

(i) The first is the sampling of the computational mesh points ($x_j = j\Delta x$, $j=0,1,..$). There is the risk of resonance between the mesh points and the oscillation. If e.g. Δx equals the wave length of the oscillation the discrete initial data may only get values from the peaks of a curve like the one in figure 1. We can never expect convergence in that case. Thus Δx can not be completely independent of the wave length. See [4] and section 1.

(ii) Another problem comes from the approximations of advection. The group velocity for the differential equation and the corresponding discretization are often very different. This means that an oscillatory pulse which is not well resolved, is not even in average transported correctly by the approximation. See figure 2. Furthermore, dissipative schemes do never advect oscillations correctly. The oscillations are damped out very fast [4].

(iii) Finally, the nonlinear interaction of different high frequency components in a solution must be modelled correctly. High frequency interaction may produce lower frequencies that influence the averaged solution. We can show that this nonlinear interaction is well approximated by certain particle schemes applied to a class of semilinear differential equations, [4], [6], section 3 . The problem is open for the approximation of more general nonlinear equations.

2. Difference approximations of linear problems

Let us consider a hyperbolic system of constant coefficient partial differential equations and review some results from [4]

$$(1)\qquad \begin{aligned} &\frac{\partial u}{\partial t} + \sum_{m=1}^{d} A_m \frac{\partial u}{\partial x_m} = 0 \qquad -\infty < x_m < \infty\,, \quad t > 0 \\ &u(x,0) = f(x) \qquad -\infty < x_m < \infty. \end{aligned}$$

Let the initial values be of the form

$$(2)\qquad f(x) = f_\varepsilon(x) = a(x, x/\varepsilon)$$

with $a(x,y) \in C^\infty$ of compact support with respect to x and 1-periodic in all components of y.

Let a corresponding difference approximation have the form

(3)
$$v_j^{n+1} = Q\, v_j^n \qquad j_m = \dots -1,0,1\dots, n = 0,1,\dots$$
$$v_j^0 = f(x_j) \qquad j_m = \dots -1,0,1\dots$$
$$m = 1,\dots,d.$$

The difference operator Q is assumed to have constant coefficients and to be defined for meshfunctions on the grid $\{(x_j,t_n)\}$, $x_j = x_0 + (j_1\Delta x_1,\dots,j_d\Delta x_d)$

$$j = (j_1,\dots,j_d), \qquad \Delta x = (\Delta x_1,\dots,\Delta x_d)$$

$$\lambda_m = \Delta t/\Delta x_m = \text{constant}$$

$$t_n = n\Delta t, \qquad n = 0,1,\dots$$

$$Qv_j = \sum_{|p|\leq s} B_p\, v_{j+p}, \; B_p \text{ are constant matrices.}$$

$$|p| = \sum_{m=1}^{d} |p_m|$$

Denote the solution operators of (1) and (3) by $S(t)$ and S^n respectively,

$$u(x,t') = S(t'-t'')u(x,t'') \quad , \qquad t' \geq t''$$
$$v_j^{n'} = S^{n'-n''}\, v_j^{n''}. \qquad n' \geq n''$$

We shall assume the system of differential equations (1) to be well posed in L_2:

(4)
$$||u(\cdot,t)|| \leq C||u(\cdot,0)|| \quad \text{or} \quad ||S(t)|| < C.$$

Let $||\cdot||$ denote the L_2 norm and the corresponding operator norm in the contionous or in the discrete sense whenever appropriate. Here and throughout the paper C denotes a generic constant which may not be the same in different formulas.

Definitions. The difference equation (3) is stable if $||S^n|| \leq C$, $t^n \leq T$. It is a consistent approximation of (1) if for smooth solutions $u(x,t)$ of (1)

$$||u(x_j,t_{n+1}) - \sum_{|p|\leq s} B_p u(x_{j+p},t_n)|| \leq C\ \Delta t^2.$$

The traditional form of an error estimate for a q-th order stable difference approximation (3) of (1) is, [12]:

(5) $$||u(\cdot,t_n) - v^n|| \leq C \ \Delta t^q \max_{|\alpha| \leq q+1} ||\partial^\alpha u||$$

where ∂^α denotes a general differential operator of order $|\alpha|$. For a stable and consistent difference approximation of our type (3) the estimate (5) is valid with $q = 1$, [12]. If the initial values are of the form (2) then

$$\max_{|\alpha|=q+1} ||\partial^\alpha u|| \sim \varepsilon^{-(q+1)}$$

and the estimate (5) is essentially useless if Δt is not small compared to ε .

Difference approximations of solutions, which oscillates with a wave length of the order of or shorter than the discretization step size, will in general have large errors. It is easy to see that the error will be substantial by studying the difference in phase velocities for the differential equation (1) and the approximation (3), (see e.g. [5],[7]). After only one timestep the L_2-error is $O(1)$ if $\varepsilon < C\Delta t$. This observation implies that a good approximation can only be expected in an average sense or after some other processing of the computed approximation.

The traditional concept of convergence is not useful for our type of problems. Even for highly oscillatory solutions the error obviously converges to zero as $C\Delta t^q$ when $\Delta t \to 0$ and ε is fixed. The trouble is the dependence of C on ε . We shall therefore introduce concepts of convergence which are relevant when the oscillations in a solution are not well resolved on the computational grid.

Definitions. The approximation v converges to u as $\Delta t \to 0$ independent of ε if for any δ , $T > 0$ there exists $\Delta t_0 > 0$ such that

(6) $$||u(\cdot,t_n) - v^n|| \leq \delta, \quad 0 < t^n < T$$

for all $\Delta t \leq \Delta t_0$ and where Δt_0 is independent of ε .

The approximation v converges to u as $\Delta t \to 0$ essentially independent of ε if for any δ , $T > 0$ there exists a set $S(\varepsilon,\Delta t_0) \subset (0,\Delta t_0]$ with measure $(S(\varepsilon,\Delta t_0)) \geq (1 - \delta)\Delta t_0$ such that (6) is valid for all $\Delta t \in S(\varepsilon,t_0)$ and where Δt_0 is independent of ε.

The averaged solution with a local continuous averaging is given by

$$\tilde{u}(x,t) = \int_{R^d} \theta(x - z)u(z,t)dz$$

$$\theta(x) = \begin{cases} C \prod_{m=1}^{d} \exp(-(1 - (x_m/\sigma)^2)^{-1}), & x_m < \sigma \\ 0, & \text{some } x_m \geq \sigma. \end{cases}$$

C is the normalization factor such that $\int_{R^d} \theta(x)dx = 1$. With θu we mean that each component of u is multiplied by θ. The discrete average is given by

$$\bar{v}_j^n = (\Delta x) \sum_k \theta_{jk} v_k^n = \prod_{m=1}^{d} \Delta x_m \sum_{k_1=-\infty}^{\infty} \cdots \sum_{k_d=-\infty}^{\infty} \theta_{jk} v_k^n$$

$$\theta_{jk} = \theta(x_j - x_k), \quad k = (k_1,\ldots,k_d).$$

The continuous average may also be applied to a meshfunction v after having defined v^0 for all x-values at $t = 0$. The stepsize Δx is regarded as fixed but x_0 variable, see[12].

Theorem 1. (a) The discretely averaged solution $\bar{v}$ of a stable and consistent difference approximation (3) converges as $\Delta t \to 0$ essentially independent of ε to the averaged solution $\tilde{u}$ of (1) if $\sigma = \Delta t^{\beta}$, $0 < \beta < 1/2$.

(b) The continuously averaged solution $\tilde{v}$ of a stable and consistent difference approximation (3) converges at $\Delta t \to 0$ independent of ε to the averaged solution $\tilde{u}$ of (1) if $\sigma = \Delta t^{\beta}$, $0 < \beta < 1/2$.

For the proof see [4], Part (b) is essentially obvious and follows since the evolution operators and the averaging commute. The proof of the more interesting part (a) is different and the convergence cannot be completely independent of ε as in (b). There is the possibility of resonance as discussed in the introduction (i). The ratio of Δx and ε must be irrational. The argument is similar to standard techniques in ergodic thoery and Monte Carlo methods, [10]. The limit $\beta < 1/2$ is sharp, [4].

The theorem can be extended to cover parabolic and other higher order equations. The trouble is that we do not know of any methods to solve the problems (ii) and (iii) from the introduction for these higher order equations.

In [4] we give simple examples of the necessity of having methods without numerical dissipation or dispersion in order to well represent the oscillations. We would at least like the average of $(v^n)^2$ to converge to the average of $u(\cdot,t_n)^2$ as $\Delta t \to 0$ essentially independent of ε. This property is needed for the class of semilinear problems studied in the following section.

An example of a method that does not work is the leap-frog scheme. It has numerical dispersion and thus different group velocities for different frequencies. If we take initial data similar to that of figure 1 and solve the scalar advection equation

$$\frac{\partial u}{\partial t} + \frac{\partial u}{\partial x} = 0 \tag{7}$$

by the leap-frog scheme with a coarse grid, the result may look like the function in figure 2. The oscillations have moved to a wrong location instead of being on top of the average pulse to the right. The oscillatory pulse has a zero local average so the averaged solution is well approximated. If however the oscillatory part through nonlinear interaction would contribute to the smooth part of the solution the scheme is useless.

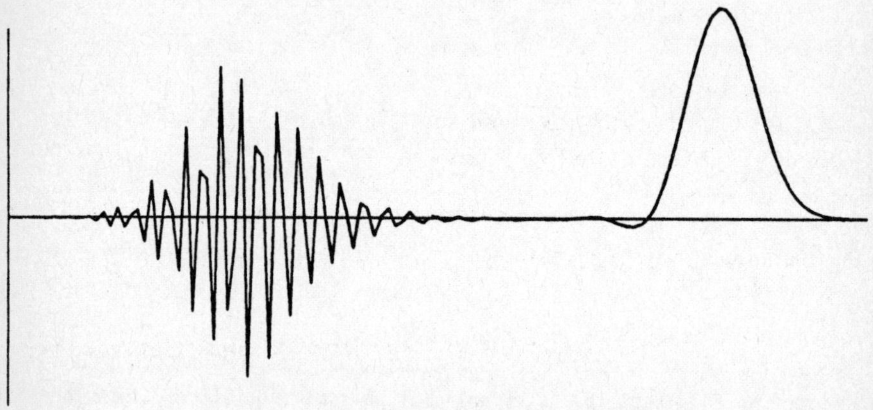

Figure 2. Leap-frog approximation of (7) with one oscillatory pulse as initial data.

3. Particle method approximations of the Carleman and Broadwell equations

Let us consider the Carleman hyperbolic system of semilinear partial differential equations. This system provides a simple example of a problem for which high frequency components of a solution can be transformed into lower frequences through nonlinear interaction and thus affect the average of the solution.

The equations have the form

$$(8)\qquad \begin{aligned} &\frac{\partial u}{\partial t} + \frac{\partial u}{\partial x} + u^2 - v^2 = 0 \\ & && -\infty < x < \infty\ , \quad t > 0 \\ &\frac{\partial v}{\partial t} - \frac{\partial v}{\partial x} - u^2 + v^2 = 0 \end{aligned}$$

and let us assume oscillatory initial values

$$(9)\qquad \begin{aligned} u(x,0) &= u_0(x) = a(x,x/\varepsilon)(=u_\varepsilon(x,0)) \\ v(x,0) &= v_0(x) = b(x,x/\varepsilon)(=v_\varepsilon(x,0)). \end{aligned}$$

Here $a(x,y)$ and $b(x,y) \in C^\infty$ are of compact support with respect to x and 1-periodic in y. General theoretical results for this simple model in kinetic theory are given in [8], [9] and [13].

High frequency components from v^2 in the first equation and from u^2 in the second will generate low frequency contributions to the solution. This is described by the following homogenization result by Tartar [13]:

$$\begin{aligned} u_\varepsilon &\to U(x,(x-t)/\varepsilon,t) \\ & && \text{as } \varepsilon \to 0 \\ v_\varepsilon &\to V(x,(x+t)/\varepsilon,t) \end{aligned}$$

where

$$(10)\qquad \begin{aligned} &\frac{\partial U}{\partial t} + \frac{\partial U}{\partial x} + U^2 - \int_0^1 V^2(x,y,t)dy = 0 \\ &\frac{\partial V}{\partial t} - \frac{\partial V}{\partial x} + V^2 - \int_0^1 U^2(x,y,t)dy = 0 \\ &U(x,y,0) = a(x,y) \\ &V(x,y,0) = b(x,y) \end{aligned}$$

We need strong convergence in L_∞, which can be proved by using the integrated form of (8) and (10) applied to the differences $u_\varepsilon - U$ and $v_\varepsilon - V$, [4].

Our numerical approximation is the simplest characteristic or particle method

$$(11)\qquad \begin{aligned} u_j^{n+1} &= u_{j-1}^n - \Delta t((u_{j-1}^n)^2 - (v_{j-1}^n)^2) \\ v_j^{n+1} &= v_{j+1}^n + \Delta t((u_{j+1}^n)^2 - (v_{j+1}^n)^2) \\ u_j^0 &= u_0(x_j), \quad v_j^0 = v_0(x_j) \\ \Delta x &= \Delta t, \quad j=\ldots-1,0,1,\ldots,n=0,1,\ldots \end{aligned}$$

The form of the approximation of the lower order terms is crucial for the convergence [4]. The principal part is here approximated exactly but we shall see below in the variable coefficient problem that this is not essential for a good approximation.

In order to illustrate the effect of more general particle methods let us modify the Carleman equations to have variable velocity

$$(12)\qquad \begin{aligned} &\frac{\partial u}{\partial t} + \alpha(x,t)\frac{\partial u}{\partial x} + u^2 - v^2 = 0 \\ &\frac{\partial v}{\partial t} - \beta(x,t)\frac{\partial v}{\partial x} - u^2 + v^2 = 0 \end{aligned}$$

$$\alpha,\beta \geq \delta > 0\,, \text{ initial values as in (10).}$$

The simple scheme (11) with $\Delta x = \Delta t$ can not be applied any longer. However, with a slight change in the scheme and in particular with a new interpretation of the grid points we get

$$(13)\qquad \begin{aligned} u_j^{n+1} &= u_{j-1}^n - (t_j^{n+1}-t_{j-1}^n)((u_{j-1}^n)^2 - (v_{j-1}^n)^2) \\ v_j^{n+1} &= v_{j+1}^n + (t_j^{n+1}-t_{j+1}^n)((u_{j+1}^n)^2 - (v_{j+1}^n)^2) \\ u_j^0 &= u_0(x_j), \quad v_j^0 = v_0(x_j) \end{aligned}$$

The grid points $\{(x_j^n,t_j^n)\}$ are now at the intersection of the characteristics starting at the locations $x_j^0 = j\Delta x$ for the initial time $(t=0)$. These grid points can also be regarded as collision points in a grid free particle method. The equations for the characteristics are as usual approximated by an ordinary differential equation solver. For details see [6].

Even if the above method introduces errors in the approximation of advection, the errors are not of dispersive or dissipative type. The situation here is similar to the approximation of advection in vortex methods. We know experimentally that vortex methods give qualitatively good results even if the solution is not well resolved by the particle distribution, [3],[11].

Our final example is the Broadwell model

(14)
$$\begin{aligned}
&\frac{\partial u}{\partial t} + \frac{\partial u}{\partial x} + uv - w^2 = 0\\
&\frac{\partial v}{\partial t} - \frac{\partial v}{\partial x} + uv - w^2 = 0\\
&\frac{\partial w}{\partial t} \qquad - uv + w^2 = 0
\end{aligned}$$

with initial data analogous to (9). High frequency oscillations in u and v may here create oscillations in w via the uv-term. See[13] for the homogenized form of the equations. As for the Carleman equations there is a simple approximation which describes the nonlinear interaction correctly in the limit as $\varepsilon \to 0$, $\Delta x = \Delta t$.

(15)
$$\begin{aligned}
u_j^{n+1} &= u_{j-1}^n - \Delta t(u_{j-1}^n v_{j-1}^n - (w_{j-1}^n)^2)\\
v_j^{n+1} &= v_{j+1}^n - \Delta t(u_{j+1}^n v_{j+1}^n - (w_{j+1}^n)^2)\\
w_j^{n+1} &= w_j^n + \Delta t(u_j^n v_j^n - (w_j^n)^2)
\end{aligned}$$

Let us summarize the convergence results in the following theorem.

Theorem 2. The solutions of the difference approximations (11) and (15) converge strongly in L_∞ to the solutions of (8) and (14) respectively as $\Delta t \to 0$ essentially independent of ε. The averaged solution of (13) with a converging approximation for the characteristics converges strongly in L_∞ to the averaged solution of (12) as $\Delta t \to 0$ essentially independent of ε.

For the proof see [4] and [6]. Maximum norm estimates of the solutions of (8), (9), (14) and their corresponding numerical approximations are needed in the proof. The techniques of [8] and [9] for discrete Boltzmann equations can be modified in order to derive estimates also for the variable coefficient and discrete cases. The homogenized form of the differential equations are used in estimates of the local truncation errors. These truncation errors are too large for a standard convergence proof. It is necessary to take cancellation between different terms into account. Ergodic mixing and a technical lemma estimating the form and number of the largest error terms are essential ingredients in this part of the proof.

In figures 3,4 and 5 are results from the Broadwell approximation (15) displayed. The computations were done by Y. Hou. The functions u, v, and w were identically zero initially but for the interval (3,5) for u and (4,6) for v. In these intervals u and v were given the values of an oscillatory pulse of the form

$$\sin^2(\pi(x-x_0)/2)(1+\sin(2\pi(x-x_0)/\varepsilon)/2.$$

The figures give the values of w at a time when the oscillations in u and v have interacted. In figure 3 a refined grid is used, $\Delta x<\varepsilon$ and in figure 4 the grid is coarse $\Delta x>\varepsilon$. Figure 5 displays the averaged solutions witch are almost the same in the two cases.

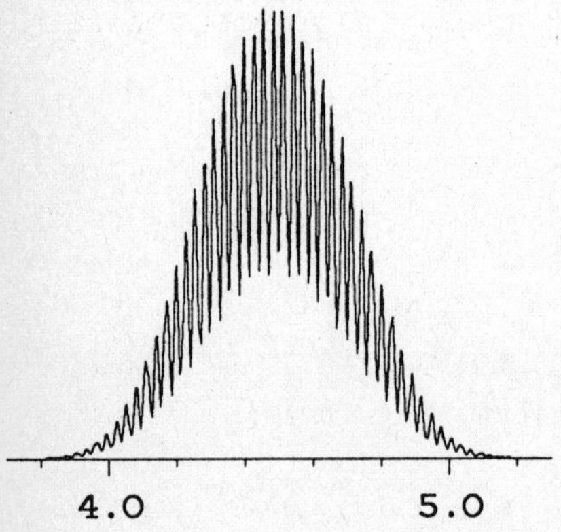

Figure 3. $w(x)$, $\Delta x<\varepsilon$

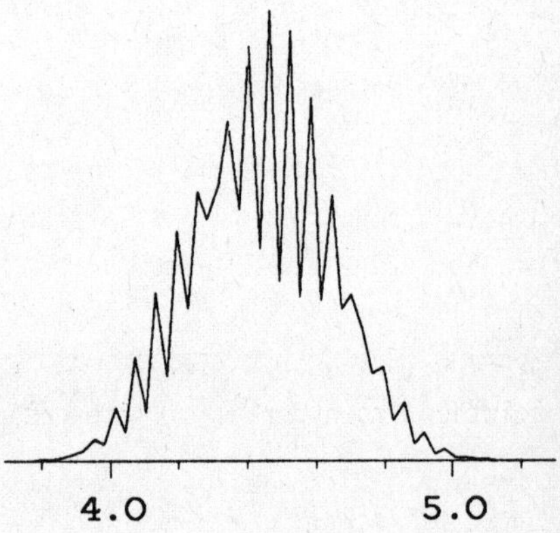

Figure 4. $w(x)$, $\Delta x>\varepsilon$

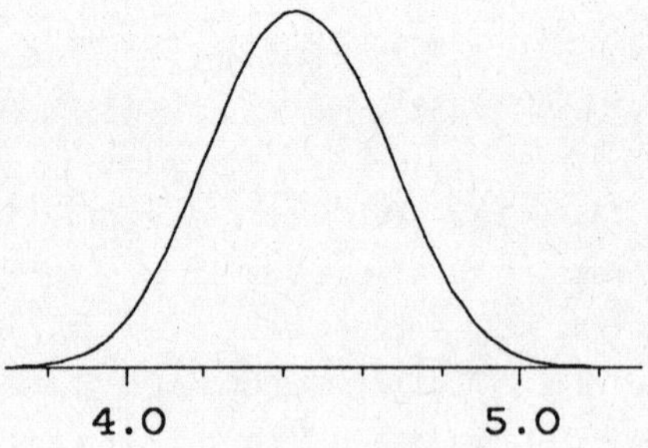

Figure 5. $\overline{w}(x)$

References

[1] A. Bensoussan, J.-L. Lions and G. Papanicolaou, "Asymptotic analysis for periodic structures," Studies in Mathematics and its applications, Vol. 5, North-Holland Publ., 1978.

[2] T. Carleman, "Problemes mathematizues dans la théorie cinétique des gaz", Publ. Sc. Inst. Mittag-Leffler, Uppsala, 1957.

[3] A. J. Chorin, "Vortex models and boundary layer instabilities," SIAM J. Sci. Comput., 1 (1980).

[4] B. Engquist, "Computation of oscillatory solutions to hyperbolic differential equations," to appear.

[5] B. Engquist, and H.O. Kreiss, "Difference and finite element methods for hyperbolic differential equations," Computer Meth. Appl. Math. Engineering, 17/18 (1979), pp. 581-596.

[6] B. Engquist, and Y. Hou, "Particle method approximations of oscillatory solutions to hyperbolic differential equations," to appear.

[7] H.-O. Kreiss, and J. Oliger, "Comparison of accurate methods for the integration of hyperbolic equations," Tellus, 24 (1972), pp. 199-215.

[8] R. Illner, "Global existence for two-velocity models of the Boltzmann equation," Math. Meth. Appl. Sc., 1, (1979), pp. 187-193.

[9] R. Illner, "Global existence results for discrete velocity models of the Boltzmann equation in several dimensions," J. Mécan. Th. Appl., 4 (1982), pp. 611-622.

[10] H. Neiderreither, "Quasi-Mote Carlo methods and pseudo-random numbers," Bull. Amer. Math. Soc., 84 (1978), pp. 957-1041.

[11] M. Perlman, "On the accuracy of vortex methods," J. Comp. Phys., 59 (1985), pp. 200-223.

[12] R. D. Richtmyer, and K. W. Morton, "Difference methods for initial-value problems," Interscience Publ., 1967.

[13] L. Tartar, "Etude des oscillations dans les équations aux dérivées partielle non linéaries," Trends and Applications of Pure Mathematics to Mechanics, Ciarlet-Roseau ed., Lecture Notes in Physics, Springer, 195 (1984), pp. 384-412.

Reserach supported by NSF Grant No. DMS85-03294, ARO Grant No. DAAG29-85-K-0190, NASA Consortium Agreement No. NCA2-IR360-403.

PRELIMINARY RESULTS ON THE EXTENSION OF ENO SCHEMES TO TWO-DIMENSIONAL PROBLEMS

Ami Harten
Department of Mathematics, UCLA
and
School of Mathematical Sciences, Tel-Aviv University

Abstract:

In this paper we present preliminary results on the extension of high-order accurate essentially non-oscillatory (ENO) schemes to the solution of hyperbolic systems of conservation laws in 2D. The design involves an essentially non-oscillatory piecewise polynomial reconstruction of the solution from its cell averages, solution in the small of the resulting piecewise polynomial initial value problem, and averaging of this solution over each cell. Unlike standard finite difference methods this procedure uses an adaptive stencil of grid points and consequently the resulting schemes are highly nonlinear.

1. Introduction.

In [9], [10], [12] we have presented an hierarchy of essentially non-oscillatory (ENO) schemes for the numerical solution of one-dimensional hyperbolic systems of conservation laws

$$u_t + f(u)_x = 0, \qquad u(x,0) = u_0(x), \tag{1.1}$$

which generalizes Godunov's scheme [3] and its second order accurate MUSCL extension [13], [2] to arbitrary order of accuracy. The design of the ENO schemes involves an essentially non-oscillatory reconstruction of the solution from its cell averages; time evolution through an approximate solution of the resulting initial value problem, and averaging of this approximate solution over each cell.

The reconstruction algorithm is derived from a new piecewise-polynomial interpolation $H_m(x;w)$ that when applied to piecewise smooth data w gives high order accuracy wherever $w(x)$ is smooth

$$\frac{d^k}{dx^k} H_m(x;w) = \frac{d^k}{dx^k} w(x) + O(h^{m+1-k}), \quad 0 \le k \le m, \tag{1.2a}$$

but avoids having a Gibbs phenomenon at discontinuities; this can be expressed by

$$TV(H_m(\cdot;w)) \le TV(w) + O(h^{m+1}), \tag{1.2b}$$

where TV denotes the total variation in x.

$H_m(x;w)$ is constructed as follows: We define

$$H_m(x;w) = q_{m,j+\frac{1}{2}}(x;w) \text{ for } \quad x_j \leq x \leq x_{j+1} \tag{1.3a}$$

where $q_{m,j+\frac{1}{2}}$ is the unique m-th degree polynomial which interpolates $w(x)$ at the $m+1$ successive points $\{x_i\}$, $i_m(j) \leq i \leq i_m(j)+m$ that include x_j and x_{j+1}, i.e.

$$q_{m,j+\frac{1}{2}}(x_i;w) = w(x_i) \text{ for } \quad i_m(j) \leq i \leq i_m(j)+m \tag{1.3b}$$

$$1-m \leq i_m(j)-j \leq 0. \tag{1.3c}$$

Clearly there are exactly m such polynomials corresponding to the m different choices of $i_m(j)$ subject to (1.3c). We achieve the properties (1.2) by using this freedom to assign to (x_j, x_{j+1}) a stencil of $m+1$ points (1.3b) in which $w(x)$ is "smoothest." The information about smoothness of $w(x)$ is extracted from a table of divided differences of w. The latter can be defined recursively by

$$w[x_i] = w(x_i) \tag{1.4a}$$

$$w[x_i, \ldots, x_{i+k}] = (w[x_{i+1}, \ldots, x_{i+k}] - w[x_i, \ldots, x_{i+k-1}])/(x_{i+k} - x_i). \tag{1.4b}$$

The stencil to be assigned to $[x_j, x_{j+1}]$ is chosen in an hierarchial fashion: We set

$$i_1(j) = j \tag{1.5a}$$

and define for $1 \leq k \leq m-1$

$$i_{k+1}(j) = \begin{cases} i_k(j)-1 & \text{if } |w[x_{i_k(j)-1}, \ldots, x_{i_k(j)+k}]| < |w[x_{i_k(j)}, \ldots, x_{i_k(j)+k+1}]| \\ i_k(j) & \text{otherwise.} \end{cases} \tag{1.5b}$$

(see [7], [9], [12] for more details).

Unlike standard finite difference schemes that use a predetermined stencil of grid points, the ENO schemes use an adaptive stencil which is selected by (1.5) based on local considerations of smoothness; consequently the ENO schemes are highly nonlinear.

In this paper we extend the ENO schemes to the two-dimensional case. We recall that Strang-type dimensional splitting [15] is only second order accurate in time and therefore is unsuitable for extending higher order accurate members of the ENO schemes to 2D. In section 2 we present the abstract formulation of the ENO schemes in 2D; in section 3 we extend the one-dimensional reconstruction via deconvolution technique of [7], [9] to 2D; in section 4 we present some preliminary results showing the performance of the ENO schemes in the scalar two dimensional constant coefficient case.

2. Formulation of ENO schemes in 2D

In this section we describe the abstract form of the ENO schemes for the solution of the initial value problem (IVP) for a two-dimensional system of conservation laws

$$u_t + f(u)_x + g(u)_y = 0, \qquad u(x,y,0) = u_0(x,y). \tag{2.1}$$

We assume that the system of conservation laws in (2.1) has an entropy function and therefore it is symmetrizable and consequently hyperbolic (see [5]). We design our numerical schemes under the assumption that the IVP (2.1) is well posed in some appropriate setting (which includes entropy considerations) and that $u(x,y,t)$ is piecewise-smooth with a finite number of discontinuity surfaces for a large class of computationally interesting initial data. We denote the evolution operator of (2.1) by $E(t)$, i.e.

$$u(\cdot,\cdot,t) = E(t)u_0. \tag{2.2}$$

Let $\{I_{ij}\times[t_n,t_{n+1}]\}$ where $I_{ij} = [x_{i-\frac{1}{2}}, x_{i+\frac{1}{2}}]\times[y_{j-\frac{1}{2}}, y_{j+\frac{1}{2}}]$, $x_\alpha = \alpha\Delta x$, $y_\beta = \beta\Delta y$, $t_n = n\tau$, be a partition of $\boldsymbol{R}^2 \times \boldsymbol{R}^+$ with $\Delta x = 0(\Delta)$, $\Delta y = 0(\Delta)$. Integrating the PDE (2.1) over $I_{ij} \times [t_n, t_{n+1}]$ we get

$$\begin{aligned} \bar{u}_{ij}^{n+1} &= \bar{u}_{ij}^n - \lambda_x[\hat{f}(x_{i+\frac{1}{2}}; y_j, t_n, u) - \hat{f}(x_{i-\frac{1}{2}}; y_j, t_n, u)] \\ &\quad - \lambda_y[\hat{g}(y_{j+\frac{1}{2}}; x_i, t_n, u) - \hat{g}(y_{j-\frac{1}{2}}; x_i, t_n, u)] \end{aligned} \tag{2.3a}$$

where $\lambda_x = \tau/\Delta x$, $\lambda_y = \tau/\Delta y$ and

$$\hat{f}(x; y, t, u) = \frac{1}{\tau\Delta y}\int_0^\tau \int_{-\Delta y/2}^{\Delta y/2} f(u(x, y+\eta, t+\mu))d\eta d\mu, \tag{2.3b}$$

$$\hat{g}(y; x, t, u) = \frac{1}{\tau\Delta x}\int_0^\tau \int_{-\Delta x/2}^{\Delta x/2} g(u(x+\xi, y, t+\mu))d\xi d\mu. \tag{2.3c}$$

Here

$$\bar{u}_{ij}^n = \bar{u}(x_i, y_j, t_n) = \frac{1}{\Delta x\Delta y}\int_{-\Delta y/2}^{\Delta y/2}\int_{-\Delta x/2}^{\Delta x/2} u(x_i+\xi, y_j+\eta, t_n)d\xi d\eta \tag{2.4}$$

is the "cell-average" of u at time t_n.

In this paper we consider schemes in the conservation form

$$v_{ij}^{n+1} = v_{ij}^n - \lambda_x(\bar{f}_{i+\frac{1}{2},j} - \bar{f}_{i-\frac{1}{2},j}) - \lambda_y(\bar{g}_{i,j+\frac{1}{2}} - \bar{g}_{i,j-\frac{1}{2}}) \equiv [\bar{\bar{E}}_\Delta(\tau)v^n]_{ij} \tag{2.5}$$

which approximates (2.3) to high-order of accuracy. Here v_{ij}^n is an approximation to $\bar{u}_{ij}^n$ and $\bar{\bar{E}}_\Delta(\tau)$ is the numerical solution operator. Setting $v^n \equiv \bar{u}^n$ in (2.5) and comparing it to (2.3) we see that if

$$\bar{f}_{i+\frac{1}{2},j} = \hat{f}(x_{i+\frac{1}{2}}; y_j, t_n, u) + c(x_{i+\frac{1}{2}}; y_j, t_n, u)(\Delta x)^r + O(\Delta^{r+1}) \tag{2.6a}$$

$$\bar{g}_{i,j+\frac{1}{2}} = \hat{g}(y_{j+\frac{1}{2}}; x_i, t_n, u) + d(y_{j+\frac{1}{2}}; x_i, t_n, u)(\Delta y)^r + O(\Delta^{r+1}), \tag{2.6b}$$

then the local truncation error in the sense of cell-averages is

$$\begin{aligned}\bar{\bar{u}}(x_i, y_j, t_{n+1}) - [\bar{\bar{E}}_\Delta(\tau)\cdot\bar{\bar{u}}^n]_{ij} = \\ = \lambda_x[c(x_{i+\frac{1}{2}}; y_j, t_n, u) - c(x_{i-\frac{1}{2}}; y_j, t_n, u)](\Delta x)^r \\ + \lambda_y[d(y_{j+\frac{1}{2}}; x_i, t_n, u)] - d(y_{j-\frac{1}{2}}; x_i, t_n, u)](\Delta y)^r + O(\Delta^{r+1});\end{aligned} \tag{2.6c}$$

the RHS of (2.6c) is $O(\Delta^{r+1})$ wherever c and d are Lipschitz continuous.

We observe that although (2.3a) is a relation between $\bar{\bar{u}}_{ij}^n$ and $\bar{\bar{u}}_{ij}^{n+1}$, the evaluation of the fluxes $\hat{f}$ and $\hat{g}$ in (2.3b), (2.3c) requires knowledge of the solution itself, not its cell-averages. In order to design numerical fluxs that satisfy (2.6) we must extract high order accurate pointwise information from the given $\{v_{ij}^n\}$, which are approximations to $\{\bar{\bar{u}}_{ij}^n\}$. For this purpose we solve the following problem in approximation of functions: Given $\{\bar{\bar{w}}_{ij}\}$, cell-averages of a piecewise-smooth function $w(x, y)$, find $R(x, y; \bar{\bar{w}})$, a piecewise-polynomial function in x and y of polynomial degree $(r-1)$, that reconstructs $w(x, y)$ in the sense that

$$R(x, y; \bar{\bar{w}}) = w(x, y) + O(\Delta^r) \tag{2.7a}$$

at all points (x, y) for which there is an open neighborhood where w is smooth, and

$$\bar{\bar{R}}(x_i, y_j; \bar{\bar{w}}) = \bar{\bar{w}}_{ij}. \tag{2.7b}$$

Solving this reconstruction problem to arbitrarily high order of accuracy r, without introducing $O(1)$ Gibbs-like spurious oscillations at discontinuities, is the most important step in the design of our new schemes.

In section 3 we present a reconstruction technique which is a generalization of the one-dimensional reconstruction via deconvolution of [7] (see also [9], [12]). In this approach we consider the given cell-averages $\{\bar{\bar{w}}_{ij}\}$ to be point values $\bar{\bar{w}}(x_i, y_j)$ of the sliding average function $\bar{\bar{w}}(x, y)$ of w

$$\bar{\bar{w}}(x, y) = \frac{1}{\Delta x \Delta y}\int_{-\Delta y/2}^{\Delta y/2}\int_{-\Delta x/2}^{\Delta x/2} w(x+\xi, y+\eta)\,d\xi d\eta \equiv (A_\Delta \cdot w)(x, y). \tag{2.8}$$

Since $\bar{\bar{w}}(x, y)$ in (2.8) is the convolution of $w(x, y)$ with the characteristic function of a cell $\psi_\Delta(x, y)$ i.e.

$$\bar{\bar{w}} = w * \psi_\Delta \tag{2.9a}$$

$$\psi_\Delta(x, y) = \begin{cases} \frac{1}{\Delta x \Delta y} & \text{for } |x| < \Delta x/2,\ |y| < \Delta y/2, \\ 0 & \text{otherwise} \end{cases} \tag{2.9b}$$

the reconstruction (2.7) can be viewed as a finite-order deconvolution of (2.9).

Using the reconstruction (2.7) we can express the abstract form of our new schemes (2.5) by

$$\bar{\bar{E}}_\Delta(\tau)\bullet\bar{\bar{w}} = A_\Delta \bullet E(\tau)\bullet R(\cdot,\cdot;\bar{\bar{w}}). \tag{2.10}$$

Here A_Δ is the cell-averaging operator which is defined by the RHS of (2.8); $E(t)$ is the exact evolution operator (2.2) and w is any piecewise smooth function of x and y. These schemes are a generalization of the second-order accurate schemes in [1] and [14]. Although the scheme $v^{n+1} = \bar{\bar{E}}_\Delta(\tau)v^n$ generates discrete values v_{ij}^n which are r-th order approximations to the cell-averages $\bar{\bar{u}}_{ij}^n$, its operation involves a globally defined pointwise approximation to $u(x,y,t)$ of the same order of accuracy, which we denote by $v_\Delta(x,y,t)$. The latter is defined for all x and y in the time-strips $t_n \le t < t_{n+1}$, with a possible discontinuity at $\{t_k\}$; we shall use the standard notation $v_\Delta(x,y,t_n \pm 0)$ to distinguish between the two possibly different values.

We define $v_\Delta(x,y,t)$ via the following algorithmic description of the scheme (2.10). We start by setting

$$v_{ij}^0 = \bar{\bar{u}}_0(x_i, y_j)$$

where u_0 is the given initial data in (2.1) and $\bar{\bar{u}}_0$ is its sliding average (2.8). Having defined $v^n = \{v_{ij}^n\}$, approximation to $\bar{\bar{u}}^n = \{\bar{\bar{u}}_{ij}^n\}$ in (2.4), we proceed to evaluate v^{n+1} by the following three steps:

(i) Reconstruction:

Define

$$v_\Delta(x,y,t_n+0) = R(x,y;v^n). \tag{2.11a}$$

Note that $v_\Delta(x,y,t_n+0)$ is a pointwise approximation to $u(x,y,t_n)$.

(ii) Solution in the small:

For $t_n \le t < t_n + \tau = t_{n+1}$, define

$$v_\Delta(\cdot,\cdot,t) = E(t-t_n)\bullet v_\Delta(\cdot,\cdot,t_n+0). \tag{2.11b}$$

(iii) Cell-averaging:

Close the time-loop of the algorithm by defining

$$v_{ij}^{n+1} = \bar{\bar{v}}_\Delta(x_i, y_j, t_{n+1}-0). \tag{2.11c}$$

We note that v_Δ, being an exact solution of (2.1) in $t_n \le t < t_{n+1}$, satisfies relation (2.3) in this time-strip. Using the conservation property (2.7b) of the reconstruction, i.e.

$$\bar{\bar{v}}_\Delta(x_i, y_j, t_n + 0) = v_{ij}^n,$$

we get from (2.3) that the scheme (2.10), (2.11) can be written in the conservation form (2.5)

$$v_{ij}^{n+1} = v_{ij}^n - \lambda_x(\bar{f}_{i+\frac{1}{2},j} - \bar{f}_{i-\frac{1}{2},j}) - \lambda_y(\bar{g}_{i,j+\frac{1}{2}} - \bar{g}_{i,j-\frac{1}{2}}) \tag{2.12a}$$

with

$$\bar{f}_{i+\frac{1}{2},j} = \hat{f}(x_{i+\frac{1}{2}}; y_j, t_n, v_\Delta), \tag{2.12b}$$

$$\bar{g}_{i,j+\frac{1}{2}} = \hat{g}(y_{j+\frac{1}{2}}; x_i, t_n, v_\Delta), \tag{2.12c}$$

where $\hat{f}$ and $\hat{g}$ are defined by (2.3b) - (2.3c). To see that the scheme (2.12) is r-th order accurate in the sense of cell-averages (2.6c) we consider a single application of (2.11) starting with $v^n = \bar{\bar{u}}^n$, where $u(x, y, t)$ is any smooth solution of (2.1). It follows from (2.7a) and (2.11a) that

$$v_\Delta(x, y, t_n + 0) = u(x, y, t_n) + 0(\Delta^r). \tag{2.13a}$$

The definition (2.11b) and our assumption of well-posedness of the IVP (2.1) imply that

$$v_\Delta(x, y, t) = u(x, y, t) + O(\Delta^r) \text{ for } \quad t_n \le t < t_{n+1}. \tag{2.13b}$$

Assuming $f(u)$ and $g(u)$ in (2.1) to be Lipschitz continuous, we get from (2.13b) that the numerical fluxes (2.12b) - (2.13b) satisfy relations (2.6a) - (2.6b); consequently the numerical scheme satisfies (2.6c).

The abstract form of our scheme calls in (2.11b) for the evaluation of the exact solution *in the small* of the IVP (2.1) with the piecewise polynomial initial data $R(x, y; v^n)$. To make our scheme a practical one we replace v_Δ in the numerical fluxes (2.12b) - (2.12c) by a closed-form expression which is a simple but adequate approximation to v_Δ; details of this approximation will be given in a future paper [11]. In the present paper we study the scalar constant coefficient case

$$f(u) = au, \qquad g(u) = bu \tag{2.14a}$$

where a and b are scalar constants. In this case

$$v_\Delta(x, y, t_n + t) = R(x - at, y - bt; v^n) \text{ for } \quad 0 \le t < \tau. \tag{2.14b}$$

and the scheme (2.10) - (2.12) takes the particularly simple form

$$v_{ij}^{n+1} = \bar{\bar{R}}(x_i - a\tau, y_j - b\tau; v^n), \quad v_{ij}^0 = \bar{\bar{u}}_0(x_i, y_j). \tag{2.14c}$$

3. Two-dimensional reconstruction

In this section we describe a two-dimensional reconstruction via deconvolution $R(x,y;\bar{w})$ of a piecewise smooth function $w(x,y)$ from its given cell-averages $\{\bar{w}_{ij}\}$. In this approach we assume the mesh to be uniform and consider $\bar{w}_{ij}$ to be $\bar{w}(x_i,y_j)$, the point values of the globally defined sliding-average function $\bar{w}(x,y)$ (2.8) of w. This enables us to obtain approximations to derivatives of $\bar{w}$ by differentiation of interpolating polynomials.

Expanding $w(x+\xi,y+\eta)$ in

$$\bar{w}(x,y)=\frac{1}{\Delta x\Delta y}\int_{-\Delta y/2}^{\Delta y/2}\int_{-\Delta x/2}^{\Delta x/2} w(x+\xi,y+\eta)d\xi d\eta \tag{3.1a}$$

around $\xi=\eta=0$, we get

$$\begin{aligned}\bar{w}(x,y)&=\sum_{l=0}^{\infty}\frac{1}{l!}\sum_{k=0}^{l}\binom{l}{k}\frac{\partial^l w(x,y)}{\partial x^k\partial y^{l-k}}\frac{1}{\Delta x\Delta y}\int_{-\Delta y/2}^{\Delta y/2}\int_{-\Delta x/2}^{\Delta x/2}\xi^k\eta^{l-k}d\xi d\eta \\ &\equiv\sum_{l=0}^{\infty}\alpha_l\sum_{k=0}^{l}\beta_{l,k}(\Delta x)^k(\Delta y)^{l-k}\frac{\partial^l w(x,y)}{\partial x^k\partial y^{l-k}}\end{aligned} \tag{3.1b}$$

where

$$\alpha_l=2^{-l}/(l+1)!,\qquad \beta_{l,k}=\binom{l+1}{k}\frac{[1+(-1)^k][1+(-1)^l]}{4(k+1)}. \tag{3.1c}$$

Next we multiply both sides of (3.1b) by $(\Delta x)^p(\Delta y)^q\frac{\partial^{p+q}}{\partial x^p\partial y^q}$ for $0\le p\le r-1$, $0\le q\le r-1-p$, and truncate the expansion in the RHS at $O(\Delta^r)$. Writing the resulting relations at (x_i,y_j) we get

$$(\bar{d}^{p,q})_{ij}=\sum_{l=0}^{r-p-q-1}\alpha_l\sum_{k=0}^{l}\beta_{l,k}(d^{k+p,l+q-k})_{ij}+O(\Delta^r) \tag{3.2a}$$

where

$$(d^{p,q})_{ij}=(\Delta x)^p(\Delta y)^q\frac{\partial^{p+q}w(x_i,y_j)}{\partial x^p\partial y^q}, \tag{3.2b}$$

$$(\bar{d}^{p,q})_{ij}=(\Delta x)^p(\Delta y)^q\frac{\partial^{p+q}\bar{w}(x_i,y_j)}{\partial x^p\partial y^q}. \tag{3.2c}$$

Our reconstruction $R(x,y;\bar{w})$ is obtained from relations (3.2) as follows: We set

$$(\bar{D}^{0,0})_{ij}=(\bar{d}^{0,0})_{ij}=\bar{w}_{ij} \tag{3.3a}$$

and use differentiation of interpolating polynomials of $\bar{w}$ to compute approximations $(\bar{D}^{p,q})_{ij}$ such that

$$(\bar{D}^{p,q})_{ij}=(\bar{d}^{p,q})_{ij}+O(\Delta^r); \tag{3.3b}$$

then we invert the system of linear equations

$$(\bar{\bar{D}}^{p,q})_{ij} = \sum_{l=0}^{r-p-q-1} \alpha_l \sum_{k=0}^{l} \beta_{l,k} (D^{k+p,l+q-k})_{ij}, \quad 0 \le p \le r-1,\ 0 \le q \le r-p-1, \tag{3.4}$$

to obtain $\{(D^{p,q})_{ij}\}$ for $0 \le p \le r-1,\ 0 \le q \le r-p-1$; clearly

$$(D^{p,q})_{ij} = (d^{p,q})_{ij} + O(\Delta^r). \tag{3.5}$$

Finally we define $R(x, y; \bar{\bar{w}})$ in the cell I_{ij} by

$$R(x,y;\bar{\bar{w}}) = \sum_{l=0}^{r-1} \frac{1}{l!} \sum_{k=0}^{l} \binom{l}{k} (D^{k,l-k})_{ij} \left(\frac{x - x_i}{\Delta x}\right)^k \left(\frac{y - y_j}{\Delta y}\right)^{l-k}, \quad (x,y) \in I_{ij}. \tag{3.6}$$

We observe that relation (2.7a) follows immediately from (3.5) and the definitions (3.2b) and (3.6). Taking the cell-average of R in (3.6) and using the equality in (3.4) for $p = q = 0$, we get as in (3.1b) that

$$\bar{\bar{R}}(x_i, y_j; \bar{\bar{w}}) = \sum_{l=0}^{r-1} \alpha_l \sum_{k=0}^{l} \beta_{l,k} (D^{k,l-k})_{ij} = (\bar{\bar{D}}^{0,0})_{ij}. \tag{3.7}$$

Hence the conservation relation (2.7b) follows from (3.3a).

The approximations $(\bar{\bar{D}}^{k,l})_{ij}$ in (3.3b) are obtained by a sequence of applications of the one-dimensional operator G_m^l

$$(G_m^l \bullet u)_j = M\left(\frac{d^l}{dz^l} H_m(z_j - 0; u), \frac{d^l}{dz^l} H_m(z_j + 0; u)\right); \tag{3.8a}$$

here u is any one-dimensional vector $\{u(z_j)\}$; $H_m(x; u)$ is the one-dimensional non-oscillatory piecewise-polynomial interpolation (1.3) - (1.5); $M(v_1, v_2)$ is the min mod function

$$M(v_1, v_2) = \begin{cases} s \cdot \min(|v_1|, |v_2|) & \text{if } \operatorname{sgn}(v_1) = \operatorname{sgn}(v_2) = s \\ 0 & \text{otherwise.} \end{cases} \tag{3.8b}$$

We recall that $H_m(z; u)$ is only continuous at $z = z_j$, and that if z_j has a neighborhood in which u is smooth, then (1.2a) implies that for $0 \le l \le m$

$$\frac{d^l}{dz^l} H_m(z_j \pm 0; u) = \frac{d^l u(z_j)}{dz^l} + O(\Delta^{m-l+1});$$

consequently

$$(G_m^l \bullet u)_j = \frac{d^l u(z_j)}{dz^l} + O(\Delta^{m-l+1}). \tag{3.9}$$

Using (3.8) and the notation convention $\bar{\bar{w}}_{\bullet,j} = \{\bar{\bar{w}}(x_i, y_j)\}_{i=1}^{K}$ we define

$$(\bar{\bar{D}}^{k,0})_{ij} = (\Delta x)^k (G_m^k \bullet \bar{\bar{w}}_{\bullet,j})_i, \qquad 1 \le k \le r-1, \tag{3.10a}$$

$$(\bar{\bar{D}}^{0,l})_{ij} = (\Delta y)^l (G_m^l \bullet \bar{\bar{w}}_{i,\bullet})_j, \qquad 1 \le l \le r-1. \tag{3.10b}$$

To obtain approximations to the mixed derivatives of $\bar{\bar{w}}$ we first evaluate

$$(\tilde{D}_1^{k,l})_{ij} = (\Delta y)^l [G_{m-k}^l \bullet (\bar{\bar{D}}^{k,0})_{i,\bullet}]_j, \qquad 1 \le l \le r-1-k,$$
$$(\tilde{D}_2^{k,l})_{ij} = (\Delta x)^k [G_{m-l}^k \bullet (\bar{\bar{D}}^{0,l})_{\bullet,j}]_i, \qquad 1 \le k \le r-1-l,$$

and then define

$$(\bar{\bar{D}}^{k,l})_{ij} = M((\tilde{D}_1^{kl})_{ij}, (\tilde{D}_2^{kl})_{ij}), \tag{3.10c}$$

where M is the min mod function (3.8b). Relations (3.3b) follow from (3.9) and the definitions (3.2c) and (3.10), provided that $m \ge r-1$ in (3.8).

4. Numerical experiments in the scalar constant coefficient case.

In this section we present some numerical experiments with the two-dimensional ENO schemes in the scalar constant coefficient case

$$u_t + au_x + bu_y = 0, \qquad u(x,y,0) = u_0(x,y) \tag{4.1a}$$

In all these experiments we define $\tilde{u}_0(x,y)$ in $U = [-1,1] \times [-1,1]$ and extend it to $\boldsymbol{R}^2$ by periodicity

$$u_0(x+2k, y+2l) = \tilde{u}_0(x,y) \text{ for } (x,y) \in U; \; k,l \text{ integers}. \tag{4.1b}$$

The ENO schemes in the scalar constant coefficient case take the particularly simple form (2.14c), i.e.

$$v_{ij}^{n+1} = \bar{\bar{R}}(x_i - a\tau, \; y_j - b\tau; v^n), \qquad v_{ij}^0 = \bar{\bar{u}}_0(x_i, y_j), \tag{4.2}$$

where R is the piecewise-polynomial reconstruction of section 3. We recall that the polynomial degree of the reconstruction (3.6) for an r-th order accurate ENO scheme (4.2) is $r-1$, and that the corresponding order of interpolation in (3.8a) is $m \ge r-1$. We observe that although the PDE (4.1a) is linear, the ENO schemes (4.2) with $r \ge 2$ are nonlinear.

In Table 1 and Table 2 we present calculations of (4.1a) with the smooth initial data

$$u_0(x,y) = (\sin \pi x)(\sin \pi y). \tag{4.3}$$

We divide U into K^2 equal cells with

$$\Delta x = \Delta y = \Delta = 2/K \tag{4.4a}$$

and define

$$x_i = -1 + i\Delta, \qquad 0 \le i \le K, \tag{4.4b}$$

$$y_j = -1 + j\Delta, \qquad 0 \le j \le K. \tag{4.4c}$$

In Table 1 and Table 2 we list the L_∞-error and the average L_1-error at $t = 2$ of a refinement sequence $K = 20,\ 40,\ 80$. In these calculations we chose $a = b = 1$ in (4.1a) and used the schemes (4.2) with a CFL number 0.8. The L_∞-error in these tables is evaluated by

$$E_\infty = \max_{1 \leq i,j \leq K} |R(x_i, y_j; v^N) - u(x_i, y_j, 2)|; \tag{4.5a}$$

the average L_1-error is calculated by

$$E_1 = \frac{\Delta^2}{4} \sum_{i=1}^{K} \sum_{j=1}^{K} |R(x_i, y_j, v^N) - u(x_i, y_j, 2)|; \tag{4.5b}$$

here $N = 2/\tau$. The value of r_c in Table 1 and Table 2 is the "computational order of accuracy" which is calculated by assuming the error to be a constant times Δ^{r_c}. As in the one-dimensional case

$$u_t + u_x = 0, \qquad u(x, 0) = \sin \pi x \tag{4.6}$$

(see [7], [9]) we find that the schemes in Table 2 with $m = r$, $2 \leq r \leq 4$, are r-th order accurate in both the L_∞ and L_1 sense. However the schemes with $m = r - 1$ in Table 1 seem to be r-th order accurate in L_1 but only $(r-1)$-th order accurate in L_∞; this is due to insufficient smoothness of the coefficients c and d (2.6c) in this case. Comparing the errors in Tables 1 and 2 to those of the corresponding one-dimensional ENO schemes for (4.6) we find that the error in 2D is about twice that of the one-dimensional case, as to be expected.

We note that the 2D reconstruction (2.6) restricted to the lines $x = x_i$ and $y = y_j$ is basically the same as the 1D reconstruction in [7], [9] applied to the one-dimensional data along these lines. Therefore we expect the two-dimensional ENO schemes to perform as well as the one-dimensional schemes, provided that the curvature of the discontinuity curves is not too large. In [10] we have pointed out that the 1D reconstruction is essentially non-oscillatory only if discontinuities are seperated by a least $r + 1$ points of smoothness. Consequently the 2D reconstruction may produce spurious oscillations in the vicinity of intersection points of discontinuity curves (such as the triple point in a Mach reflection).

In order to evaluate the severity of this problem we apply the ENO schemes to the IVP (4.1) with

$$\tilde{u}_0(x, y) = \begin{cases} 1 & (x, y) \in S \\ 0 & (x, y) \in U - S \end{cases} \tag{4.7a}$$

where S is a square centered at the origin and rotated by 45°; the sides of this square are of unit length, i.e.

$$S = \{(x, y) : |x - y| < 2^{-1/2}, |x + y| < 2^{-1/2}\}. \tag{4.7b}$$

In figures 1-5 we present calculations of the periodic IVP (4.1), (4.7) with $a = b = 1$. These calculations were performed with K = 20 in (4.4) and a CFL number of 0.8. Each of these figures contains 6 plots. Plots (a), (b) and (c) show the solution after 1 period in time which corresponds to $t = 2$ and $n = 25$ time-steps: (a) Level curves of v^{25}; (b) $R(x_i, 0; v^{25})$ and $u(x, 0, 2)$; (c) $R(x_i, -0.4; v^{25})$ and $u(x, -0.4, 2)$. Plots (d), (e) and (f) show the same information as (a), (b) and (c) after 4 periods in time, which corresponds to $t = 8$ and $n = 100$ time-steps.

In Figure 1 we show the results of the first order accurate upwind scheme ($r = 1, m = 0$). This scheme is monotone and therefore the 2D total variation of its solution

$$TV(v^n) = \Delta x \Delta y \sum_i \sum_j \left[|v^n_{i+1,j} - v^n_{i,j}|/\Delta x + |v^n_{i,j+1} - v^n_{i,j}|/\Delta y \right] \tag{4.8a}$$

is diminishing in time, i.e.

$$TV(v^{n+1}) \leq TV(v^n). \tag{4.8b}$$

In figure 2 we show the results of the ENO scheme (4.2) with $r = 2$, $m = 1$; this scheme is a 2D extension of the one-dimensional TVD scheme of [6]. Goodman and LeVeque have shown in [4] that this scheme, as any other scheme which is not first order accurate and monotone, cannot be total variation diminishing in the sense of (4.8). Since the results in figure 2 do not show any spurious oscillations, it seems that the Goodman-LeVeque statement has more to do with the inadequacy of the definition (4.8) than with the actual performance of the scheme.

In figure 3 we show the results of the second order ENO scheme (4.2) with $r = 2$, $m = 2$; this scheme is a 2D extension of the (strictly) non-oscillatory scheme of [8]. We remark that this scheme differs from the "second order TVD" scheme ($r = 2$, $m = 1$) only by a modified expression for the slopes $D^{0,1}$ and $D^{1,0}$ in (3.5). Comparing figure 3 to figure 2, and the corresponding results in Tables 1 and 2, we see that this simple modification results in a considerable inprovement in performance.

In figures 4 and 5 we present the results of the third order ENO scheme ($r = 3$, $m = 3$) and the fourth order ENO scheme ($r = 4$, $m = 4$), respectively. We observe that the level of spurious oscillations is sufficiently small to be computationally acceptable; contour plots in (a) and (d) look "noisy" only because the plotting routine finds several contours of value 1 inside the square, and several contours of value 0 outside the square; the small negative values result primarily from the periodicity of the problem. We note that the third and fourth order accurate schemes have a much better resolution than the second order ones.

Acknowledgment

I would like to thank Sukumar Chakravarthy, Bjorn Engquist and Stan Osher for many stimulating discussions and for various contributions to this research (which is part of a joint project). This research was supported by NSF Grant No. DMS85-03294, ARO Grant No. DAAG29-85-K-0190 and NASA Consortium Agreement No. NCA2-IR390-403.

References

[1]. P. Colella, "Multidimensional upwind methods for hyperbolic conservation laws," Lawrence Berkeley Laboratory Preprint #17023, May 1984.

[2]. P. Colella and P.R. Woodward, "The piecewise-parabolic method (PPM) for gas-dynamical simulations," J. Comp. Phys., V. 54 (1984), 174-201.

[3]. S.K. Godunov, "A finite difference method for the numerical computation of discontinuous solutions of the equations to fluid dynamics, " Mat. Sb., 47 (1959), pp. 271-290.

[4]. J.B. Goodman and R. LeVeque 76, "On the accuracy of stable schemes for 2D scalar conservation laws," Math. Comp., V. 45 (1985) pp. 15-21.

[5]. A. Harten, "On the symmetric form of systems of conservation laws with entropy," J. Comp. Phys., V. 49 (1983), pp. 151-169.

[6]. A. Harten, "High resolution schemes for hyperbolic conservation laws," J. Comp. Phys., V. 49 (1983), pp. 357-393.

[7]. A. Harten, "On high-order accurate interpolation for non-oscillatory shock capturing schemes," MRC Technical Summary Report #2829, University of Wisconsin, (1985).

[8]. A. Harten and S. Osher, "Uniformly high-order accurate non-oscillatory schemes, I.," MRC Technical Summary Report #2823, May 1985, to appear in SINUM.

[9]. A. Harten, B. Engquist, S. Osher and S.R. Chakravarthy, "Uniformly high-order accurate non-oscillatory schemes, II." (in preparation).

[10]. A. Harten, B. Engquist, S. Osher and S.R. Chakravarthy, "Uniformly high-order accurate non-oscillatory schemes, III." submitted to J. Comp. Phys.

[11]. A. Harten, B. Engquist, S. Osher and S.R. Chakravarthy, "Uniformly high-order accurate non-oscillatory schemes, IV." (in preparation).

[12]. A. Harten, S. Osher, B. Engquist and S.R. Chakravarthy, "Some results on uniformly high-order accurate essentially non-oscillatory schemes," to appear in J. App. Num. Math.

[13]. B. van Leer, "Towards the ultimate conservative difference schemes V. A second order sequel to Godunov's method," J. Comp. Phys., V. 32 (1979) pp. 101-136.

[14]. B. van Leer, "Multidimensional explicit difference schemes for hyperbolic conservation laws," ICASE Report 172254, November 1983.

[15]. G. Strang, "On the construction and comparison of difference schemes," SINUM, V. 5 (1968) pp. 506-517.

Table 1. L_∞-error and L_1-error for solutions of (4.3) by the schemes (4.2) with $m = r - 1$.

	K \ order	$r = 1$, $m = 0$	r_c	$r = 2$, $m = 1$	r_c	$r = 3$, $m = 2$	r_c	$r = 4$, $m = 3$	r_c
E_∞	20	3.317×10^{-1}		1.630×10^{-1}		1.106×10^{-2}		4.147×10^{-3}	
			0.88		1.21		2.97		3.17
	40	1.808×10^{-1}		7.029×10^{-2}		1.409×10^{-3}		4.622×10^{-4}	
			0.94		1.26		2.48		3.18
	80	9.445×10^{-2}		2.935×10^{-2}		2.519×10^{-4}		5.113×10^{-5}	
E_1	20	1.333×10^{-1}		3.057×10^{-2}		4.708×10^{-3}		6.083×10^{-4}	
			0.87		1.75		2.91		3.69
	40	7.314×10^{-2}		9.076×10^{-3}		6.274×10^{-4}		4.728×10^{-5}	
			0.94		1.72		2.94		3.75
	80	3.826×10^{-2}		2.757×10^{-3}		8.199×10^{-5}		3.502×10^{-6}	

Table 2. L_∞-error and L_1-error for solutions of (4.3) by the schemes (4.2) with $m = r$.

	K \ order	$r = 2$, $m = 2$	r_c	$r = 3$, $m = 3$	r_c	$r = 4$, $m = 4$	r_c
E_∞	20	1.863×10^{-2}		4.348×10^{-3}		2.083×10^{-4}	
			2.29		3.15		4.67
	40	3.814×10^{-3}		4.883×10^{-4}		8.158×10^{-6}	
			2.18		3.16		4.34
	80	8.417×10^{-4}		5.464×10^{-5}		4.020×10^{-7}	
E_1	20	7.891×10^{-3}		6.233×10^{-4}		9.347×10^{-5}	
			2.29		3.77		4.71
	40	1.615×10^{-3}		4.584×10^{-5}		3.577×10^{-6}	
			2.17		3.53		4.58
	80	3.599×10^{-4}		3.959×10^{-6}		1.494×10^{-7}	

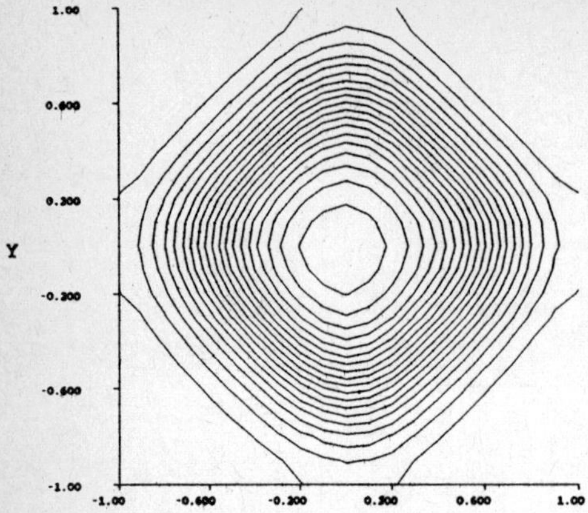

(a) Level curves of v^{25} $(t = 2)$

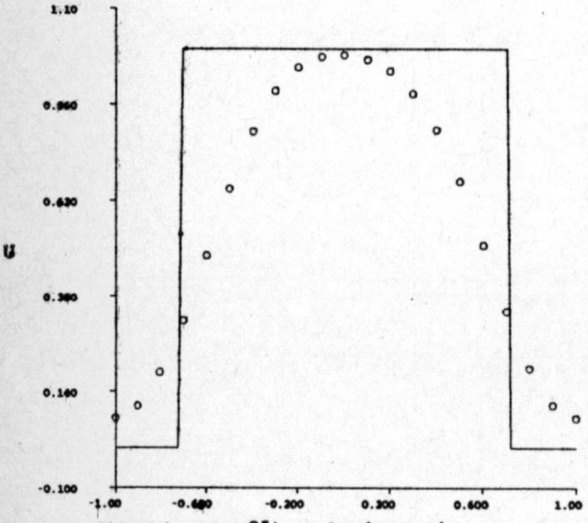

(b) $R(x_i, 0; v^{25})$ and $u(x, 0, 2)$

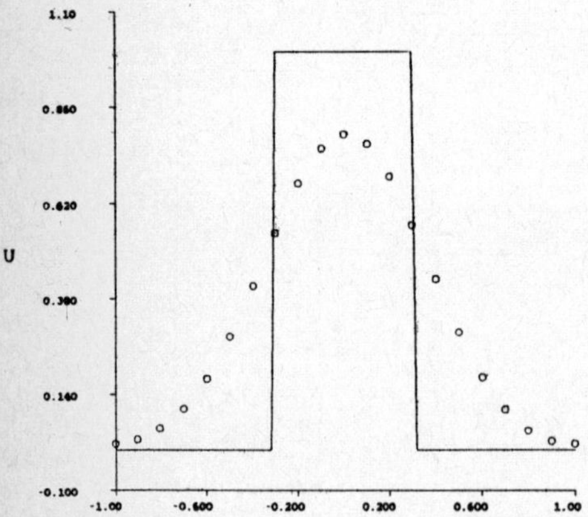

(c) $R(x_i, -0.4, v^{25})$ and $u(x, -0.4, 2)$

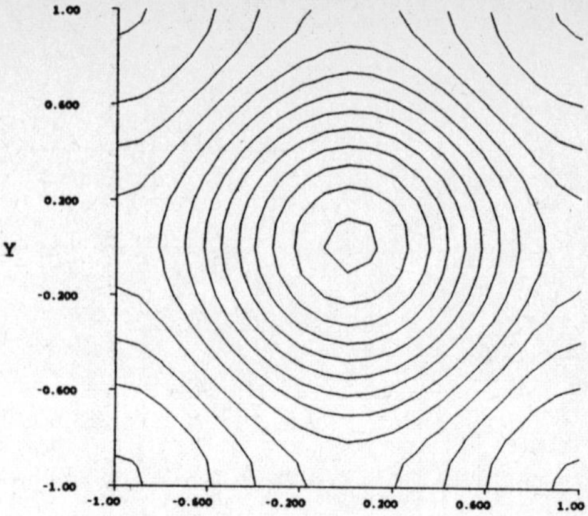

(d) level curves of v^{100} $(t = 8)$

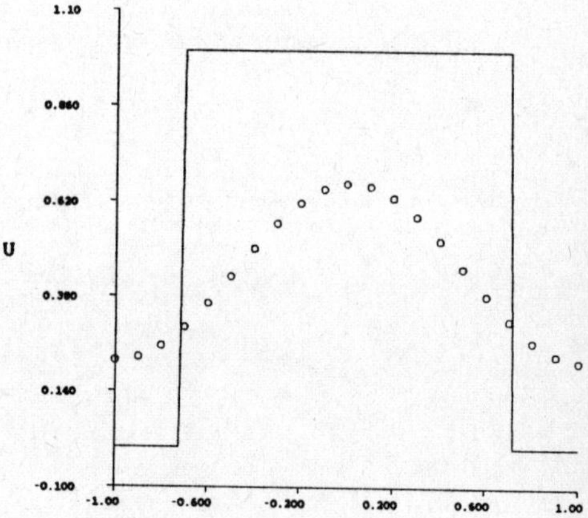

(e) $R(x_i, 0; v^{100})$ and $u(x, 0, 8)$

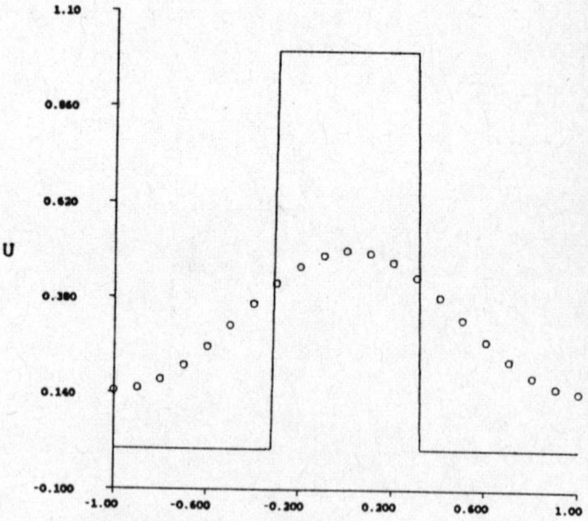

(f) $R(x_i, -0.4; v^{100})$ and $u(x, -0.4, 8)$

Figure 1. First order upwind scheme $(r = 1, m = 0)$

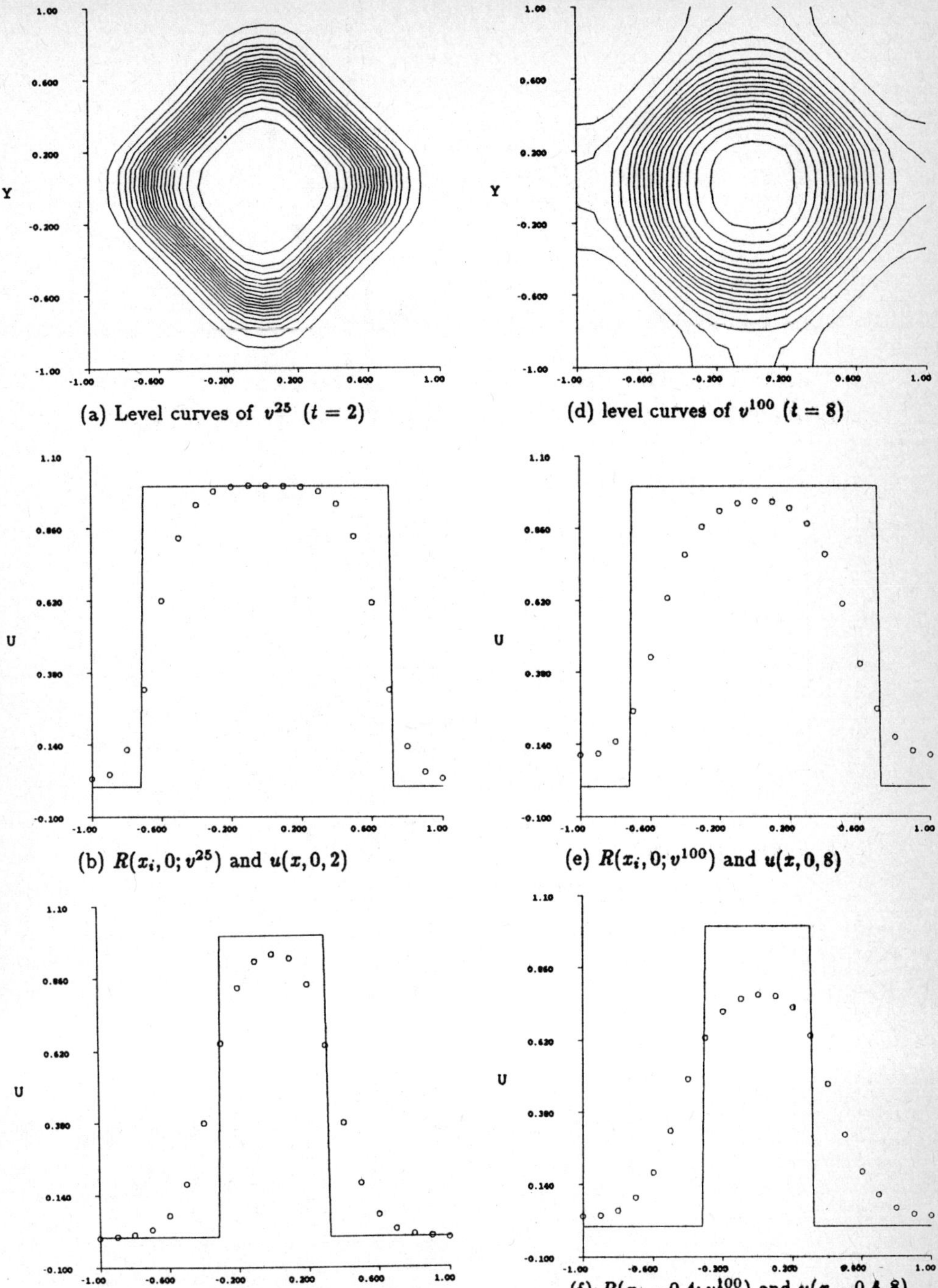

(a) Level curves of v^{25} $(t = 2)$

(d) level curves of v^{100} $(t = 8)$

(b) $R(x_i, 0; v^{25})$ and $u(x, 0, 2)$

(e) $R(x_i, 0; v^{100})$ and $u(x, 0, 8)$

(c) $R(x_i, -0.4, v^{25})$ and $u(x, -0.4, 2)$

(f) $R(x_i, -0.4; v^{100})$ and $u(x, -0.4, 8)$

Figure 2. Second order "TVD" scheme $(r = 2, m = 1)$

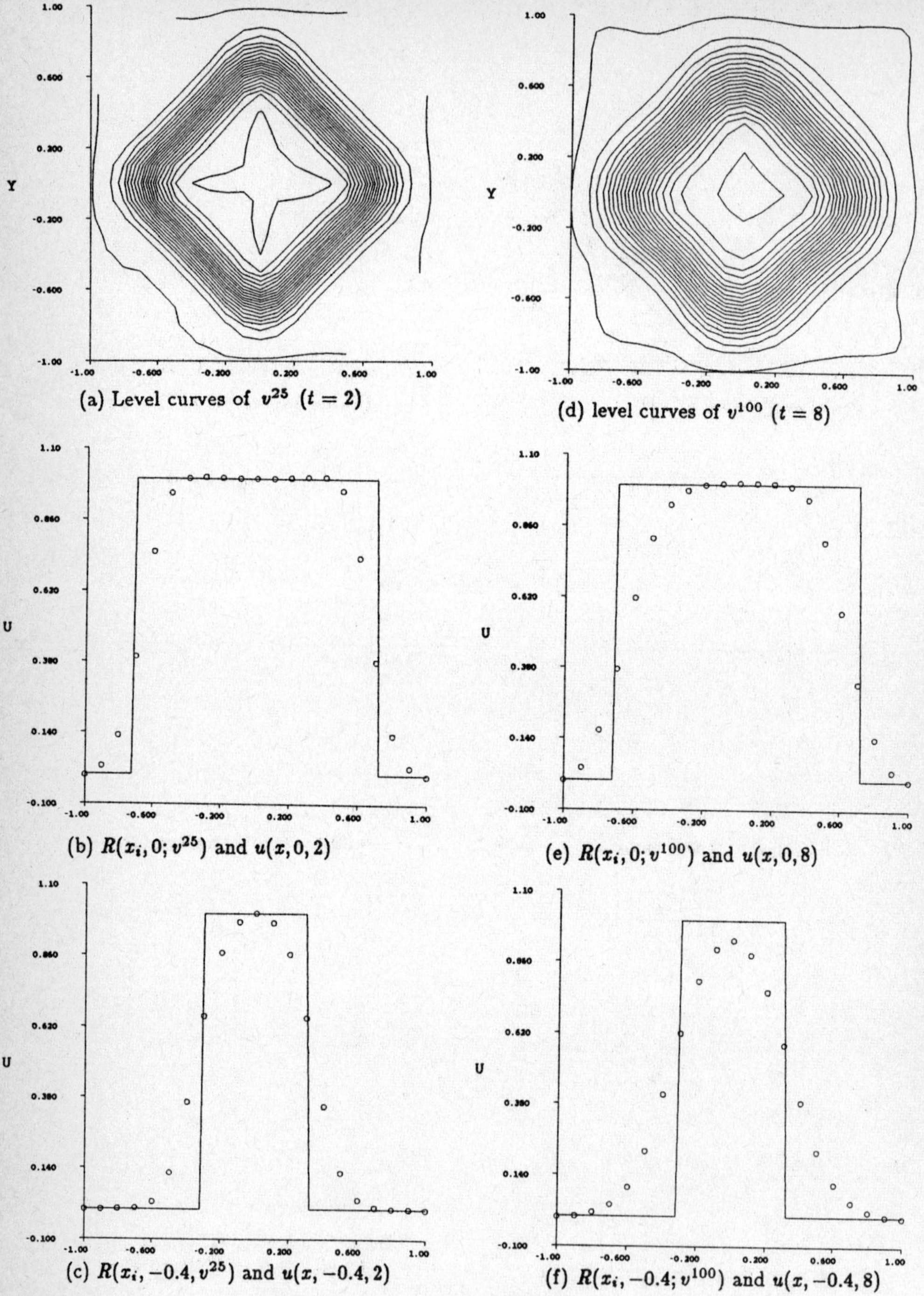

(a) Level curves of v^{25} $(t = 2)$

(d) level curves of v^{100} $(t = 8)$

(b) $R(x_i, 0; v^{25})$ and $u(x, 0, 2)$

(e) $R(x_i, 0; v^{100})$ and $u(x, 0, 8)$

(c) $R(x_i, -0.4, v^{25})$ and $u(x, -0.4, 2)$

(f) $R(x_i, -0.4; v^{100})$ and $u(x, -0.4, 8)$

Figure 3. Second order ENO scheme $(r = 2, m = 2)$

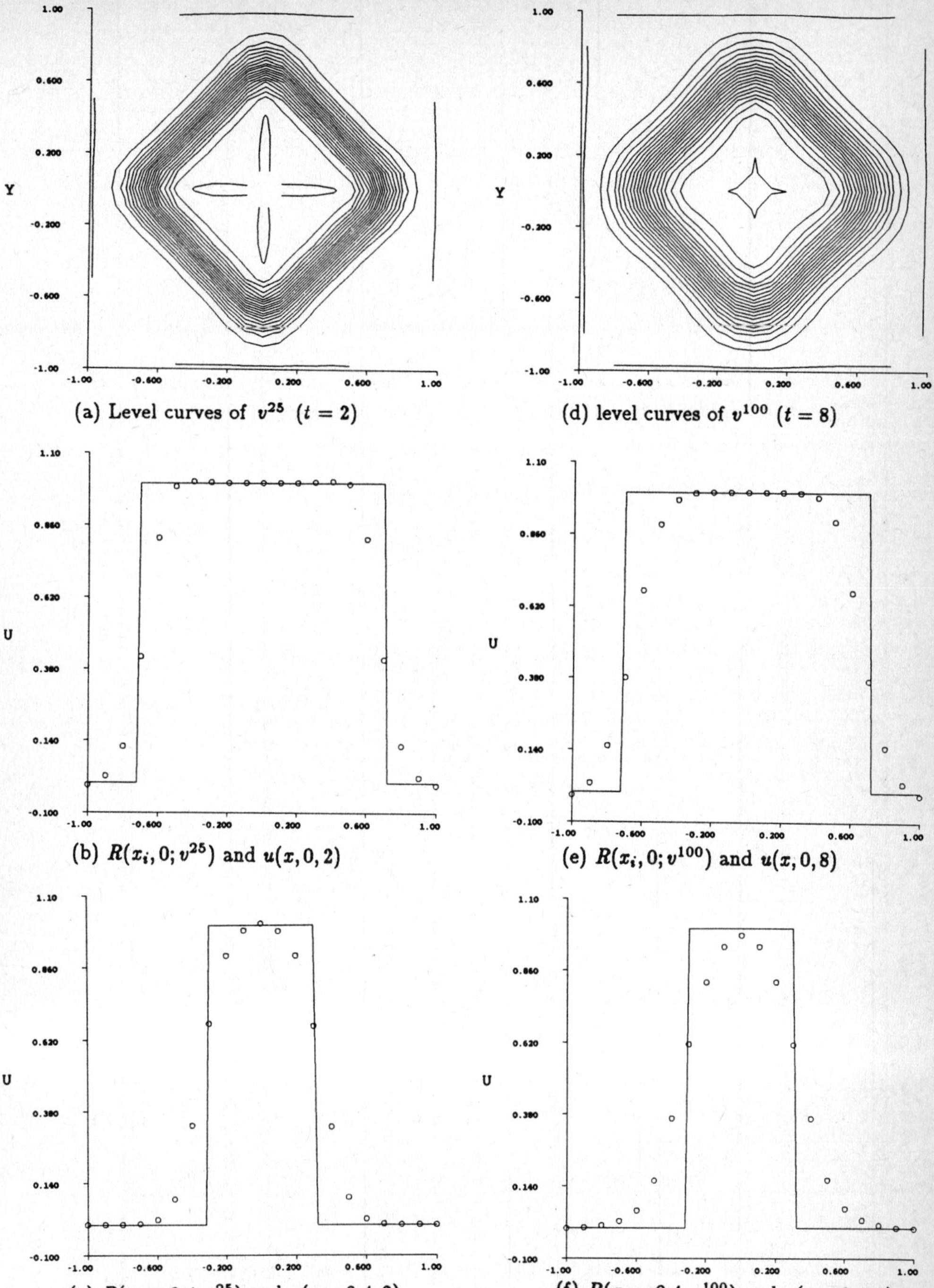

(a) Level curves of v^{25} $(t = 2)$

(d) level curves of v^{100} $(t = 8)$

(b) $R(x_i, 0; v^{25})$ and $u(x, 0, 2)$

(e) $R(x_i, 0; v^{100})$ and $u(x, 0, 8)$

(c) $R(x_i, -0.4, v^{25})$ and $u(x, -0.4, 2)$

(f) $R(x_i, -0.4; v^{100})$ and $u(x, -0.4, 8)$

Figure 4. Third order ENO scheme $(r = 3, m = 3)$

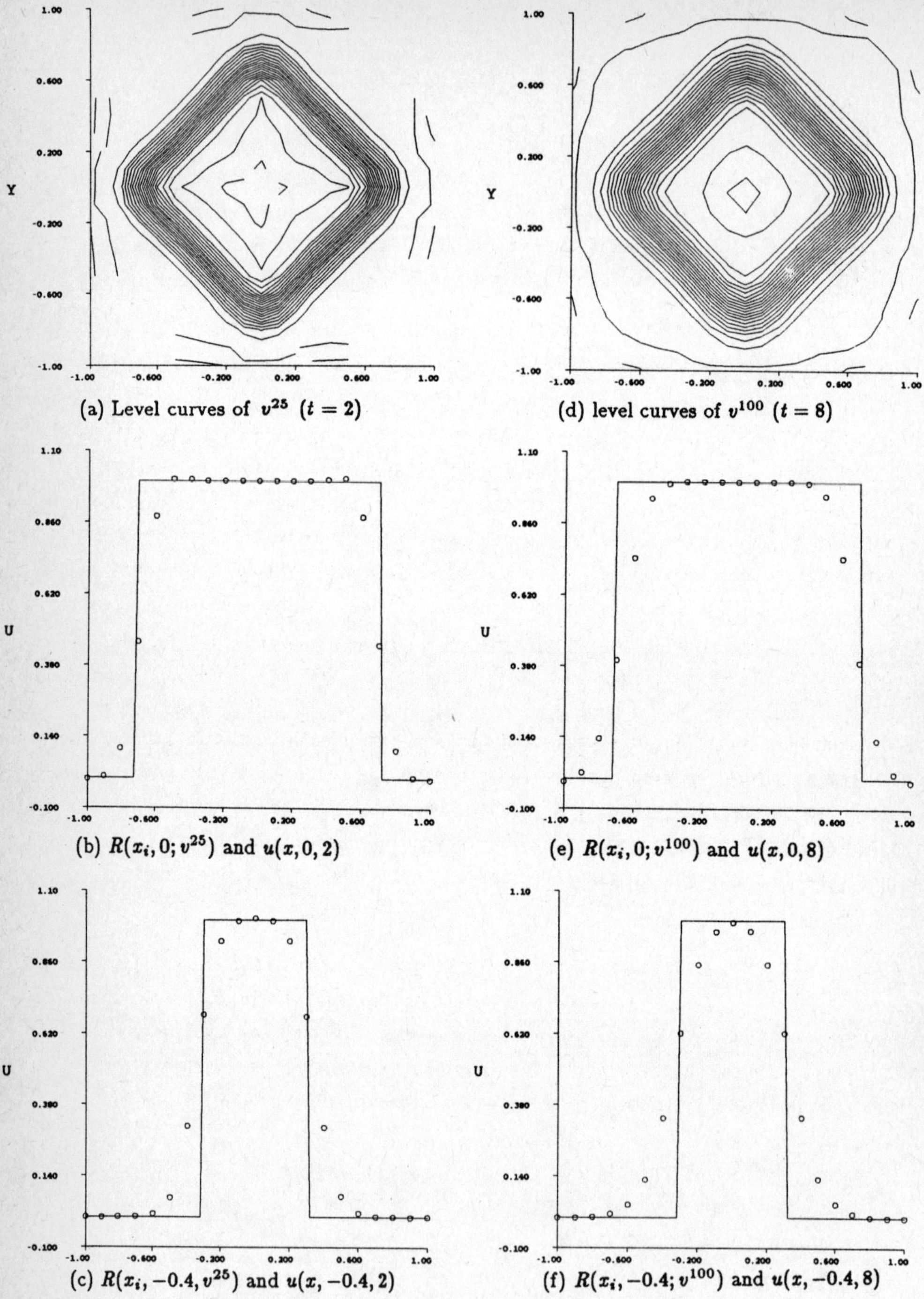

(a) Level curves of v^{25} $(t = 2)$

(d) level curves of v^{100} $(t = 8)$

(b) $R(x_i, 0; v^{25})$ and $u(x, 0, 2)$

(e) $R(x_i, 0; v^{100})$ and $u(x, 0, 8)$

(c) $R(x_i, -0.4, v^{25})$ and $u(x, -0.4, 2)$

(f) $R(x_i, -0.4; v^{100})$ and $u(x, -0.4, 8)$

Figure 5. Fourth order ENO scheme $(r = 4, m = 4)$

UPWIND DIFFERENCING SCHEMES FOR HYPERBOLIC CONSERVATION LAWS WITH SOURCE TERMS

P.L. Roe
College of Aeronautics
Cranfield Institute of Technology
Cranfield, Bedford MK43 OAL, U.K.

1. Introduction

Numerical schemes for hyperbolic conservation laws have advanced rapidly in recent years. Naturally the theoretical investigations are most complete in the simplest case. Homogeneous problems (i.e. without source terms) in one space dimension can now be solved very accurately with cleanly captured discontinuities, and improvements in technique are still being made; see the contributions by Osher and Harten to these proceedings. Nevertheless, most problems of technological interest are either non-homogeneous, or multidimensional, or both, and here the foundations for numerical work are not yet so well-established.

The present work contributes to the study of non-homogeneous problems in one dimension. We particularly have in mind problems governed by the Euler equations with non-vanishing source terms on the right hand side. These could describe the effects on the flow of area variation, chemical reaction, mass or energy release, or interaction with other material, as in a dusty gas. In all cases the equations to be solved are

$$\underset{\sim}{u}_t + \underset{\sim}{F}_x = \underset{\sim}{Q} \qquad (1)$$

where $\underset{\sim}{u}$, $\underset{\sim}{F}$ are the vectors representing respectively the conserved quantities and the fluxes, and $\underset{\sim}{Q} = (\underset{\sim}{u},x)$ is the source term. (The more general $\underset{\sim}{Q}(\underset{\sim}{u},x,t)$ is avoided because explicit dependence on time would be unusual in any real application.) If several materials or phases are simultaneously present, then each material must be allowed its own density, momentum and (less obviously) pressure [1].

In Section 2 some numerical aspects of eqn (1) are discussed in the simplest case, that of a linear scalar equation. The discussion is extended to systems in Section 3, and to non-linear systems in Section 4. High-order schemes are derived in Section 5, and Section 6 demonstrates an example.

2. The Linear Scalar Case

The simplest problem relevant to our interests is the scalar equation

$$u_t + au_x = q(x) \tag{2}$$

where a is a constant, which is assumed positive. If we consider the initial-value problem for (2) with data $u = u_0(x)$, the general solution is

$$u(x,t) = u_0(x - at) + \frac{1}{a}\int_{x-at}^{x} q(x)dx \tag{3}$$

To be slightly more specific, if the solution is sought on a regular grid $x = i\Delta x$, $t = n\Delta t$, with $i \varepsilon Z$, $n \varepsilon N$, we have

$$u(i\Delta x,(n + 1)\Delta t) = u((i - \nu)\Delta x,n\Delta t) + \frac{1}{a}\int_{(i-\nu)\Delta x}^{i\Delta x} q(x)dx \tag{4}$$

where we have introduced the Courant number $\nu = a\Delta t/\Delta x$. The appearance in eqn (4) of the coordinate $(i - \nu)\Delta x$, which is not in general part of the grid, introduces the numerical problem, how such values are to be approximated when ν is not an integer.

Godunov [2] was the first author to show the potency of such simple problems in inspiring the solution to more complex problems, and to indicate the technique by which they may be generated. The best numerical scheme for making the approximations in (4) will depend upon the Courant number ν, and in the scalar case the best approximations (in various senses) are easy to find. In realistic problems, with several unknowns, and therefore several different wave speeds, each wave will have its own Courant number. Therefore we analyse the data to reveal the effect of each system, and treat each such effect in its own optimal manner. Systematic ways to do this were presented by the present author in [3,4].

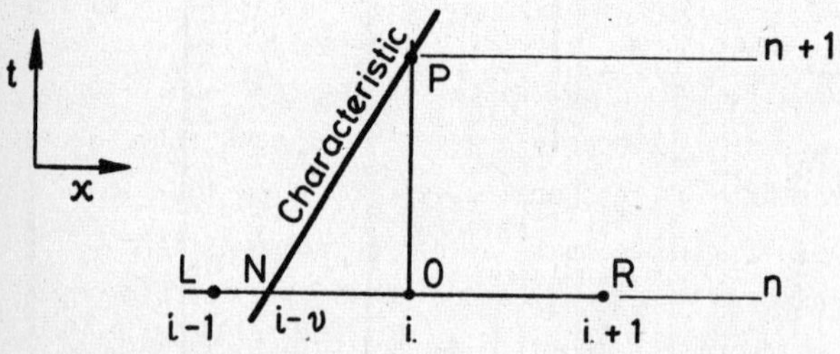

Fig. 1 Illustration of equation (4)

Returning to eqn (4) we see that each term on the RHS exhibits an upwind domain of dependence. Fig. 1 illustrates the case when $0 < \nu < 1$.

We need to interpolate at N, and to integrate between N and O. Strictly speaking no data except that at L and O is physically relevant, since all the required information lies in the interval LO. In pursuit of greater accuracy, it is permissible to use other data, provided some measure of smoothness guarantees that extrapolation will not be misleading. However, for the time being we restrict attention to LO. There is then only one reasonable way to approximate the interpolation term in (4). It must be

$$u((i - \nu)\Delta x, u\Delta t) = (1 - \nu)u_i^n + \nu u_{i+1}^n + O(\Delta x^2) \tag{5}$$

because no other formula is consistent with the p.d.e.. Nevertheless, the formula is a good one, because it has least truncation error amongst the class of all formulae (with wider support) which have positive coefficients depending only on ν [2]. Positivity ensures freedom from spurious oscillations near shockwaves.

The treatment of the integral term presents at least two contrasts. Firstly, even with only two-point support there is a choice of approximations. The formula

$$\frac{1}{a}\int_{(i-\nu)\Delta x}^{i\Delta x} q(x)dx = \Delta t[(1 - \alpha)q_i + \alpha q_{i+1}] + O(\Delta x\Delta t) \tag{6}$$

holds for arbitrary α. For any α between 0 and 1 we have positive coefficients, which seems desirable because in the solution (4) depends monotonically on the values of q.

Secondly, by making the particular choice $\alpha = \frac{1}{2}\nu$, we obtain a second-order accurate formula which still has positive coefficients. Thus positivity does not restrict the integral term to first-order accuracy.

To digress for a moment to schemes with wider support, it is easily shown that if the integral term is represented as

$$\frac{1}{a}\int_{(i-\nu)\Delta x}^{i\Delta x} q(x)dx \simeq \Delta t \sum_k \beta_k q_{i+k}, \tag{7}$$

then the condition for this expression to be n^{th}-order accurate is that

$$\sum_k k^p \beta_k = \frac{(-\nu)^p}{p+1} \tag{8}$$

for $p = 0,1,...n$. It is not easy to see a simple argument which excludes positive solutions of the system (8) for $n > 2$. However the author confesses that he has not been able to find any such solutions even if $n = 3$.

Returning to two-point schemes, the choice of $\alpha = \frac{1}{2}$ in (6) also has a distinguished property. It produces second-order accuracy in the steady state. The scheme becomes

$$u_i^{n+1} = u_i^n - \nu(u_i^n - u_{i-1}^n) + \tfrac{1}{2}\Delta t(q_i + q_{i-1}); \tag{9}$$

hence, if $u_i^{n+1} = u_i^n$, for all i,

$$a(u_i^n - u_{i-1}^n) - \Delta x(q_i + q_{i-1}) = 0 \tag{10}$$

which is a second-order approximation to

$$au_x - q = 0 \tag{11}$$

because it is centrally differenced at $(i - \frac{1}{2})$. This scheme will reappear later.

To obtain second-order accuracy in the transient solution the interpolation term in (4) has to be given more than two-point support. One way to derive a formula is to follow van Leers MUSCL approach [5]. We approximate the data u(x) by a piecewise linear function; that is, within the range $(i - \frac{1}{2})\Delta x < x < (i + \frac{1}{2})\Delta x$, we take

$$u(x) = u_i + \frac{(x - i\Delta x)S_i}{2\Delta x} \tag{12}$$

so that the cell average value is u_i, and the slope (yet to be specified) is S_i. (see Fig. 2). In general u(x) may be discontinuous between cells.

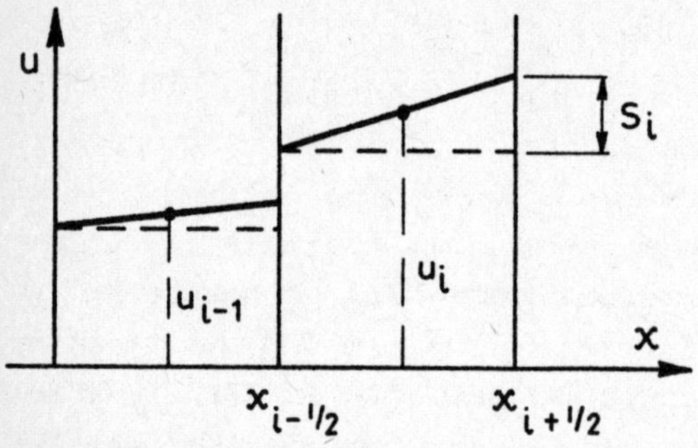

Fig. 2 Piecewise linear data

If we further suppose that q(x) is piecewise constant (which is consistent in that both the convection and production terms are approximated to the same accuracy), we can find the exact solution for $\Delta t < a\Delta x$, from eqn (3)

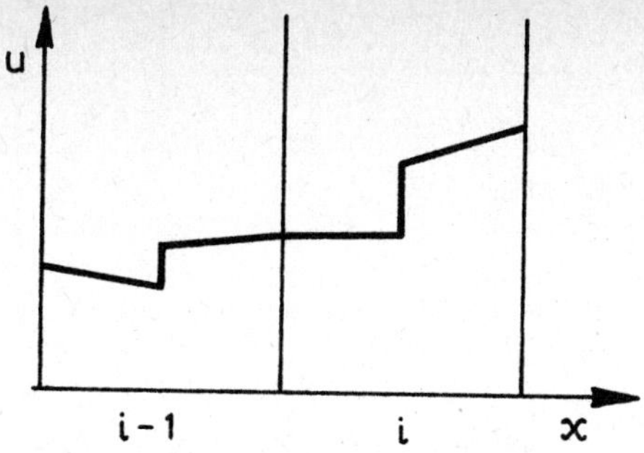

Fig. 3 Exact evolution of data in Fig. 2

What we require is the new cell averages of u, which are found to be

$$u_i^{n+1} = u_i^n - \nu(u_i^n - u_{i-1}^n) + \tfrac{1}{2}\nu(1 - \nu)S_{i-1} - \tfrac{1}{2}\nu(1 - \nu)S_i + [(1 - \tfrac{1}{2}\nu)q_i + \tfrac{1}{2}\nu q_{i-1}]\Delta t \quad (13)$$

It is now the choice of S_i which determines the scheme. The physically illogical choice

$$S_i = (u_{i+1} - u_i)$$

(that is, downwind differencing!) yields second-order accuracy, treating the convective term as in the Lax-Wendroff scheme. We will also get second-order accuracy from any choice

$$S_i = (u_{i+1} - u_i) + O(\Delta x) \quad (14)$$

There is an alternative motivation for choosing the $\{S_i\}$. In considering the full problem (non-linear systems) several authors [6,7,8] have felt the attraction of considering data which is in piecewise equilibrium. That is, the data is projected into a representation such that the steady flow equations are satisfied within each cell. In our simple model equation, that means choosing

$$S_i = \frac{q_i \Delta x}{a} = \frac{q_i \Delta t}{\nu} \quad (15)$$

With that choice, (13) becomes

$$u_i^{n+1} = u_i^n - \nu(u_i^n - u_{i-1}^n) + \tfrac{1}{2}\Delta t(q_{i-1} + q_i) \quad (16)$$

which is the same as (9).

This particular formula could be motivated in yet another way. If we begin by considering pairs of cells, such as (i - 1, i), we can measure the extent to which they are out of equilibrium (with each other now, now internally) by the quantity

$$\phi_{i-\frac{1}{2}} = a(u_i - u_{i-1}) - \tfrac{1}{2}\Delta x(q_{i-1} + q_i) \tag{17}$$

and if this quantity is not zero we may change the value of n in the downwind cell of the pair by $\phi_{i-\frac{1}{2}} \times (\Delta t/\Delta x)$.

3. Linear Systems

All the discussion of the previous secton goes over immediately to the linear system

$$\underset{\sim}{u}_t + A\underset{\sim}{u}_x = \underset{\sim}{Q}(x) \tag{18}$$

by introducing the left and right eigenvectors $\{\underset{\sim}{l}_k\}$, $\{\underset{\sim}{r}_k\}$ of the constant matrix A. Then

$$\underset{\sim}{l}_k \cdot \underset{\sim}{u}_t + \lambda_k \underset{\sim}{l}_i \cdot \underset{\sim}{u}_x = \underset{\sim}{l}_k \cdot \underset{\sim}{Q} \tag{19}$$

where λ_k is the eigenvalue corresponding to $\underset{\sim}{l}_k$, is a scalar equation with source term $\underset{\sim}{l}_k \cdot \underset{\sim}{Q}$, and can be solved by one of the methods in Section 2. Since these depend upon the sign of the wavespeed (here λ_k), the subproblems corresponding to various i may be solved by different methods, and this applies just as much to the source term as to the convection term. If we wish to work in the original variables $\underset{\sim}{u}$ we reconstruct them from the characteristic variables $(\underset{\sim}{l}_k \cdot \underset{\sim}{u})$ by means of the changes in time

$$\delta_t \underset{\sim}{u} = \sum_k \delta_t (\underset{\sim}{l}_k \cdot \underset{\sim}{u}) \underset{\sim}{r}_k \tag{20}$$

where it is assumed that the eigenvectors are normalised in the sense that

$$\underset{\sim}{l}_i \cdot \underset{\sim}{r}_j = \delta_{ij}$$

4. Non-Linear Systems

In a realistic problem, the matrix A in (17) is not a constant. Usually it depends upon u rather than x or t, and this is what we assume here. To be able to apply the analysis of Sections 2 and 3 we assume that A is a piecewise constant matrix, constant in each interval (i,i+1), and depending only on those two states, thus

$$A_{i+\frac{1}{2}} = A_{i+\frac{1}{2}}(\underset{\sim}{u}_i, \underset{\sim}{u}_{i+1}) \tag{21}$$

For the Euler equations of gasdynamics, it is possible to find a matrix $A_{i+\frac{1}{2}}$ having the following properties

$$A_{i+\frac{1}{2}}(\underset{\sim}{u}_{i+1} - \underset{\sim}{u}_i) = \underset{\sim}{F}_{i+1} - \underset{\sim}{F}_i \tag{22}$$

$$\lim_{\underset{\sim}{u}_i \to \underset{\sim}{u}_{i+1}} A_{i+\frac{1}{2}}(\underset{\sim}{u}_i, \underset{\sim}{u}_{i+1}) = A(\underset{\sim}{u}_i) \tag{23}$$

Specific expressions for $A_{i+\frac{1}{2}}$ in the case of an ideal gas can be found in, for example [3,4,9]. An important recent development is that Glaister [10] has extended the analysis to arbitrary equations of state.

Now consider the following analogue of equation (17)

$$\underset{\sim}{\phi}_{i+\frac{1}{2}} = \underset{\sim}{F}_{i+1} - \underset{\sim}{F}_i - \tfrac{1}{2}\Delta x(\underset{\sim}{Q}_i + \underset{\sim}{Q}_{i+1}) \tag{24}$$

Because of (22) this can be written

$$\underset{\sim}{\phi}_{i+\frac{1}{2}} = A_{i+\frac{1}{2}}(\underset{\sim}{u}_{i+1} - \underset{\sim}{u}_i) - \tfrac{1}{2}\Delta x(\underset{\sim}{Q}_i + \underset{\sim}{Q}_{i+1}) \tag{25}$$

so that

$$\underset{\sim}{l}_{k,i+\frac{1}{2}}\cdot \underset{\sim}{\phi}_{i+\frac{1}{2}} = \lambda_{k,i+\frac{1}{2}}\underset{\sim}{l}_{k,i+\frac{1}{2}}\cdot(\underset{\sim}{u}_{i+1} - \underset{\sim}{u}_i) - \tfrac{1}{2}\Delta x\underset{\sim}{l}_{k,i+\frac{1}{2}}\cdot(\underset{\sim}{Q}_i + \underset{\sim}{Q}_{i+1}) \tag{26}$$

For each k, (26) describes a scalar quantity measuring the fluctuation in the solution due to the kth wave. If this is multiplied by $(\Delta t/\Delta x)\underset{\sim}{r}_{k,i+\frac{1}{2}}$ it gives a vector to be subtracted from $\underset{\sim}{u}_i$ if $\lambda_{k,i+\frac{1}{2}} < 0$, or from $\underset{\sim}{u}_{i+1}$ if $\lambda_{k,i+\frac{1}{2}} > 0$. It is important to note that when this switching takes place (i.e. when $\lambda_{k,i+\frac{1}{2}} = 0$) the first (convection) term in (26) vanishes, but the second may not, so that the effect of the source term switches discontinuously. For some physical problems this seems realistic. For example, sonic flow in a nozzle is unstable unless it occurs at a place where the area is locally constant, so that the source term vanishes. Whether it is realistic in all situations is something that remains to be investigated.

It is easy to see that the changes induced by the first term of (26), summed over k, are merely $\frac{\Delta t}{\Delta x}(\underset{\sim}{F}_{i+1} - \underset{\sim}{F}_i)$, making that part of the scheme conservative.

The most familiar example of fluid flow involving source terms is that of compressible flow in a duct of variable area, say S(x). We can include source-free flow with cylindrical or spherical symmetry by the device of setting $S = x$ or x^2 respectively. Then the conserved quantities are $S\underset{\sim}{u}$, where $\underset{\sim}{u}$ are the usual variables $(\rho, \rho u, \rho e)$, and the equations of motion are

$$(S\underset{\sim}{u})_t + (S\underset{\sim}{F})_x = \underset{\sim}{Q}$$

where $\underset{\sim}{Q} = (0, p dS/dx, 0)^T$. Carrying through the algebra gives the following expressions for the fluctuations (26) due to each wave.

$$\phi_1 = \frac{u - a}{2a^2}[\Delta p - \rho a\Delta u] + \frac{\rho u}{2}\frac{\Delta S}{S} \tag{27a}$$

$$\phi_2 = \frac{u}{a^2}[a^2\Delta\rho - \Delta p] \tag{27b}$$

$$\phi_3 = \frac{u + a}{2a^2}[\Delta p + \rho a\Delta u] + \frac{\rho u}{2}\frac{\Delta S}{S} \tag{27c}$$

where $\Delta(.) = (.)_{i+1} - (.)_i$, and the coefficients ρ, u, a are those occuring in the averaged matrix $A_{i+\frac{1}{2}}$ [3,9].

In this particular case the terms arising from the area variation are extremely simple, and add a negligible overhead to the calculation. The reader is warned that a general source may produce additional terms that are rather lengthy.

5. Higher-Order Schemes

Higher-order schemes can always be derived by ensuring that enough time derivatives of the solution are added as corrections. The time derivatives can be related to the space derivatives in a systematic way. Here we will only concern ourselves with second-order effects, so we need to evaluate $\underset{\sim}{u}_{tt}$.

$$\begin{aligned}\underset{\sim}{u}_{tt} &= (\underset{\sim}{Q} - \underset{\sim}{F}_x)_t \\ &= \underset{\sim}{Q}_t - \underset{\sim}{F}_{tx} \\ &= Q_u\underset{\sim}{u}_t + (Au_t)_x \\ &= Q_u(\underset{\sim}{Q} - \underset{\sim}{F}_x) + (A(\underset{\sim}{Q} - \underset{\sim}{F}_x))_x \qquad (28)\end{aligned}$$

where Q_u is the Jacobian of the source term with respect to the conserved variables. Hence the change of $\underset{\sim}{u}$ in one time step is

$$\begin{aligned}\underset{\sim}{u}^{n+1} - \underset{\sim}{u}^n &= \Delta t\underset{\sim}{u}_t + \frac{\Delta t^2}{2}\underset{\sim}{u}_{tt} \\ &= \Delta t(\underset{\sim}{Q} - \underset{\sim}{F}_x) + \frac{\Delta t^2}{2}[Q_u(\underset{\sim}{Q} - \underset{\sim}{F}_x) + (A(\underset{\sim}{Q} - \underset{\sim}{F}_x))_x] \qquad (29)\end{aligned}$$

If the terms in this expression are evaluated straightforwardly, the familiar 'wiggles' associated with classical second-order schemes will of course appear. The familiar remedy in the homogeneous case is of course to progressively discard the second-order terms in non-smooth regions of the flow, reverting toward a first-order scheme known to be robust and free from wiggles. The scheme described in Section 4 seems a good candidate for this. There is, however, a gap in the theoretical argument at this stage. In the homogeneous case, we know that the total variation of the solution (for non-linear scalar problems) or of the Riemann invariants (for linear systems) is a non-increasing function of time, and this result has proved a reliable guide, even for non-linear systems. In the non-homogeneous case, the author knows of no corresponding results sharp enough to be of use in algorithm design. Therefore any limiting of the second-order terms is rather more empirical.

Looking now to see how the second-order scheme may be implemented, let us observe how (29) differs from the homogeneous case. There, the second term would be absent, and in the other terms $(\underset{\sim}{Q} - \underset{\sim}{F}_x)$ would simply read $(-F_x)$. Now the second term does not seem to be a source of trouble. It produces positive coefficients in the linear case. The chief problem with it is that the matrix Q_u could be expensive to evaluate and to multiply by. However, the meaning of the second term is merely that $\underset{\sim}{Q}$ should be evaluated at time $t = (n + \frac{1}{2})\Delta t$. In fact, if we define

$$\underset{\sim}{Q}^* = \underset{\sim}{Q}^n + \tfrac{1}{2}\Delta t Q_u \underset{\sim}{u}_t \tag{30}$$

then (29) can be written

$$\underset{\sim}{u}^{n+1} - \underset{\sim}{u}^n = \Delta t(\underset{\sim}{Q}^* - \underset{\sim}{F}_x) + \frac{\Delta t^2}{2}(A(\underset{\sim}{Q}^* - \underset{\sim}{F}_x))_x \tag{31}$$

to an approximation involving $(\Delta x)^3$.

We can therefore propose the following second-order two-step scheme.

A. Use the scheme of Section 4 to advance by $\frac{1}{2}\Delta t$

B. Re-evaluate $\underset{\sim}{Q}_i^* = \underset{\sim}{Q}(\underset{\sim}{u}_i^{n+\frac{1}{2}})$

C. Using the quantities $(\underset{\sim}{Q}^* - \underset{\sim}{F}_x)_{i+\frac{1}{2}}$ as modified flux differences, advance by Δt using a second-order flux-difference splitting method [4,11,12].

6. Results

Some results have been given by Glaister, using the above ideas [13]. He obtains excellent results for several hard problems in cylindrical and spherical symmetry. Here we show some computations using a test problem recently devised [14] to discriminate between various methods intended for use in combustion problems.

The conditions of the problem are so contrived that an analytical solution is possible. The Euler equations with source term take the form

$$\underset{\sim}{u}_t + \underset{\sim}{F}_x = G(x,\underset{\sim}{u})\underset{\sim}{R} \tag{32}$$

where $\underset{\sim}{R}$ is the vector $(\rho,\rho u,\rho h)^T$. It can be shown that this corresponds to the isentropic addition of mass at a rate proportional to G. The specific choice

$$G(x,\underset{\sim}{u}) = H(-x)(C/a)$$

where H is the Heaviside function, C is a constant with the dimensions of velocity, and a is sound speed, allows the equations to be integrated analytically along the characteristics, so that an exact solution is available.

Fig. 4 (a),(b),(c) show respectively the density, velocity and pressure at three times. At the latest time, the analytical solution has become multi-valued on the far right, so that a shock should be inserted. A singularity is just about to form at x = 0. Very small overshoots can be observed in a few places. It is thought that these may reflect the rather empirical nature of limiting devices for non-homogenous problems already mentioned. One hundred grid points are used to define the flow, and the "superbee" limiter function was employed [4]. More details can be found in [15].

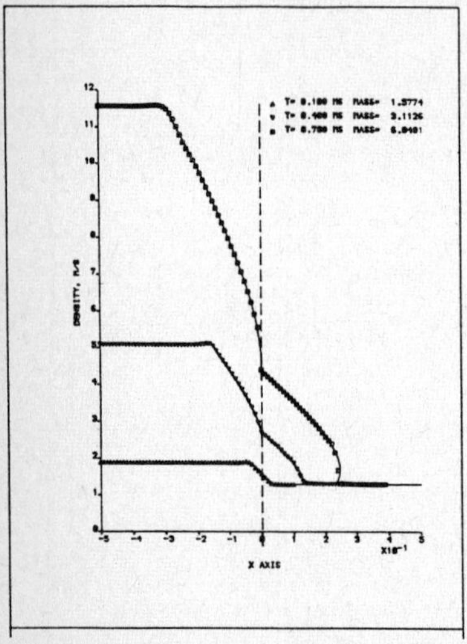

Fig. 4a Density distribution in test problem

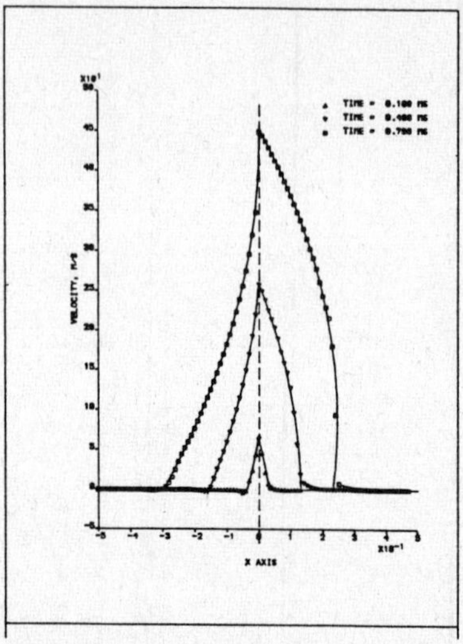

Fig. 4b Velocity distribution in test problem

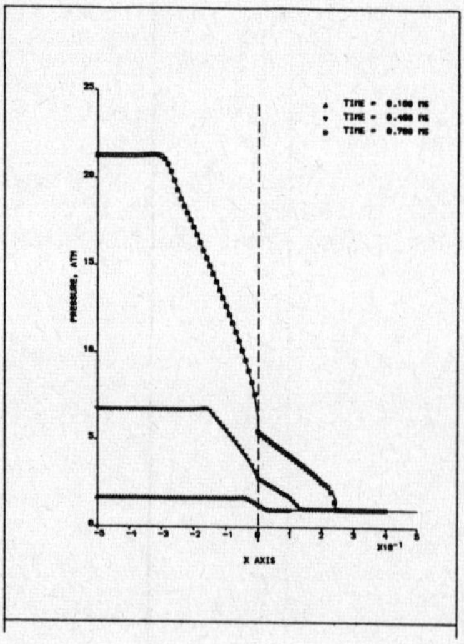

Fig. 4c Pressure distribution in test problem

Acknowledgements

I am very grateful to Mr Bram van Leer for interesting discussions on this topic, especially on the relationship between the present ideas and those of reference 6. Thanks are also due to Dr Tito Toro for helpful comments and for supplying the computations shown in Section 6.

References

1. V.H. Ransom, D.L. Hicks, 1984. Hyperbolic two-pressure models for two-phase flow. J. Comp. Phys. 53; 124-151.

2. S.K. Godunov, 1959. A finite difference method for the numerical computation of discontinous solutions of the equations of gas dynamics. Mat. Sb. 47; 357-393.

3. P.L. Roe, 1980. The use of the Riemann problem in finite-difference schemes. Lect. Notes. Phys. 141; 354-359.

4. P.L Roe, 1985. Some contributions to the modelling of discontinuous flows. In Lectures in Applied Mathematics 22, part 2. Am. Math. Soc; 163-193.

5. B. van Leer, 1979. Towards the ultimate conservative difference scheme, V. A second-order sequel to Godunov's method. J. Comp. Phys. 14; 361-376.

6. B. van Leer, 1981. On the relationship between the upwind differencing schemes of Godunov, Engquist-Osher, and Roe. SIAM J. Sci. Stat. Comput. 5; 1-20.

7. H. Glaz, T-P. Lon, 1984. The asymptotic analysis of wave interactions and numerical calculations of transonic nozzle flow. Adv. Appl. Math. 5; 111-146.

8. J. Glimm, G. Marshall, B. Plohr, 1984. A generalised Riemann problem for quasi-one-dimensional gas flows. Adv. Appl, Math. 5; 1-30.

9. P.L. Roe, J. Pike, 1984. Efficient construction and utilisation of approximate Riemann solutions, in Computing Methods in Applied Science and Engineering VI. Eds R. Glowinski, J-L. Lions, North Holland; 499-518.

10. P. Glaister. An approximate Riemann solver for the Euler equations in one dimension with a general equation of state. Num. Anal. Rept 7/86. University of Reading.

11. A. Harten, 1983. High resolution schemes for hyperbolic conservation laws. J. Comp. Phys. 49; 357-393.

12. P.K. Sweby, 1985. High resolution TVD schemes using flux limiters, in Lectures in Applied Mathematics 22, part 2 Am. Math. Soc; 289-309.

13. P. Glaister, 1985. Flux-difference splitting techniques for the Euler equations in non-Cartesian geometry, Num. Anal. Rept 8/85.

14. J.F. Clarke, E.F. Toro, 1985. Flows generated by burning of solid propellants. Lect. Notes. Phys. 241; 192-205.

15. E.F. Toro, 1986. Roe's method in gas dynamical problems associated with the combustion of high-energy solids in a closed tube. Cranfield Inst. of Tech. Rept CoA No NFP 86/18.

THE ENTROPY DISSIPATION BY NUMERICAL VISCOSITY IN NONLINEAR CONSERVATIVE DIFFERENCE SCHEMES

Eitan Tadmor*
School of Mathematical Sciences, Tel-Aviv University
and
Institute for Computer Applications in Science and Engineering

ABSTRACT

We study the question of entropy stability for discrete approximations to hyperbolic systems of conservation laws. We quantify the amount of numerical viscosity present in such schemes, and relate it to their entropy stability by means of comparison. To this end, two main ingredients are used: the entropy variables and the construction of certain entropy conservative schemes in terms of piecewise-linear finite element approximations. We then show that conservative schemes are entropy stable, if they contain more numerical viscosity than the above mentioned entropy conservative ones.

1. THE ENTROPY VARIABLES

We consider semi-discrete schemes of the form

$$\frac{d}{dt} u_\nu(t) = -\frac{1}{\Delta x_\nu} [f_{\nu+1/2} - f_{\nu-1/2}], \tag{1}$$

which are consistent with the system of conservation laws

$$\frac{\partial}{\partial t} u + \frac{\partial}{\partial x} [f(u)] = 0, \qquad (x,t)\in R\times[0,\infty) \tag{2}$$

Here, $f \equiv f(u) = (f_1,\dots,f_N)^T$ is a smooth flux function of the conservative variables $u \equiv u(x,t) = (u_1,\dots,u_N)^T$, $u_\nu(t)$ denote the discrete solution along the gridline (x_ν,t) with $\Delta x_\nu \equiv \frac{1}{2}(x_{\nu+1} - x_{\nu-1})$ being the variable meshsize,

*Research was supported in part by NASA Contract Nos. NAS1-17070 and NAS1-18107 while the author was in residence at ICASE, NASA Langley Research Center, Hampton, VA 23665. Additional support was provided by NSF Grant No. DMS85-03294 and ARO Grant No. DAAG29-85-K-0190 while in residence at the University of California, Los Angeles, CA 90024. The author is a Bat-Sheva Foundation Fellow.

and $f_{\nu+1/2}$ is the Lipschitz continuous numerical flux consistent with differential one[(1)]

$$f_{\nu+1/2} = f(u_{\nu-p+1},\dots,u_{\nu+p}), \qquad f(u,u,..,u) = f(u). \tag{3}$$

We are concerned here with the entropy stability question of such schemes. To this end, let $(U \equiv U(u),\ F \equiv F(u))$ be an entropy pair associated with the system (2), such that

$$U_u^T f_u = F_u^T, \qquad U_{uu} > 0. \tag{4}$$

We ask whether the scheme (1) is entropy stable w.r.t. such pair, in the sense that it satisfies a discrete entropy inequality of the form

$$\frac{d}{dt} U(u_\nu(t)) + \frac{1}{\Delta x_\nu} [F_{\nu+1/2} - F_{\nu-1/2}] \leq 0, \tag{5}$$

with $F_{\nu+1/2}$ being a consistent numerical entropy flux

$$F_{\nu+1/2} = F(u_{\nu-p+1},\dots,u_{\nu+p}), \qquad F(u,u,\dots,u) = F(u). \tag{6}$$

If, in particular, equality takes place in (5), we say that the scheme (1) is entropy conservative. We note in passing that if it holds for a large enough class of entropy pairs, such a discrete entropy inequality is intimately related to both questions of convergence toward a limit solution as well as this limit solution being the unique physically relevant one, e.g. [1], [2], [3].

Making use of the entropy pair (4), Mock [4] (see also [5]), has suggested the following procedure to symmetrize the system (2).

Define the entropy variables

$$v = v(u) = \frac{\partial U}{\partial u}(u). \tag{7}$$

---

(1) The same notations are used for differential and discrete fluxes; the distinction between the two is by the number of their arguments.

Thanks to the convexity of $U(u)$ the mapping $u \to v$ is one-to-one. Hence, one can make the change of variables $u = u(v)$ which puts the system (2) into its equivalent symmetric form

$$\frac{\partial}{\partial t}[u(v)] + \frac{\partial}{\partial x}[g(v)] = 0, \qquad g(v) \equiv f(u(v)). \tag{8}$$

The system (8) is symmetric in the sense that the Jacobians of its temporal and spatial fluxes are

$$H \equiv H(v) = \frac{\partial}{\partial v}[u(v)] > 0, \qquad B \equiv B(v) = \frac{\partial}{\partial v}[g(v)]. \tag{9}$$

Indeed, if we introduce the so-called potential functions

$$\begin{aligned} \phi &\equiv \phi(v) = v^T u(v) - U(u(v)) \\ \psi &\equiv \psi(v) = v^T g(v) - F(u(v)), \end{aligned} \tag{10}$$

then making use of (4) we find

$$\begin{aligned} u(v) &= \frac{\partial \phi}{\partial v}, \\ g(v) &= \frac{\partial \psi}{\partial v}, \end{aligned} \tag{11}$$

and hence the Jacobians $H(v)$ and $B(v)$ in (9), are the symmetric Hessians of $\phi(v)$ and $\psi(v)$, respectively.

Example 1.1. Consider the Euler equations

$$\frac{\partial}{\partial t}\begin{bmatrix} \rho \\ m \\ E \end{bmatrix} + \frac{\partial}{\partial x}\begin{bmatrix} m \\ \frac{m^2}{\rho} + p \\ \frac{m}{\rho}(E+p) \end{bmatrix} = 0, \qquad p = (\gamma - 1)\cdot[E - \frac{m^2}{2\rho}], \tag{12}$$

asserting the conservation of the density ρ, momentum m and energy E. Harten [6] has noted that this system is equipped with a family of entropy pairs; Godunov [7] and Hughes et al. [8] have studied the canonical choice

$$(U = -\rho S,\ F = -mS), \qquad S = \ln(p\rho^{-\gamma}), \tag{13}$$

which leads to the entropy variables

$$v \equiv \begin{bmatrix} v_1 \\ v_2 \\ v_3 \end{bmatrix} = \frac{1-\gamma}{p} \begin{bmatrix} E + \frac{p}{\gamma-1}(S - \gamma - 1) \\ -m \\ \rho \end{bmatrix} \tag{14}$$

The inverse mapping $v \rightarrow u$ can be found in [8]. We call attention to the fact that the corresponding potential pair in this case is given by $(\phi = (\gamma - 1)\cdot\rho$, $\psi = (\gamma - 1)\cdot m)$, and hence, in view of (11), Euler equations can be rewritten in the intriguing form [10]

$$\frac{\partial}{\partial t}[\mathrm{grad}_v \rho] + \frac{\partial}{\partial x}[\mathrm{grad}_v m] = 0. \tag{15}$$

Returning to our question of entropy stability, the answer provided in [9], [10], consists of two main ingredients: the use of the entropy variables described above, and the comparison with appropriate entropy conservative schemes. To this end we proceed as follows.

We use the entropy variables--rather than the conservative ones--as our primary dependent quantities, by making the change of variables $u_\nu = u(v_\nu)$, e.g., [11], [12]. The scheme (1) is now equivalently expressed as

$$\frac{d}{dt}u_\nu(t) = -\frac{1}{\Delta x_\nu}[g_{\nu+1/2} - g_{\nu-1/2}], \qquad u_\nu = u(v_\nu(t)), \tag{16}$$

with a numerical flux

$$g_{\nu+1/2} = g(v_{\nu-p+1},\dots,v_{\nu+p}), \qquad g(\dots,v,\dots) = f(\dots,u(v),\dots), \tag{17}$$

consistent with the differential one

$$g(v,v,\dots,v) = g(v), \qquad g(v) = f(u(v)). \tag{18}$$

Defining

$$F_{\nu+1/2} = v^T_{\nu+1}g_{\nu+1/2} - \psi(v_{\nu+1}), \tag{19}$$

then multiplication on the left of (16) by v^T_ν gives

$$\frac{d}{dt}U(u_\nu(t)) + \frac{1}{\Delta x_\nu}[F_{\nu+1/2} - F_{\nu-1/2}] = \frac{1}{\Delta x_\nu}[\Delta v^T_{\nu+1/2}\, g_{\nu+1/2} - \Delta\psi_{\nu+1/2}], \tag{20}$$

$$\Delta\psi_{\nu+1/2} \equiv \psi(v_{\nu+1}) - \psi(v_\nu).$$

In view of (10), $F_{\nu+1/2}$ is a consistent numerical entropy flux, and this brings us to (e.g., [13], [9, Theorem 5.2])

Theorem 1.2: The conservative scheme (16) is entropy stable (respectively, entropy conservative), if the following inequality (21a) (respectively, equality (21b)) holds

$$\Delta v^T_{\nu+1/2}\, g_{\nu+1/2} \leq \Delta\psi_{\nu+1/2}, \tag{21a}$$

$$\Delta v^T_{\nu+1/2}\, g_{\nu+1/2} = \Delta\psi_{\nu+1/2}. \tag{21b}$$

2. THE SCALAR PROBLEM

We discuss the entropy stability of scalar conservative schemes, $N = 1$. Defining

$$Q_{\nu+1/2} = \frac{f(u_\nu) + f(u_{\nu+1}) - 2g_{\nu+1/2}}{\Delta v_{\nu+1/2}}, \qquad \Delta v_{\nu+1/2} = v_{\nu+1} - v_\nu \tag{22}$$

then our scheme recast into the more convenient viscosity form

$$\frac{d}{dt} u_\nu(t) = -\frac{1}{2\Delta x_\nu} [f(u_{\nu+1}) - f(u_{\nu+1})] + \frac{1}{2\Delta x_\nu} [Q_{\nu+1/2}\Delta v_{\nu+1/2} - Q_{\nu-1/2}\Delta v_{\nu-1/2}], \tag{23}$$

thus revealing the role of $Q_{\nu+1/2}$ as the numerical viscosity coefficient [14].

According to (21), the scalar entropy conservative schemes are uniquely determined by $g_{\nu+1/2} = g^*_{\nu+1/2}$ where

$$g^*_{\nu+1/2} = \frac{\Delta\psi_{\nu+1/2}}{\Delta v_{\nu+1/2}} = \int_{\xi=0}^{1} g(v_{\nu+1/2}(\xi))d\xi, \quad v_{\nu+1/2}(\xi) = v_\nu + \xi\Delta v_{\nu+1/2}. \tag{24}$$

Writing

$$g^*_{\nu+1/2} = \int_{\xi=0}^{1} \frac{d}{d\xi}(\xi - \tfrac{1}{2})\, g(v_{\nu+1/2}(\xi))d\xi, \tag{25}$$

and integrating by parts, these entropy conservative schemes assume the viscosity form (23), with viscosity coefficient $Q_{\nu+1/2} = Q^*_{\nu+1/2}$, where

$$Q^*_{\nu+1/2} = \int_{\xi=0}^{1} (2\xi - 1)\, g'(v_{\nu+1/2}(\xi))d\xi. \tag{26}$$

We are now ready to characterize entropy stability, by comparison with the above entropy conservative schemes.

Theorem 2.1: <u>The conservative scheme (23) is entropy stable, if it contains more viscosity than the entropy conservative one (26), i.e.,</u>

$$Q^*_{\nu+1/2} \le Q_{\nu+1/2}. \tag{27}$$

<u>Proof:</u> Multiplying by $\Delta v_{\nu+1/2}$ on both sides, the inequality (27) reads

$$\Delta v_{\nu+1/2} Q^*_{\nu+1/2} \Delta v_{\nu+1/2} \le \Delta v_{\nu+1/2} Q_{\nu+1/2} \Delta v_{\nu+1/2}, \tag{28}$$

or equivalently, consult (22),

$$\Delta v_{\nu+1/2} [f(u_\nu) + f(u_{\nu+1}) - 2g^*_{\nu+1/2}] \le \Delta v_{\nu+1/2} [f(u_\nu) + f(u_{\nu+1}) - 2g_{\nu+1/2}]. \tag{29}$$

In view of (24), this yields

$$\Delta v_{\nu+1/2}\, g_{\nu+1/2} \le \Delta v_{\nu+1/2}\, g^*_{\nu+1/2} \equiv \Delta\psi_{\nu+1/2}, \tag{30}$$

and entropy stability follows by Theorem 1.1.

We note in passing that the entropy conservative viscosity, $Q^*_{\nu+1/2}$, is of order $O(|\Delta v_{\nu+1/2}|)$, and consequently the entropy conservative schemes (24) are second order accurate. Indeed, change of variables $\xi \to 1-\xi$ in (26) yields

$$Q^*_{\nu+1/2} = -\int_{\xi=0}^{1} (2\xi-1)\, g'(v_{\nu+1/2}(1-\xi))d\xi, \tag{31}$$

and by averaging of (26) and (31) we find

$$\begin{aligned} Q^*_{\nu+1/2} &= \int_{\xi=0}^{1} (\xi-1/2)\cdot \int_{\eta=0}^{1} \frac{d}{d\eta} g'[\eta v_{\nu+1/2}(\xi) + (1-\eta)\, v_{\nu+1/2}(1-\xi)]d\eta d\xi = \\ &= 2\int_{\xi=0}^{1} (\xi-1/2)^2 \cdot \int_{\eta=0}^{1} g''(v_{\nu+1/2}[\xi\eta + (1-\xi)(1-\eta)])d\eta d\xi \cdot \Delta v_{\nu+1/2}, \end{aligned} \tag{32}$$

as asserted. Second order accuracy then follows in view of

Lemma 2.2: Consider the conservative schemes (25) with viscosity coefficient, $Q_{\nu+1/2}$, such that $\dfrac{Q_{\nu+1/2}}{\Delta v_{\nu+1/2}}$ is Lipschitz continuous. Then these schemes are second-order accurate, in the sense that their truncation is of the order

$$O[|x_{\nu+1} - x_\nu|^2 + |x_\nu - x_{\nu-1}|^2 + |x_{\nu+1} - 2x_\nu + x_{\nu-1}|]. \tag{33}$$

The proof of Lemma 2.2 is straightforward and therefore omitted.

Theorem 2.1 together with Lemma 2.2 enable us to verify the entropy stability of first- as well as second-order accurate schemes. We consider a couple of examples.

Example 2.3: Using the simple upper bound, see (32),

$$Q^*_{\nu+1/2} \leq \frac{1}{6} \operatorname*{Max}_{v} |g''(v)| \cdot |\Delta v_{\nu+1/2}|, \tag{34}$$

we obtain, on the right of (34), a viscosity coefficient which according to Theorem 2.1 and Lemma 2.2 maintains both, entropy stability and second-order accuracy. Similar viscosity terms were previously derived in a number of special cases, e.g., [15], [16], [17]. We remark that the careful lengthy calculations required in those derivations is due to the delicate balance between the cubic order of entropy loss and the third-order dissipation in this case.

Example 2.4: Consider the genuinely nonlinear case where $f(u)$ is, say, convex. The quadratic entropy function, $U(u) = \frac{1}{2} u^2$, leads to entropy variables which coincide with the conservative ones, $g(v) = f(u)$. Thus, according to (32), viscosity is required only at rarefactions where $\Delta u_{\nu+1/2} > 0$, since $\operatorname{sign}(Q^*_{\nu+1/2}) = \operatorname{sign}(\Delta u_{\nu+1/2}) < 0$ otherwise. A simple second-order accurate entropy stable flux of this kind is given by

$$f_{\nu+1/2} = \begin{bmatrix} f(\frac{1}{2}(u_\nu + u_{\nu+1})) & \Delta u_{\nu+1/2} > 0 \\ \\ \frac{1}{2}(f(u_\nu) + f(u_{\nu+1})) & \Delta u_{\nu+1/2} < 0 \end{bmatrix}. \tag{35}$$

3. SYSTEMS OF CONSERVATION LAWS

We generalize the construction of the scalar entropy conservative schemes (24), to systems of conservation laws, using piecewise linear finite-elements.

To this end, consider the weak formulation of (8)

$$\int_\Omega w^T \frac{\partial}{\partial t}[u(v)]dxdt = \int_\Omega \frac{\partial w^T}{\partial x} g(v)dxdt. \tag{36}$$

Let the trial solutions $w \to \hat{w}(x,t) = \sum_k w_k(t)\hat{H}_k(x)$ be chosen out of the typical finite-element set spanned by the C^0 "hat functions"

$$\hat{H}_k(x) = \begin{bmatrix} \dfrac{x - x_{k-1}}{x_k - x_{k-1}} & x_{k-1} \le x \le x_k \\ \\ \dfrac{x_{k+1} - x}{x_{k+1} - x_k} & x_k \le x \le x_{k+1} \end{bmatrix}. \tag{37}$$

The spatial part on the right of (36) yields--after change of variables,

$$\begin{aligned} &\int_{x_{\nu-1}}^{x_{\nu+1}} \frac{\partial \hat{H}_\nu(x)}{\partial x} g[\hat{v}(x,t) = \sum_{\nu-1}^{\nu+1} v_k(t)\hat{H}_k(x)]dx = \\ &= -[\int_{\xi=0}^{1} g(v_{\nu+1/2}(\xi))d\xi - \int_{\xi=0}^{1} g(v_{\nu-1/2}(\xi))d\xi], \quad v_{\nu+1/2}(\xi) = v_\nu + \xi\Delta v_{\nu+1/2}. \end{aligned} \tag{38}$$

A second-order mass lumping on the left of (36) leads to

$$\int_{x_{\nu-1}}^{x_{\nu+1}} \hat{H}_\nu(x) \frac{\partial}{\partial t} u[\hat{v}(x,t) = \sum_{\nu-1}^{\nu+1} v_k(t)H_k(x)] = \Delta x_\nu \frac{\partial}{\partial t}[u(v_\nu(t))] + O(|\Delta v_{\nu+1/2}|^2) \tag{39}$$

Equating (38) and (39) while neglecting the quadratic error terms, we end up with, compare (24),

$$\frac{\partial}{\partial t} u_\nu(t) = -\frac{1}{\Delta x_\nu}[g^*_{\nu+1/2} - g^*_{\nu-1/2}], \quad g^*_{\nu+1/2} = \int_{\xi=0}^{1} g(v_{\nu+1/2}(\xi))d\xi. \tag{40}$$

The resulting scheme (40) is entropy conservative since, consult (11),

$$\Delta v^T_{\nu+1/2} g^*_{\nu+1/2} = \int_{\xi=0}^{1} \Delta v_{\nu+1/2} g(v_\nu + \xi\Delta v_{\nu+1/2})d\xi = \int_{v_\nu}^{v_{\nu+1}} dv^T g(v) = \Delta\psi_{\nu+1/2}, \tag{41}$$

in agreement with Theorem 1.2. Integration by parts along the lines which follow (25) shows that the entropy conservative schemes (40) can be also rewritten in the viscosity form

$$\frac{d}{dt} u_\nu(t) = \frac{-1}{2\Delta x_\nu} [f(u_{\nu+1}) - f(u_{\nu-1})] + \frac{1}{2\Delta x_\nu} [Q_{\nu+1/2} \Delta v_{\nu+1/2} - Q_{\nu-1/2} \Delta v_{\nu-1/2}] \quad (42)$$

with a numerical viscosity coefficent matrix, $Q_{\nu+1/2} = Q^*_{\nu+1/2}$, where

$$Q^*_{\nu+1/2} = \int_{\xi=0}^{1} (2\xi - 1)\cdot B(v_{\nu+1/2}(\xi))d\xi, \qquad B(v) \equiv \frac{\partial}{\partial v} [g(v)]. \quad (43)$$

Once the entropy conservative schemes were identified for systems of conservation laws, we can repeat our previous arguments concerning the entropy stability of (42); in analogy with Theorem 2.1, we now have

Theorem 3.1: The conservative scheme (42) is entropy stable, if it contains more viscosity than the entropy conservative one (40), i.e.,

$$\Delta v^T_{\nu+1/2} Q^*_{\nu+1/2} \Delta v_{\nu+1/2} \leq \Delta v^T_{\nu+1/2} Q_{\nu+1/2} \Delta v_{\nu+1/2}. \quad (44)$$

If, in particular, $Q_{\nu+1/2}$ is symmetric, then a sufficient entropy stability criterion is

$$Q^*_{\nu+1/2} \leq Q_{\nu+1/2}, \quad (45)$$

where the inequality is understood in the usual sense of order among symmetric matrices.

Example 3.2: Consider the conservative scheme (42) with a numerical viscosity coefficient given by

$$Q_{\nu+1/2} = \int_{\xi=0}^{1} |B(v_{\nu+1/2}(\xi)| d\xi. \quad (46)$$

Here the absolute value of a symmetric matrix is evaluated in the usual fashion from its spectral representation, $B = U^*\Lambda U$,

$$|B(v_{\nu+1/2}(\xi))| = U^*(v_{\nu+1/2}(\xi))|\Lambda(v_{\nu+1/2}(\xi))|U(v_{\nu+1/2}(\xi)). \tag{47}$$

Since

$$(2\xi-1)\Lambda(v_{\nu+1/2}(\xi)) \leq |\Lambda(v_{\nu+1/2}(\xi))|, \tag{48}$$

(45) holds and entropy stability follows.

Away from sonic points, (46) amounts to the usual upwind differencing, e.g., [18]. In the neighborhood of such sonic points, however, an exact evaluation of $Q_{\nu+1/2}$ in (46) may turn out to be a difficult task. Yet, in view of Theorem 3.1, one can use instead simpler upper bounds. In [10] this is achieved using the construction of a whole family of entropy conservative schemes which take into account the characteristic directions associated with the system (2).

REFERENCES

1. DiPerna, R. J., "Convergence of Approximate Solutions to Conservation Laws," Arch. Rational Mech. Anal., Vol. 82, 1983, pp. 27-70.

2. Lax, P. D., Hyperbolic Systems of Conservation Laws and the Mathematical Theory of Shock Waves, SIAM Regional Conference Lectures in Applied Mathematics, No. 11, 1972.

3. Harten, A., Hyman, J. M., and Lax, P. D., "On Finite Difference Approximations and Entropy Conditions for Shocks," Comm. Pure Appl. Math., Vol. 29, 1976, pp. 297-322.

4. Mock, M. S., "Systems of Conservation Laws of Mixed Type," J. Differential Equations, Vol. 37, 1980, pp. 70-88.

5. Harten, A., and Lax, P. D., "A Random Choice Finite-Difference Schemes for Hyperbolic Conservation Laws," SINUM, Vol. 18, 1981, pp. 289-315.

6. Harten, A., "On the Symmetric Form of Systems of Conservation Laws with Entropy," J. Comp. Physics, Vol. 49, 1983, pp. 151-164.

7. Godunov, S. K., "The Problem of a Generalized Solution in the Theory of Quasilinear Equations and in Gas Dynamics," Russian Math. Surveys, Vol. 17, 1962, pp. 145-156.

8. Hughes, T. J. R., Franca, L. P., and Mallet, M., "Symmetric Forms of the Compressible Euler and Navier-Stokes Equations and the Second Law of Thermodynamics," Applied Mechanics and Engineering, in press.

9. Tadmor, E., "The Numerical Viscosity of Entropy Stable Schemes for Systems of Conservation Laws. I.," ICASE Report No. 85-51, 1985, NASA Langley Research Center, Hampton, VA.

10. Tadmor, E., "The Numerical Viscosity of Entropy Stable Schemes for Systems of Conservation Laws. II.," ICASE Report, to appear.

11. Tadmor, E., "Skew-Selfadjoint Form for Systems of Conservation Laws," J. Math. Anal. Appl., Vol. 103, 1984, pp. 428-442.

12. Hughes, T. J. R., Mallet, M., and Franca, L. P., "Entropy-Stable Finite Element Methods for Compressible Fluids; Applications to High Mach Number Flows with Shocks," Finite Element Methods for Nonlinear Problems, Springer-Verlag, to appear.

13. Osher, S., "Riemann Solvers, the Entropy Condition and Difference Approximations," SINUM, Vol. 21, 1984, pp. 217-235.

14. Tadmor, E., "Numerical Viscosity and the Entropy Condition for Conservative Difference Schemes," Math. Comp., Vol. 43, 1984, pp. 369-382.

15. Majda, A., and Osher, S., "Numerical Viscosity and the Entropy Condition," Comm. Pure Appl. Math., Vol. 32, 1979, pp. 797-838.

16. Osher, S., "Convergence of Generalized MUSCL Schemes," SINUM, Vol. 22, 1984, pp. 947-961.

17. Osher, S., and Tadmor, E., "On the Convergence of Difference Approximations to Scalar Conservation Laws," ICASE Report No. 85-28, 1985, NASA Langley Research Center, Hampton, VA.

18. Roe, P. L., "Approximate Riemann Solvers, Parameter Vectors, and Difference Schemes," J. Comp. Physics, Vol. 43, 1981, pp. 357-372.

2-D AND 3-D EULER COMPUTATIONS WITH FINITE ELEMENT METHODS IN AERODYNAMICS

BILLEY V.*, PERIAUX J.*, PERRIER P.*, STOUFFLET B.*

I. INTRODUCTION

Since the prediction on supercomputers of supersonic and hypersonic inviscid 3-D flows has become a challenging and determinant target in Aerospace Engineering, a great deal of work has been done already in the scientific community for the numerical solution of the Euler equations. [01], [05], [07], [11], [14], [15], [18], [24], [25]...

The main goal of this paper is to show how the general finite element methods discussed in [11] and [05] and using unstructured meshes can be applied to the solution of complicated non linear problems of industrial interest in two and three dimensions, namely the numerical simulation of Euler flows around/inside inlets and also around complete space shuttle configurations.

We recall in Section 2 the governing equations with the appropriate boundary conditions and also the general definitions of discrete spaces to be used in the sequel. Section 3 and 4 are devoted to spatial approximations : a first and second order accurate upwinded scheme based on a approximated Riemann solver of Osher type and also linear hermitian or directional interpolation is described in Section 3 while a second order accurate centered scheme using a Galerkin variational formulation plus an artificial viscosity model is discussed in Section 4.

Efficiency and robustness of the solution obtained from delta schemes using block relaxation and linearized first order preconditionner as left hand side and the above second order approximations as right hand side are discussed in Section 5.

In order to improve the solution where accuracy is needed-discontinuities or stagnation lines -self adaptive mesh refinement techniques using physical criteria are introduced in Section 6. [05], [18], [20], [28],...

Finally in Section 7, various numerical experiments illustrate the possibility of the two methods and further comments on alternative adaptative full multigrid methods for the Euler equations as discussed in [06], [08], [21], [23] are suggested.

I.1. Mathematical modelling

Let us consider a compressible inviscid fluid. If Ω and Γ denote the region of the flow ($\Omega \subset R^N$, N = 2,3 in practice) and its boundary, respectively, the non dimensional conservative law form of the Euler equations modelling this flow is

*Avions Marcel Dassault-Bréguet Aviation, 78 quai Carnot 92214 Saint Cloud (France)

given below :

$$\frac{\partial W}{\partial t} + \frac{\partial}{\partial x} F_1(W) + \frac{\partial}{\partial y} F_2(W) + \frac{\partial}{\partial z} F_3(W) = 0 \qquad (1.1)$$

$$\text{with } W = \begin{pmatrix} \rho u \\ \rho v \\ \rho w \\ e \end{pmatrix} = (W^{(k)})_{k=1,m} \qquad (1.2)$$

$$\text{and } F_1(W) = \begin{pmatrix} \rho u \\ \rho u^2+p \\ \rho uv \\ \rho uw \\ u(e+p) \end{pmatrix}, \ F_2(W) = \begin{pmatrix} \rho v \\ \rho uv \\ \rho v^2+p \\ \rho vw \\ v(e+p) \end{pmatrix}, \ F_3(W) = \begin{pmatrix} \rho w \\ \rho uw \\ \rho vw \\ \rho w^2+p \\ w(e+p) \end{pmatrix}, \qquad (1.3)$$

where ρ is the density, $\underset{\sim}{U} = (u,v,w)$ is the velocity vector, e is the total specific energy and p is the pressure. We suppose that the fluid satisfies the perfect gas law

$$p = (\gamma-1)\ (e - \frac{1}{2} \rho|\underset{\sim}{U}|^2). \qquad (1.4)$$

Using a symbolic notation for the non linear vectors $(F_i)_{i=1,3}$ in (1.3),

$$\underset{\sim}{F}(W) = \begin{pmatrix} F_1(W) \\ F_2(W) \\ F_3(W) \end{pmatrix}, \qquad (1.5)$$

then the conservative system (1.1) becomes :

$$\frac{\partial W}{\partial t} + \underset{\sim}{\nabla}.\underset{\sim}{F}(W) = 0 \qquad (1.6)$$

I.2. Boundary conditions

In the sequel, we consider domains of computation occuped by external/internal Euler flows around airfoils and inside inlets ; for exemple in the case of the airfoil A of figure 1.1

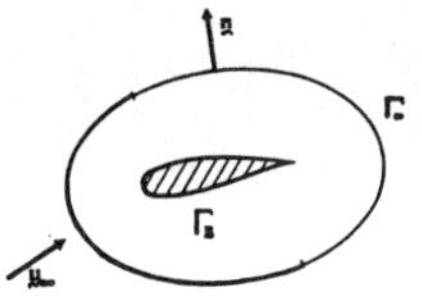

Figure 1.1

the following boundary conditions have to be added.

We assume the flow to be uniform at infinity and the variables to be non-dimensionalized by the free-stream vector W_∞, and we prescribe :

$$\rho_\infty = 1 \qquad (1.7)$$

$$\underset{\sim}{U}_\infty = \begin{pmatrix} \cos\alpha \cos\beta \\ \sin\beta \\ \sin\alpha \cos\beta \end{pmatrix} \qquad (1.8)$$

where α is the angle of attack and β the yaw angle

$$p_\infty = \frac{1}{\gamma M_\infty^2} \qquad (1.9)$$

where M_∞ denotes the freestream Mach number.

On the wall Γ_B, we use the following condition :

$\underset{\sim}{U}.\underset{\sim}{n} = 0$, slip condition. (1.10)

Finally, for the time dependent problem (equation (1.1)-(1.5)) an initial condition such as

$\underset{\sim}{W}(x,0) = \underset{\sim}{W}_0(x)$ on Ω, (1.11)

is prescribed.

II. DEFINITIONS

We assume that Ω is a polygonal bounded domain of $\mathbb{R}^3$. Let $\mathcal{T}_h$ a standard tetrahedrization of Ω and h the maximal length of the edges of the tretrahedra of $\mathcal{T}_h$. We need to introduce the following notations as in [05] :

For every vertex S_i $(i=1,\ldots,n_h)$ of $\mathcal{T}_h$, we define a cell $\hat{S}_i$ as follows (the approach is different from [16]) :

- every tetrahedron having S_i as a vertex is subdivided in 24 subtetrahedra by means of the median planes.
- the cell $\hat{S}_i$ is the union of the resulting subtetrahedra having S_i as a vertex. The boundary of $\hat{S}_i$ is denoted by $\partial\hat{S}_i$.
- K(i) is the set of numerous of nodes neighbour of S_i.
- $\underset{\sim}{\nu}_i = (\nu_{ix},\nu_{iy},\nu_{iz})$ is the unit vector of the outward normal to $\partial\hat{S}_i$.

We introduce the following discrete spaces (with P_k = space of the polynomials of degree $\le k$) :

$$\mathcal{V}_h = \{v_h : \mathbb{R}^3 \to \mathbb{R}^5,\ \ v_{h|\hat{S}_i} = v_{h_i} = \text{constant},\ i = 1,\ldots,n_h\} \tag{2.1}$$

$$S_h = \{\phi \in C^0(\bar{\Omega}),\ \ \phi_{h|T} \in P_1,\ \ \forall T \in \mathcal{T}_h\} \tag{2.2}$$

$$\mathcal{W}_h = \{v_h \in (C^0(\bar{\Omega}))^5\ ,\ v_{h|T} \in (P_1)^5\ ,\ \forall T \in \mathcal{T}_h\} \tag{2.3}$$

The P_1 - interpolation operator will be denoted π_h.
A base of W_h is defined by the set of functions $\{\phi_i\}_{i=1,\ldots,n_h}$ of W_h such that $\phi_i(S_j) = \delta_{ij}$, $j = 1,\ldots,n_h$.
Finally, a mass-lumping operator Σ_0 is defined by

$$\begin{cases} \forall\ v_h \in W_h\ ,\ \Sigma_0\ v_h \in V_h \text{ and} \\ \Sigma_0\ v_{h|\hat{S}_i} = v_h\ (S_i)\quad ,\quad \forall S_i \text{ node of } \mathcal{T}_h \end{cases} \tag{2.4}$$

For every domain $\mathcal{O} \subset \Omega$, we recall that $\chi(\mathcal{O})$ is the caracteristic function of $\mathcal{O}$.

III. UPWIND SECOND ORDER ACCURATE APPROXIMATIONS

III.1. First-order approximations

We consider the Cauchy problem in one dimension

$$\frac{\partial W}{\partial t} + \frac{\partial F(W)}{\partial x} = 0 \quad , \quad W \in \mathbb{R}^m \quad , \tag{3.1}$$

where the flux function $F \in C^1(\mathbb{R}^m)$. The Jacobian matrix $\frac{\partial F(W)}{\partial W}$ is denoted $A(W)$. Whe know that any conservative Finite Difference three-points scheme is characterized by its numerical flux function $\phi_F(U,V)$ which verifies the consistency condition [13] :

$$\phi_F(U,U) = F(U). \tag{3.2}$$

The tridimensional extension of the class of three-points upwind schemes is made as follows [4].

Using the Green formula, the Finite-Volume formulation of (1.1) with an approximation $W_h \in \mathcal{V}_h$ of W can be written

$$\mathrm{Vol}(\hat{S}_i) \; \frac{W_{h_i}^{n+1} - W_{h_i}^{n}}{\Delta t} + \int_{\partial S_i} \underset{\sim}{F}(W_h^n).\underset{\sim}{\nu}_i \, d\sigma = 0 \tag{3.3}$$

To compute the integral arising in (3.5), we split the surface $\partial\hat{S}_i$ in panels ∂S_{ij} $(j \in K(i))$, separating node S_i and node S_j (See Fig. 2.1).

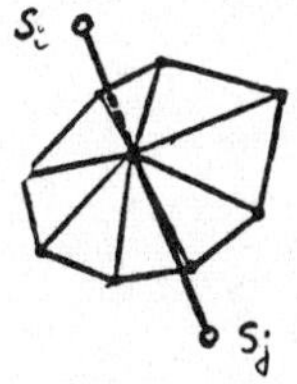

Figure 2.1

Let introduce the notations

$$F_{ij}(U) = \underset{\sim}{F}(U). \int_{\partial S_{ij}} \underset{\sim}{\nu}_i \, d\sigma \tag{3.4}$$

$$P_{ij}(U) = F_1'(U) \int_{\partial S_{ij}} \nu_{ix} d\sigma + F_2'(U) \int_{\partial S_{ij}} \nu_{iy} d\sigma + F_3'(U) \int_{\partial S_{ij}} \nu_{iz} d\sigma \tag{3.5}$$

Equality (3.3) has the following formulation :

$$\left\{ \begin{aligned} &\mathrm{Vol}(\hat{S}_i) \; \frac{W_{h_i}^{n+1} - W_{h_i}^{n}}{\Delta t} + \sum_{j \in K(i)} \overbrace{\underset{<1>}{\int_{\partial S_{ij}} \underset{\sim}{F}(W_h^n).\underset{\sim}{\nu}_i \, d\sigma}}^{H_{ij}^n} \\ &+ \underset{<2>}{\int_{\partial\hat{S}_i \cap \Gamma_B} \underset{\sim}{F}(\bar{W}_h^n)\underset{\sim}{n} \, d\sigma} + \underset{<3>}{\int_{\partial\hat{S}_i \cap \Gamma_\infty} \underset{\sim}{F}(W_h^n).\underset{\sim}{n} \, d\sigma} = 0 \end{aligned} \right. \tag{3.6}$$

We now specify the computation of the integral H_{ij}^n ; actually, the evaluation of this term corresponds to the one-dimensional calculation of the flux through ∂S_{ij} along the direction S_iS_j. Using the numerical flux function ϕ, the approximation system becomes :

$$\mathrm{Vol}(\hat{S}_i)\ \frac{W_i^{n+1} - W_i^n}{\Delta t} + \sum_{j\in K(i)} H_{ij}^{(1)} + \int_{\partial\hat{S}_i\cap\Gamma_B} \underset{\sim}{F}(\bar{W}_h^n).\underset{\sim}{n}d\sigma + \int_{\partial\hat{S}_i\cap\Gamma_\infty} \underset{\sim}{F}(\bar{\bar{W}}_h^n).\underset{\sim}{n}d\sigma = 0 \qquad (3.7)$$

$$H_{ij}^{(1)} = \phi_{F_{ij}}\ (W_i^n,\ W_j^n) \qquad (3.8)$$

We recall three first-order accurate upwind schemes defined by their numerical flux functions (or Riemann solvers) :

- Steger and Warming's splitting [26] :

$$\phi^V(U,V) = A^+(U)\ U + A^-(V)\ V$$

- The simplified splitting of Vijayasundaram [31] :

$$\phi(U,V) = A^+\left(\frac{U+V}{2}\right) U + A^-\left(\frac{U+V}{2}\right) V$$

- Osher's scheme [19] :

$$\phi^0(U,V) = \frac{F(U)+F(V)}{2} - \frac{1}{2}\int_U^V |A(W)|dW$$

Terms <2> and <3> contain the physical boundary condition. They are taken into account through the vector $\bar{W}_h^n$ which is computed from quantities depending on the interior value W_h^n and quantities determined by physical boundary conditions. For a wall boundary, the vector $\bar{W}_h^n$ is assumed to satisfy the slip conditions and thus term <2> becomes

$$\int_{\partial\hat{S}_i\cap\Gamma_B} \underset{\sim}{F}(\bar{W}_h^n).\underset{\sim}{n}\ d\sigma = \int_{\partial\hat{S}_i\cap\Gamma_B} \begin{pmatrix} 0 \\ \bar{p}n_x \\ \bar{p}n_y \\ \bar{p}n_z \\ 0 \end{pmatrix} d\sigma\ , \qquad (3.9)$$

where $\bar{p}$ is to be equal to the interior pressure $p(W_h^n)$.

With this procedure, the condition is thus applied in a weak variational way.

For inflow and outflow boundaries, we have to select a precise set of exterior datas, depending on the flow regime and direction of velocity ; for this purpose a plus-minus flux splitting is applied between exterior datas and interior values. So the integral <3> is evaluated with the use of the following decomposition :

$$\int_{\partial\hat{S}_i\cap\Gamma_\infty} \underset{\sim}{F}(\bar{W}_h^n).\underset{\sim}{n}d\sigma = P_{i\infty}^+(W_{h_i}^n)W_{h_i}^n + P_{i\infty}^-(W_{h_i}^n)W_\infty \qquad (3.10)$$

where $P_{i\infty}(U) = \underset{\sim}{F}'(U).\int_{\partial\hat{S}_i\cap\Gamma_\infty} \underset{\sim}{n}\ d\sigma$.

III.2. Two approaches for second-order accurate approximations

With constant by cell dependant variables, the numerical integration with flux splitting as described before will lead to schemes which are only first order accurate. Higher-order accurate schemes provide well-accurate solutions in smooth regions, but produce oscillatory approximations to discontinuities. The ingredients to build a second-order accurate approximation are : a dissipative first-order accurate (quasi-monotone) scheme ; a second-order scheme derived from the previous one by using linear interpolations or adding an antidiffusive term ; a limiting procedure acting to reduce the oscillations of the solution.

A procedure of this type is developped in the present context and intends to extend Van Leer idea's [29] to non-structured meshes [12].

The approximation discrete space is the following :

$$\mathcal{V}_h^{(1)} = \{v_h : \mathbb{R}^3 \to \mathbb{R}^5 , v_h|_{\hat{S}_i} \text{ is linear}, i = 1,\dots,n_h\} \quad (3.11)$$

Clearly, a function v_h of $\mathcal{V}_h^{(1)}$ is defined by its value at S_i and the values of its slopes in $\hat{S}_i$ denoted $(\underset{\sim}{\nabla} v_h)_i$.

In the following, the gradient $(\underset{\sim}{\nabla} w_h)_i$ will be derived from a Galerkin approximation using the P_1 - interpolation operator π_h :

$$(\underset{\sim}{\nabla} v_h)_i = \frac{\int_{\hat{S}_i} \underset{\sim}{\nabla}(\pi_h v_h)dx}{Vol(\hat{S}_i)} .$$

<u>A second-order accurate approximation using Hermitian interpolation</u> [12].

This scheme is a direct extension of the first order accurate scheme (3.7) and its formulation is :

$$Vol(\hat{S}_i)\ \frac{W_i^{n+1}-W_i^n}{\Delta t} + \sum_{j\epsilon K(i)} H_{ij}^{(2)} + \int_{\partial \hat{S}_i \cap \Gamma_B} \underset{\sim}{F}(\bar{W}_h^n).\underset{\sim}{n}\ d\sigma + \int_{\partial \hat{S}_i \cap \Gamma_\infty} \underset{\sim}{F}(\bar{\bar{W}}_h^n).\underset{\sim}{n}\ d\sigma = 0 \quad (3.12)$$

$$H_{ij}^{(2)} = \phi_{F_{ij}} (\tilde{W}_{ij}^-, \tilde{W}_{ij}^+) \quad (3.13)$$

$$W_{ij}^+ = W_i^n + \frac{1}{2} (\underset{\sim}{\nabla} W)_i . \underset{\sim}{S_i S_j} \quad (3.14a)$$

$$W_{ij}^- = W_i^n - \frac{1}{2} (\underset{\sim}{\nabla} W)_j . \underset{\sim}{S_i S_j} \quad (3.14b)$$

where the limiting procedure acts to preserve monotonicity in defining $\tilde{W}_{ij}^-$ and $\tilde{W}_{ij}^+$ as follows

$$\tilde{W}_{ij}^{-(k)} = \text{Max } \{\text{Min}[W_{ij}^{-(k)}, \text{Max}(W_i^{n(k)}, W_j^{n(k)})], \text{Min}(W_i^{n(k)}, W_j^{n(k)})]\} \quad (3.15a)$$

$$\tilde{W}_{ij}^{+(k)} = \text{Max } \{\text{Min}[W_{ij}^{+(k)}, \text{Max}(W_i^{n(k)}, W_j^{n(k)})], \text{Min}(W_i^{n(k)}, W_j^{n(k)})]\} \quad (3.15b)$$

for each component k = 1, m (m=5).

This construction results in a Fromm-like half-upwind scheme. The boundary integrals are computed as previously.

<u>A Second order accurate approximation using directional interpolation</u>

This approximation is related to the general upwind biased MUSCL-type scheme introduced by Van-Leer and discussed in [09], [29], [01].

For this construction, we will define for each segment S_iS_j, directional derivatives in introducing fictive points S_i^- and S_j^+ such that

$\underset{\sim}{S_jS_j^+} = \underset{\sim}{S_i^-S_i} = \underset{\sim}{S_iS_j}$ as in Figure 3.1.

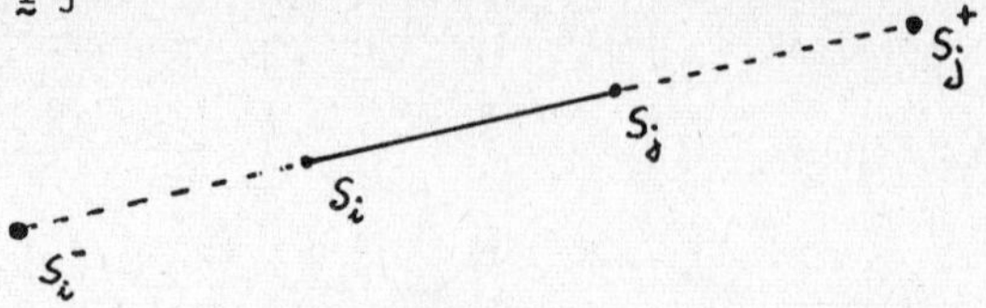

Figure 3.1

Using these points, we can compute extrapolated values of the function W to define the upwind differences

$$"W_i-W_{i+1}" = (\Delta_-)_i = 2((\underset{\sim}{\nabla} w)_i . \underset{\sim}{S_iS_j}) - [W_j-W_i] \quad (3.16a)$$

$$"W_{j+1}-W_j" = (\Delta_+)_j = 2((\underset{\sim}{\nabla} w)_j . \underset{\sim}{S_iS_j}) - [W_j-W_i] \quad (3.16b)$$

The approximation has the following formulation

$$\mathrm{Vol}(\hat{S}_i)\,\frac{W_i^{n+1}-W_i^n}{\Delta t} + \sum_{j\epsilon K(i)} H_{ij}^{(2)} + \int_{\partial\hat{S}_i\cap\Gamma_B} \underset{\sim}{F}(\bar{W}_h^n).\underset{\sim}{n}\, d\sigma + \int_{\partial\hat{S}_i\cap\Gamma_\infty} \underset{\sim}{F}(\bar{\bar{W}}_h^n).\underset{\sim}{n}\, d\sigma = 0 \quad (3.17)$$

$$H_{ij}^{(2)} = \phi_{F_{ij}}(\tilde{W}_{ij}^-, \tilde{W}_{ij}^+) \quad (3.18)$$

$$\tilde{W}_{ij}^- = W_i^n + \frac{s_i}{4}\,[(1-\chi\, s_i)(\Delta_-)_i + (1+\chi\, s_i)(W_j-W_i)] \quad (3.19a)$$

$$\tilde{W}_{ij}^+ = W_j^n - \frac{s_j}{4}\,[(1-\chi\, s_j](\Delta_+)_j + (1+\chi\, s_j)(W_j-W_i)] \quad (3.19b)$$

where the limiter s acts in a continuously differentiable manner [01] and is given by

$$s_i = \frac{2(\Delta_-)_i(W_j-W_i)+\epsilon}{(\Delta_-)_i^2+(W_j-W_i)^2+\epsilon} \quad \text{and} \quad s_j = \frac{2(\Delta_+)_j(W_j-W_i)+\epsilon}{(\Delta_+)_j^2+(W_j-W_i)^2+\epsilon},$$

where ϵ is a small number preventing division by zero.

We notice that the so-called upwind-biased scheme (χ=0) is closed to the Hermitian scheme (3.12)-(3.14).

IV. A GALERKIN SPATIAL APPROXIMATION

We consider secondly a Galerkin variational formulation of the 3-D Euler equations.

An explicit Richtmyer-Galerkin scheme is constructed ; it consists of a predictor-corrector sequence [02]. We recall the two steps of this scheme where the predictor is taken constant on each tetrahedron :

A. **Predictor step** : for each T_h tetrahedron of $\mathcal{T}_h$

$$\hat{W}(T_h) = \frac{1}{\mathrm{Volume}(T_h)}\left(\int_{T_h} W^n d + \alpha\, \Delta t \int_{\partial T_h} \underset{\sim}{\nabla} F(W^n).\underset{\sim}{n}(T_h)\, d\sigma\right) \quad (4.1)$$

where $\underset{\sim}{n}\,(T_h)$ is the unit vector of the outwed normal to ∂T_h.

B. **Corrector step** : $W^{n+1} \in \mathcal{W}_h$, $\forall \phi \in S_h$,

$$\left\{\begin{aligned} &\int_{\Omega_h} \Sigma_o \frac{W^{n+1}-W^n}{\Delta t} \Sigma_o \phi \, dx = \beta_1 \int_{\Omega_h} [F_1(W^n)\phi_x + F_2(W^n)\phi_y + F_3(W^n)\phi_z] \, dx \\ &+ \beta_2 \int_{\Omega_h} [F_1(\hat{W})\, \phi_x + F_2(\hat{W})\phi_y + F_3(\hat{W})\phi_z] dx + \int_{\Gamma_B} \Phi . [F_1(\bar{W})n_x + F_2(\bar{W})n_y + F_3(\bar{W})n_z] d\sigma \quad (4.2) \\ &+ \chi \int_{\Omega_h} f(W^n) \,\underset{\sim}{\nabla}\, W^n . \underset{\sim}{\nabla}\phi \, dx + \int_{\Gamma_\infty} \Phi . [F_1(\bar{\bar{W}})n_x + F_2(\bar{\bar{W}})n_y + F_3(\bar{\bar{W}})n_z] d\sigma \end{aligned}\right.$$

with $\alpha = 1 + \frac{\sqrt{5}}{2}$, $\beta_1 = \frac{2\alpha - 1}{2\alpha}$, $\beta_2 = \frac{1}{2\alpha}$.

The above two-step explicit scheme is stable under a classical Courant condition ; for general non-structured meshes and scalar linear case, a sufficient condition has been exhibited [05].

The parameters α, β_1, β_2 correspond to an optimal choice of the length of the predictor time step [17]. The last term of the right hand side is an artificial viscosity one with a tensor $f(W^n)$ and a parameter χ. For further details we refer to Angrand and Dervieux [02]. Using the $\mathcal{W}_h$ basis $\{\phi_i\}$, the corrector step is rewritten as

$$\left\{\begin{aligned} \frac{W_i^{n+1} - W_i^n}{\Delta t} \, Vol(\hat{S}_i) &= \beta_1 \int_{\Omega_h} \underset{\sim}{F}(W^n) . \underset{\sim}{\nabla}\phi_i dx + \beta_2 \int_{\Omega_h} \underset{\sim}{F}(\hat{W}) . \underset{\sim}{\nabla}\phi_i dx \\ &+ \int_{\Gamma_B} \phi_i \, \underset{\sim}{F}(\bar{W}) . \underset{\sim}{n} \, d\sigma + \int_{\Gamma_\infty} \phi_i \, \underset{\sim}{F}(\bar{\bar{W}}) . \underset{\sim}{n} \, d\sigma \qquad (4.3) \\ &+ \chi \int_{\Omega_h} f(W^n) \, \underset{\sim}{\nabla} W^n . \underset{\sim}{\nabla}\phi_i \, dx \end{aligned}\right.$$

The boundary integrals are identical to the integrals appearing in the previous section noticing that

$$\int_{\Gamma_B (\text{or } \Gamma_\infty)} \phi_i \underset{\sim}{n} \, d\sigma = \int_{\partial \hat{S}_i \cap \Gamma_B (\text{or} \Gamma_\infty)} \underset{\sim}{n} \, d\sigma$$

A mathematical analysis of the method can be found in [02] and [03].

V. IMPLICIT DELTA SCHEMES

Implicit delta schemes combined with relaxation have gained interest in the last years for the construction of faster solvers [10], [22], [27], [30]. They are much more reliable for the construction of faster solvers flow configurations and lead to faster convergence rates. They are, for the moment, more easily derived for unstructured grids than multigrid methods and have been studied quite early in combination with a Galerkin approximation [27]. We present in this section a general procedure to build such solvers from an upwind first order accurate approximation.

V.1 First order implicit schemes

Firstly, we consider the 1-D case for simplicity. We study the following class of upwind explicit schemes :

$$\begin{cases} \dfrac{W_i^{n+1} - W_i^n}{\Delta t} + \dfrac{1}{\Delta x}\left(\Phi^n_{i+\frac{1}{2}} - \Phi^n_{i-\frac{1}{2}}\right) = 0 \\ \Phi^n_{i+\frac{1}{2}} = \Phi(W_i^n, W_{i+1}^n) \\ \Phi(U,V) = H_1(U,V)U + H_2(U,V)V \text{ for } U,V \in \mathbb{R}^3 \\ H_1(U,U) + H_2(U,U) = A(U) \end{cases} \tag{5.1}$$

where the third line is a tentative linearization and where the fourth line is a consistency property.

The so-called Steger-Warming or Vijayasundaram's schemes belong to this class of upwind schemes.

These schemes do not satisfy the following assumption of differentiability of fluxes :

$$\phi(U,V) \text{ is differentiable w.r.t. } (U,V). \tag{5.2}$$

Anyway, let us assume that (5.2) is true ; so we can construct the Newton-like linearized implicit version of scheme (5.1) :

$$\begin{cases} \dfrac{W_{h_i}^{n+1} - W_{h_i}^n}{\Delta t} + \dfrac{1}{\Delta x}\left(\Phi^N_{i+\frac{1}{2}} - \Phi^N_{i-\frac{1}{2}}\right) = 0 \\ \Phi^N_{i+\frac{1}{2}} = \Phi^{nN}(W_i^n, W_{i+1}^n, W_i^{n+1}, W_{i+1}^{n+1}) \\ \Phi^N(U,V,W,Z) = \Phi(U,V) + \dfrac{\partial\Phi}{\partial U}(U,V)(W-U) + \dfrac{\partial\Phi}{\partial V}(U,V)(Z-V) \end{cases} \tag{5.3}$$

For Δt tending to infinity, scheme (5.5) becomes Newton's method for finding the stationary solution of scheme (5.1).

The derivatives of Φ may be very expensive to compute ; so we introduce the simplified linearized implicit version of scheme (5.1) :

$$\begin{cases} \dfrac{W_i^{n+1} - W_i^n}{\Delta t} + \dfrac{1}{\Delta x}\left(\Phi^S_{i+\frac{1}{2}} - \Phi^S_{i-\frac{1}{2}}\right) = 0 \\ \Phi^S(U,V,W,Z) = H_1(U,V)W + H_2(U,V)Z \end{cases} \tag{5.4}$$

It can be shown that under the assumptions (5.1) and (5.2), schemes (5.3) and (5.4) have the same equivalent system up to the second order [27].

We cannot derive that scheme (5.4) will become a quadratically converging method ("quasi-Newton") for Δt tending to infinity, but we may expect a similar efficiency for the two schemes for large Δt if the unknowns do not vary much as it is the case at convergence to steady state.

Returning to schemes, which do not satisfy condition (5.2) we cannot expect anything ; we just note that nondifferentiability arise quite rarely in the physical domain, namely in shocks, sonic points and stagnation points. We recall also that the existence of these singularities make the convergence of Newton methods hazardous in the continuous (non discretized) context.

We can rewrite (5.6) with a delta formulation ($\sigma = \frac{\Delta t}{\Delta x}$),

$$\sigma\, H^n_{2,i+\frac{1}{2}}\, \delta W^{n+1}_{i+1} + [I+\sigma\, H^n_{1,i+\frac{1}{2}} - \sigma\, H^n_{2,i-\frac{1}{2}}]\delta W^{n+1}_i - \sigma\, H^n_{1,i-\frac{1}{2}}\, \delta W^{n+1}_{i-1} = - \sigma\, [\phi^n_{i+\frac{1}{2}} - \phi^n_{i-\frac{1}{2}}] \tag{5.5}$$

V.2. Second Order implicit schemes

We consider the delta formulation (5.5). An efficient way to get second-order accurate stationary solutions while keeping the interesting properties of the first-order accurate upwind matrix is to replace the R.H.S. of (5.5) by a second-order accurate spatial approximation ; it is convenient to present the resulting scheme as a two phases scheme :

Suppose given any second order accurate approximation $E^2_h(W)$ built as in chapter III or IV, the resulting scheme can be written as :

<1> Physical / explicit / second order accurate phase

$$\delta\hat{W}_i = - \sigma\, E^2_h(W^n_h)$$

<2> Mathematical / implicit / first order accurate phase

$$\sigma H^n_{2,i+\frac{1}{2}}\, \delta W^{n+1}_{i+1} + [I + \sigma\, H^n_{1,i+\frac{1}{2}} - \sigma\, H^n_{2,i-\frac{1}{2}}]\, \delta W^{n+1}_i - \sigma\, H^n_{1,i-\frac{1}{2}}\, \delta W^{n+1}_{i-1} = \delta\hat{W}_i \tag{5.6}$$

$$W^{n+1}_i = W^n_i + \delta W^{n+1}_i$$

In all the applications, we have built the mathematical part from "Steger Warming's" flux decomposition. With such a choice, we can give the complete formulation (including linearization of boundary terms) of the linear system in three space dimensions :

$$\text{<1>} \quad \delta\hat{W}_i = - \frac{\tilde{\Delta t}_i}{\mathrm{Vol}(\hat{S}_i)}\, E^2_h(W^n)_i \;\; ; \text{ (with local time step } \tilde{\Delta t}_i) \tag{5.7}$$

$$\text{<2>} \quad \delta W^n_i + \frac{\tilde{\Delta t}_i}{\mathrm{Vol}(\hat{S}_i)} [\sum_{j\in K(i)} P^+_{ij}(W^n_i)]\, \delta\, W^n_i + \frac{\tilde{\Delta t}_i}{\mathrm{Vol}(\hat{S}_i)} \sum_{j\in K(i)} P^-_{ij}(W^n_j)\, \delta\, W^n_j$$

$$+ \frac{\tilde{\Delta t}_i}{\mathrm{Vol}(\hat{S}_i)}\, \chi(\partial\hat{S}_i \cap \Gamma_\infty)\, [P^+_{i\infty}(W^n_i)\, \delta W^n_i] + \frac{\tilde{\Delta t}_i}{\mathrm{Vol}(\hat{S}_i)} \int_{\partial\hat{S}_i\cap\Gamma_B} H(W^n)\delta W^n = \delta\hat{W}_i \;, \tag{5.8}$$

$$W^{n+1}_i = W^n_i + \delta W^n_i \;, \tag{5.9}$$

where $$H(W^n)=(\gamma-1)\begin{pmatrix} 0 & 0 & 0 & 0 \\ \frac{q_n^2}{2}n_x - u^n n_x & -v^n n_x & n_x \\ \frac{q_n^2}{2}n_y - u^n n_y & -v^n n_y & n_y \\ 0 & 0 & 0 & 0 \end{pmatrix},$$

coming from the linearization of the body boundary conditions [27].

VI. SELF ADAPTIVE MESH REFINEMENTS TECHNIQUES

The arising of various flow singularities in the Euler simulations is a strong motivation to use mesh adaptation especially for 3-D problems in which limiting the number of nodes is crucial.

In fact, with non structured meshes, local enrichment algorithms are rather easy to handle, as shown in the references, using finite elements.

A physical criterion is chosen in order to decide the regions where the density of nodes has to be increased. In those regions each tetrahedron will be splitted into 8 sub-tetrahedra ; the refined region and the non refined one are matched by splitting along the interface an extra strip of elements.

Remark 1 : In this procedure, the nodes are not moved but extra nodes are added.

Remark 2 : Even starting from a structured grid, we obtain after refinement a non structured one, on which the scheme should provide a more accurate solution.

VII. NUMERICAL EXPERIMENTS FOR 2-D AND 3-D FLOWS

In this section, we present the results of some tests whose purpose is the numerical solution of 2-D and 3-D transonic, supersonic and hypersonic Euler flows using the methods discussed in the previous sections. In fact, most of calculations have been performed on IBM 3090 using implicit schemes presented in Section V and accurate solutions on adapted meshes obtained with ingredients of Section VI. It appears from the experiments that the methods discussed in the paper are well suited to the solution of complicated problems.

VII. 1 Simulation of transonic flows around a NACA 0012 airfoil

As a first example, we have considered a transonic flow around a NACA0012 at angle of attack α=1° and Mach number at infinity M_∞=.85. We have selected this problem since it is a quite classical and significant test problem for Euler solvers [32].

Fig. 7.1 shows an enlargment of the final adapted triangulation near the profile used to solve the test problem. Then Fig. 7.2 (a)(b)(c) (resp. 7.3(a)(b)(c)) show, on one hand the pressure lines and on the other hand the pressure and entropy distributions on the airfoil obtained with the implicit second-order upwind approximation (UA) of Section III (resp implicit second-order Galerkin approximation (GA)). We can observe the similar shocks locations obtained with the two schemes implying only a small production of numerical entropy in the vicinity of the stagnation point.

VII.2 Simulation of a supersonic flow around and inside and inlet

In military aircraft design, numerical Euler simulations are intensively used to increase the performance of fitted configurations of conical inlets. With supersonic regime at infinity, complex features of the flow such as lambda shock and angular stagnation lines have to be accurately captured.

2-D Simulation :

We have used the (GA) scheme with the value of the viscosity parameter $\chi=.8$ introduced in the Section IV. The angle of attack is $\alpha=0°$ so that only the half geometry is considered. The Mach number at infinity is $M_\infty=2.$. While along the engin outflow boundary, the Mach number has a prescribed value, M(engine) = .27. For this case, a lambda shock occurs with a strong oblique shock and a weak normal one. The normal shock does not reach the upper part of the inlet.

To obtain an accurate solution, the self adaptive procedure described in section VI is applied. Starting from the computation on a regular coarse mesh, a refined one given on Fig. 7.4(a) with 2818 nodes is obtained. The Mach and pressure contours of Fig. 7.5(a)(b) show that the weak normal shock merges with the oblique one accurately.

3-D Simulation :

A 2-D coarse mesh is used to generate by means of rotation, the tetrahedrization around a 3-D axisymetric inlet. For the same supersonic Mach number at infinity $M_\infty=2.$, we consider a subcritical case in prescribing at the engine outflow a value of the Mach number M(engine) = .45. For these values, the strong shock is now the normal one. As in the 2-D experiments, a 3-D self-adaptive algorithm is used to refine the mesh in the vicinity of both shocks and stagnation lines and generate an adapted tetrahedrization shown in Fig. 7.6(a)(b). Improvement of the accuracy of the solution is shown with the pressure lines of Fig. 7.7.

VIII.3 Simulation of a supersonic flow around a 2-D space vehicle

The self adaptive mesh refinements procedure described in Section VI has been applied to identify the main feactures of a supersonic (M_∞=2.5) flow around a shuttle, at large angle of attack (α=18°), with the Galerkin approximation. A convenient threshold value of a physical criterion detects the main zones of interest of the flow ; discontinuities such as the bow shock and the canopy shock ; and also high curvature regions generating non physical numerical entropy, are progressively identified via a sequence of refinement as shown in Fig. 8. Corresponding pressure and entropy contours are presented on Fig. 9 (a)(b).

VII.4 3-D calculations around a complete space aircraft

In order to show the possibilities of the methods described in this paper for solving problems of industrial interest, we have used the implicit upwind scheme (UA) for the simulation of a 3-D supersonic and hypersonic flow around a complete space shuttle, namely the US Orbiter. Fig. 10 shows the trace on the space vehicle of the three dimensional finite element mesh used for the calculation. Two different computations have been performed, a supersonic case at M_∞=1.3 with an angle of attack α=15° using a second-order accurate scheme. Fig. 11(a) shows the Mach number distribution on the space aircraft, a hypersonic case at M_∞=25. with an angle of attack of 40°, whose Mach number contours on the body are presented on Fig. 11(b). This last computation provides an Euler initial solution to compute numerically thermal fluxes in order to predict or identify high temperature levels on the body including chemical effects of the reacting flow, due to dissociation and ionization phenomena. The method described in this paper is also used to design the European Hermes shuttle.

VIII CONCLUSION

We have presented in this paper several finite element methods using upwind or Galerkin approximations for the accurate numerical solution of Euler equations modelling transonic, supersonic and hypersonic flows. With these methods we can simulate 3-D flows around complex geometries such as complete space aircraft.

Variants of these methods can be extended to more complicated situations such as non equilibrium reacting flows for real gas effects or the coupling of two mathematical modellings using the Euler and Navier-Stokes equations.

Acknowledgements : This work was partly supported by DRET, under contract 84/104. We would like to thank for helpful discussions A. Dervieux and also F. Angrand, J.A. Desideri, L. Fezoui and B. Palmerio from INRIA, who have been our principal collaborators in developping the methods.

REFERENCES

[01] ANDERSON W.K., THOMAS J.L., VAN LEER B., A comparison of Finite volume Flux vector splitting for the Euler Equations, AIAA 23rd Aerospace Sciences Meeting, January 1985, Reno, Nevada.

[02] ANGRAND F., DERVIEUX A., Some explicit triangular Finite Element Schemes for the Euler Equation, I.J. For Num. Meth. in Fluids, Vol. 4, pp 749-764 (1984).

[03] ANGRAND F., DERVIEUX A., BOULARD V., PERIAUX J., VIJAYASUNDARAM G., Transonic Euler Simulations by means of Finite Element Explicit Schemes, 6th AIAA Conference on Computational Fluid Dynamics (Danvers, Mass. USA) and AIAA paper 83-1924.

[04] ANGRAND F., BOULARD V., DERVIEUX A., PERIAUX J., VIJAYASUNDARAM G., Triangular Finite Element Methods for the Euler Equations, 6th International Conference on Computing Methods in Applied Sciences and Engineering, Glowinski R., Lions J.L., Eds., North Holland (1984).

[05] ANGRAND F., BILLEY V., DERVIEUX A., PERIAUX J., POULETTY C., STOUFFLET B., 2-D and 3-D Euler flow calculations with a second order Galerkin Finite Element method, AIAA 18th Fluid Dynamics and Plasmodynamics and Lasers Conference, July 16-18, 1985, Cincinatti.

[06] BANK R.E., SHERMAN A.H., An adaptive Multi-Level Method for Elliptic BVP, Computing 26, pp 91-105 (1981).

[07] BAKER T.J., JAMESON A., WEATHERILL N.P., Calculation of Inviscid Transonic Flow Over a Complete Aircraft, AIAA 86-0103, AIAA 24th Aerospace Sciences Meeting, January 1986, Reno, Nevada.

[08] BERGER M.J., JAMESON A., An Adaptive Multigrid Method for the Euler Equations IC9NMFD, Lecture Notes in Physics, 218, Springer (1985).

[09] BORREL M., MONTAGNE J.L., Numerical Study of a Non-Centered Scheme with Application to Aerodynamics, AIAA 18th Fluid Dynamics and Plasmadynamics and Lasers Conference, July 16-18, 1985, Cincinatti.

[10] CHAKRAVARTHY S.R., Relaxation Methods for Unfactored Implicit Upwind Schemes, AIAA 2nd Aerospace Sciences Metting, January 9-12, 1984, Reno, Nevada.

[11] EBERLE A., MISEGADES K., Euler Solution for a Complete Aircraft at sub-and Supersonic Speed, 58th Meeting on the Fluid Dynamics Panel Symposium on Applications of Computational Fluid Dynamics in Aeronautics (AGARD) April 7-10 1986, Aix en Provence, France.

[12] FEZOUI F., STOUFFLET B., PERIAUX J., DERVIEUX A., Implicit High Order Upwind Finite Element Schemes for the Euler Equations, GAMNI Conference, Atlanta, March 1986.

[13] HARTEN A., LAX P.D., VAN LEER B., On upstream differencing and Godunov-type schemes for hyperbolic conservation laws, ICASE report n° 82-5, March 17, 1982.

[14] HUGHES T.J.R., MALLET M., FRANCA L.P., New Finite Elements Methods for the compressible Euler and Navier Stokes Equations, 7th International Conference on Computing Methods in Applied Sciences and Engineering, December 9-13, 1985, Versailles, France.

[15] JAMESON A., SCHMIDT W., TURKEL E., Steady state solution of the Euler equations for transonic flow, AIAA paper 81-1259, 1981.

[16] JAMESON A., MAVRIPLIS D., Finite volume solution of 2-D Euler equation on a regular triangular mesh, AIAA paper 85-0435, 1985, Aerospace Science Meeting, January 14-17, 1985, Reno, Nevada.

[17] LERAT A., PEYRET R., Sur le choix de schémas aux différences du second ordre fournissant des profils de chocs sans oscillation, Comptes Rendus Acad. Sc. Paris, Serie A, 277, 363-366 (1973).

[18] LOHNER R., MORGAN K., PERAIRE J., ZIENKIEWICZ O.C., KONG L., Finite Element Methods for Compressible Flows, Proc. of the ICFD Conference on Numerical Methods for Fluid DYNAMICS, April 1-4 1985, Reading, UK.

[19] OSHER S., CHAKRAVARTHY S., Upwind difference schemes for hyperbolic systems of conservation laws, Journal Math. Computation, April 1982.

[20] PALMERIO B., BILLEY V., DERVIEUX A., PERIAUX J., Self Adaptive Mesh Refinements and Finite Element Methods for Solving the Euler Equations, Proc. of the ICFD Conference on Numerical Methods for Fluid Dynamics, April 1-4, 1985, Reading, UK.

[21] PEREZ E., PERIAUX J., ROSENBLUM J.P., STOUFFLET B., DERVIEUX A., LALLEMAND M.H., Adaptive Full-Multigrid Finite Element Methods for Solving the Two-Dimensional Euler Equations, IC1ONMFD, 23-27 June 1986, Beying, China.

[22] RAI M.M., Patched-grid calculations with the Euler and Navier-Stokes equations : Theory and Applications, Von Karman Institute for FLuid Dynamics, lecture Series 1986-04, Mach 3-7, 1986.

[23] RIVARA M.C., Algorithms for Refining Triangular Gride Suitable for Adaptive and Multigrid Techniques, International Journal for Numerical Methods in Engineering, Vol. 20, pp 745-756, 1984.

[24] RIZZI A., Cyber 205 Dense Mesh Solutions to the Euler Equations for Flows around the M6 and Dillher Wings, 9th ICNMFD, Lecture Notes in Physics 218, Springer Verlag 1985.

[25] SATOFUKA N., Unconditionally Stable Explicit Method for the Numerical solutions of the Compressible Navier-Stokes Equations, Proceed. of 5th GAMM-Conf. On Num. Meth. in Fluid Mechanics, V7, Vieweg, 1984.

[26] STEGER J., WARMING R.F., Flux Vector splitting for the inviscid gas dynamic equations with applications to finite difference methods, Journal Comp. Physics, Vol. 40 n°2, pp 263-293, 1981.

[27] STOUFFLET B., Implicit Finite Element Methods for the Euler Equations, Proceedings of the INRIA Workshop on "Numerical Methods for Compressible Inviscid Fluids", 7-9 dec. 1983, Rocquencourt (France), Glowinski R., Ed., SIAM.

[28] USAB W.J., Jr., MURMAN E.M., Embedded Mesh Solutions of the Euler Equation Using a Multiple-Grid Method, 6th AIAA Computational Fluid Dynamics Conference, July 13-15, 1983, Danvers, Massachusetts.

[29] VAN LEER B., Computational Methods for Ideal Compressible Flow, Von Karman Institute for Fluid Dynamics, Lectures series 1983-04 Computational Fluid Dynamics, March 7-11 1983.

[30] VAN LEER B., MULDER W.A., Relaxation methods for hyperbolic equations, Proceedings of the INRIA workshop on "Numerical methods for compressible inviscid fluids", dec. 7-9 1983, Glowinski R. Ed., to be published by SIAM.

[31] VIJAYASUNDARAM G., Transonic Flow Simulations Using an Upstream Centered Scheme of Godunov in Finite Elements, Journal Comp. Physics, Vol 63 n°2, pp 416-433, 1986.

[32] GAMM Workshop on the Numerical Simulation of Compressible Euler Flows, Rocquencourt, France, 10-13 June 1986.

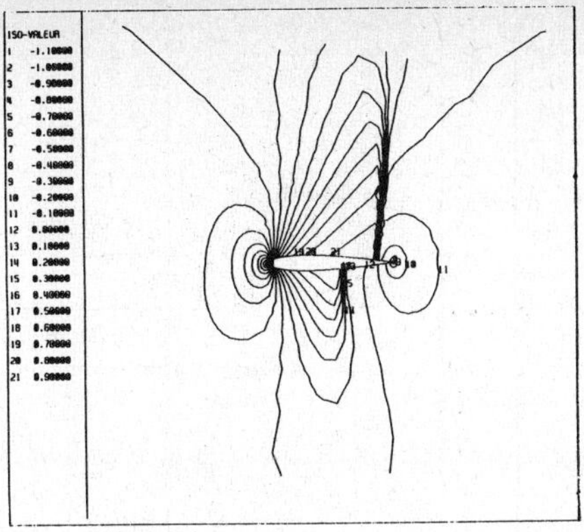

Figure 7.2.(a) : Iso-C_p lines

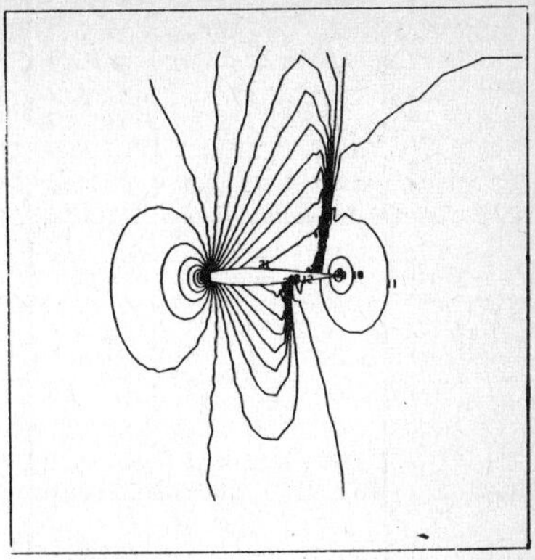

Figure 7.3.(a) : Iso-C_p lines

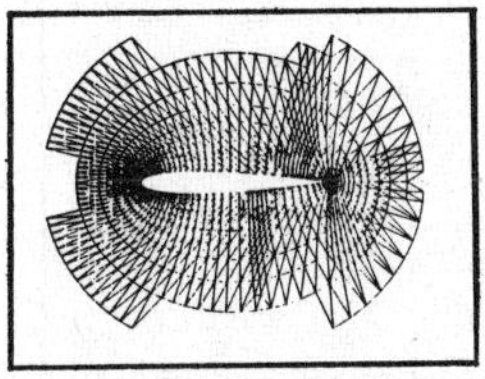

Figure 7.1.:Adapted mesh

NACA 0012 airfoil
$M_\infty = .85, \alpha = 1°$

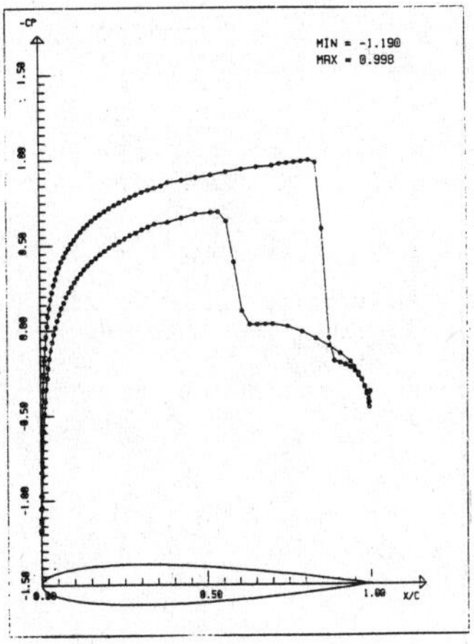

Figure 7.2.(b) : Pressure distribution

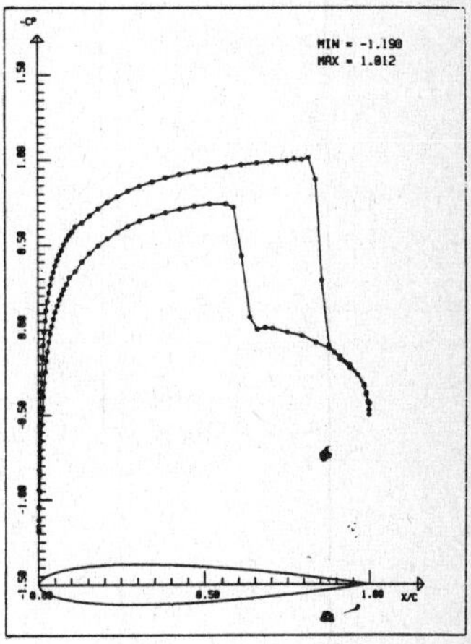

Figure 7.3.(b) : Pressure distribution

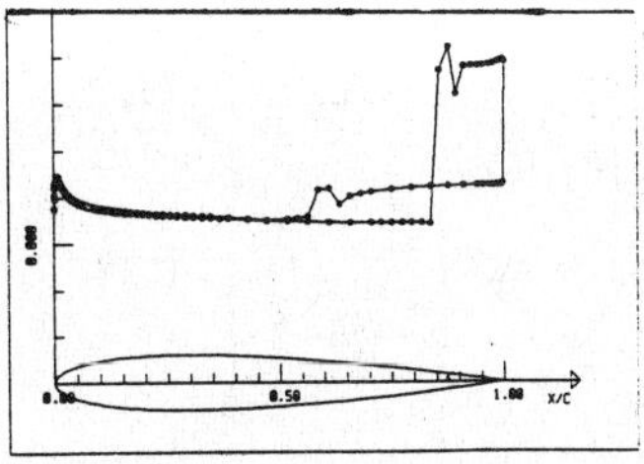

Figure 7.2.(c) : Entropy deviation

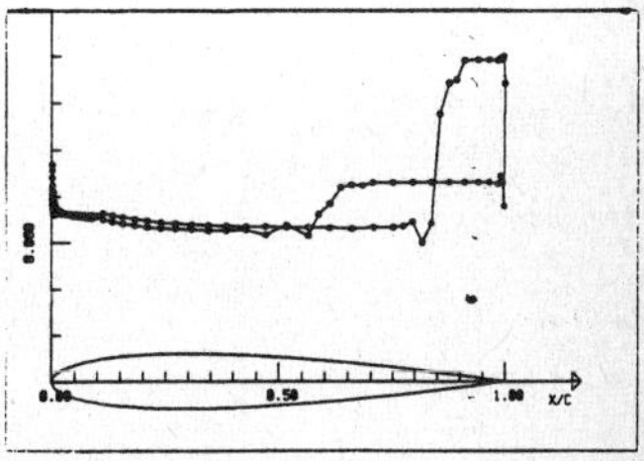

Figure 7.3.(c) : Entropy deviation

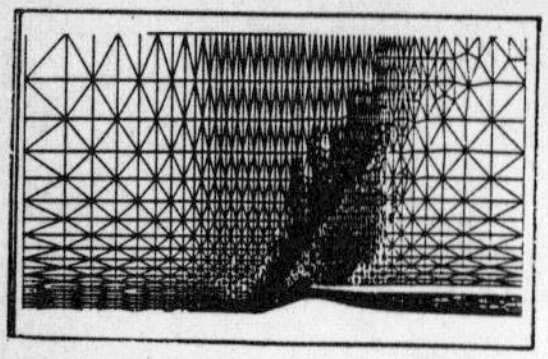

Figure 7.4.(a):Adapted mesh

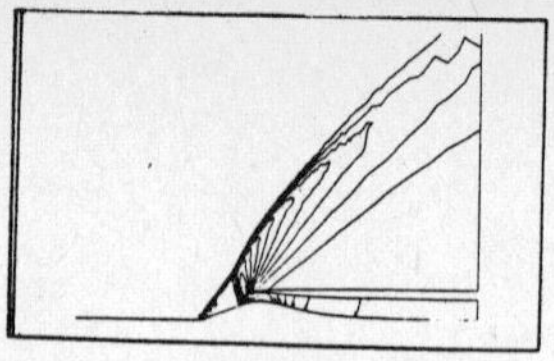

Figure 7.5.(b):Iso-C_p contours

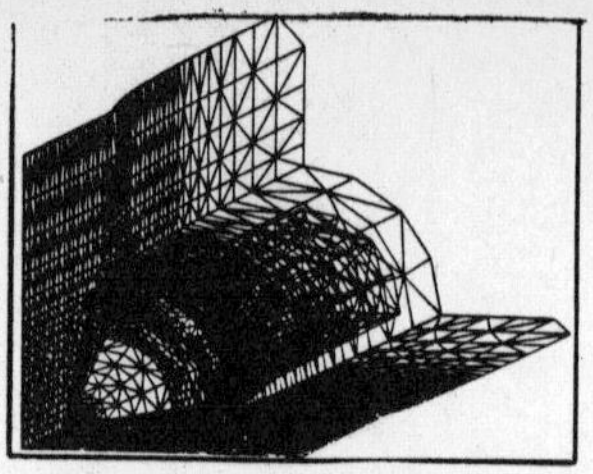

Figure 7.6.(a):Adapted mesh around the inlet

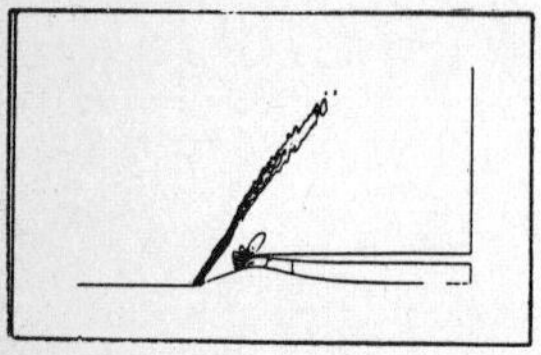

Figure 7.4.(b):Mach number gradient

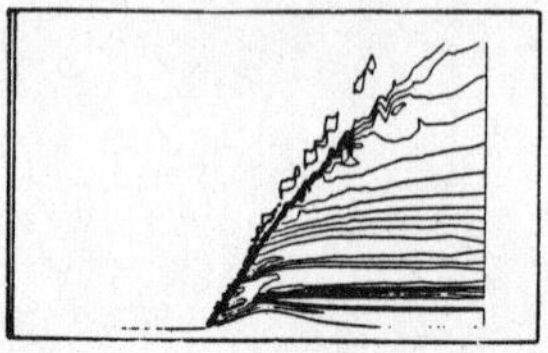

Figure 7.5.(c):Iso-entropy contours

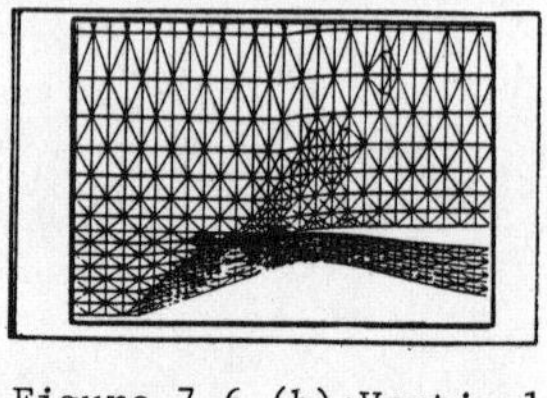

Figure 7.6.(b):Vertical section of the adapted mesh

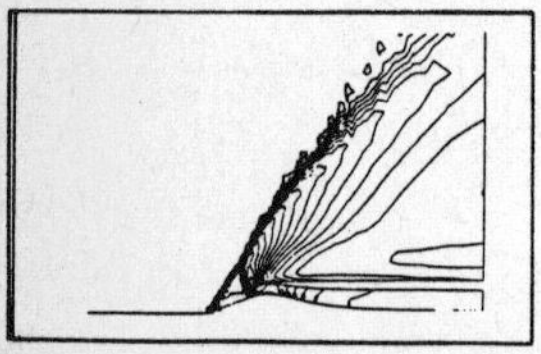

Figure 7.5.(a):Isomach contours

2-D and 3-D inlets

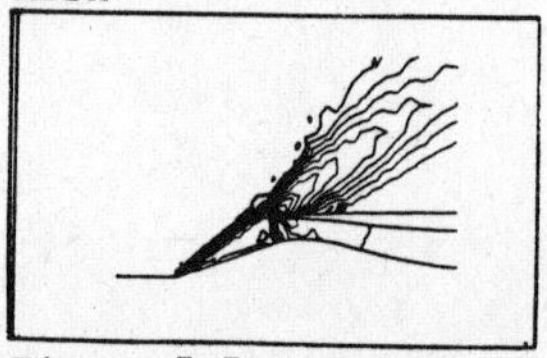

Figure 7.7.:Pressure contours

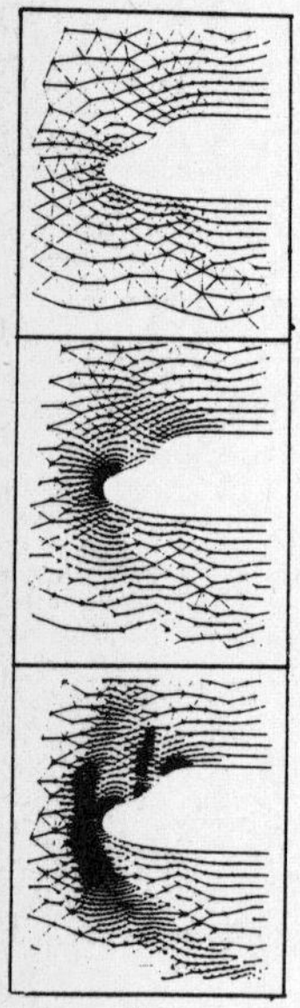

initial mesh
number of nodes : 704
number of elements : 1262

refined mesh
number of nodes : 1109
number of elements : 2070

twice refined mesh
number of nodes : 1836
number of elements : 3502

Figure 8:Adapted meshes

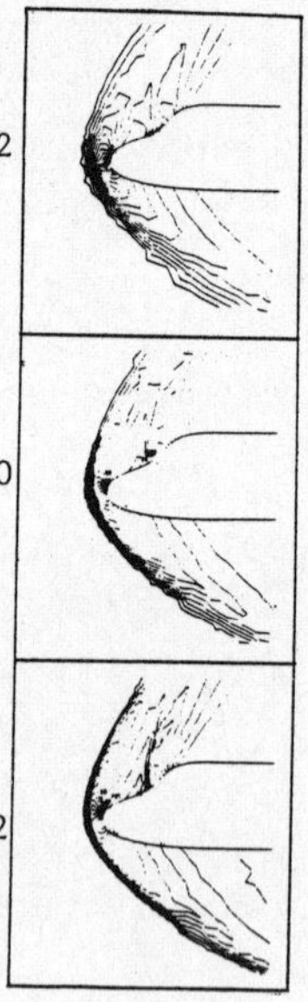

Figure 9.(a): Pressure contours

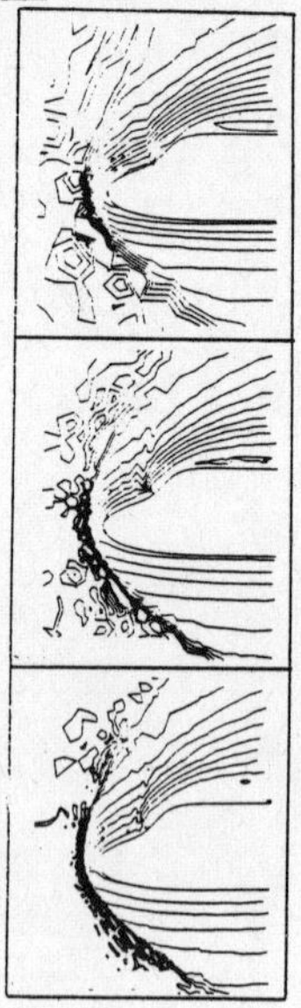

Figure 9.(b): Entropy contours

2-D Shuttle

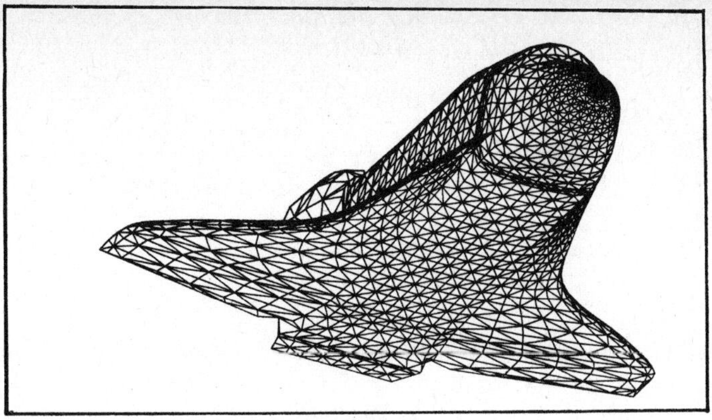

Figure 10:3-D mesh on the US orbiter

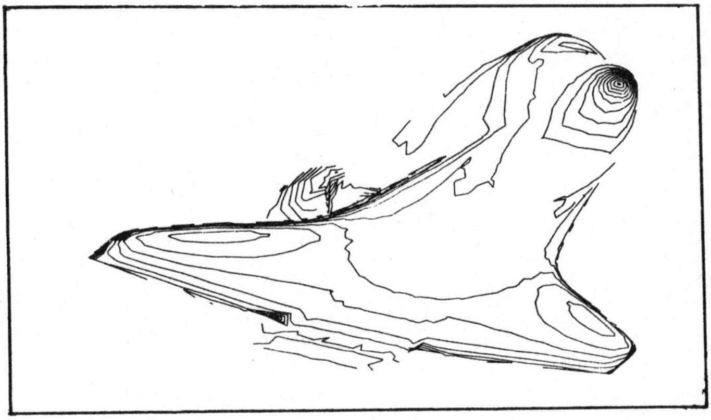

Figure 11.(a):Supersonic flow simulation by finite elements using a second order accurate upwind scheme. Isomach number distribution on the US orbiter (M_∞ = 1.3, α= 15°).

Figure 11.(b):Hypersonic flow simulation by finite elements using a first-order accurate upwind scheme. Isomach number distribution on the US orbiter (M_∞ = 25,α = 40°).

A "BOX-SCHEME" FOR THE EULER EQUATIONS

Jean-Jacques Chattot
MATRA, 37 avenue Louis Bréguet, 78140 VELIZY

Sylvie Malet
Université Pierre et Marie Curie (Paris 6)
4 Place Jussieu, 75230 - PARIS CEDEX 05

Abstract

For the solution of first order partial differential equations with boundary conditions a box scheme is introduced based on a compact discretization in space and the use of the characteristic directions for the integration in time. The scheme is first developped for a non-linear scalar conservation law. Then it is presented for the equations of gas dynamics in a domain of varying area. Applications to the shock tube and to a steady flow in a nozzle exhibit the major features of the scheme. Preliminary results in two-dimensions seem to indicate that the extension is worthy of interest.

Introduction

Most of the large three-dimensional codes used in the Industry for practical simulations of complex flows rely on classical and well studied numerical schemes which are robust and cheap to run on scalar and vector computers. Such is the case of Mac Cormack, Lax-Wendroff and Runge-Kutta type schemes, developped and used for the last two decades. Improvements have come steadily since their first introduction : the conservative or finite volume discretization contributes to better satisfaction of the conservation laws, the progress in mesh construction allows to benefit in practice of the second order accuracy in complex geometries, and the local time stepping and multiple grid or implicit techniques yield fast convergence in the search of steady solutions.

The drawbacks of these centered, type-independant schemes are the occurence of wiggles when strong gradients are present in the flow, and the need for artificial viscosity for the captures of true shocks and the damping of oscillations.

The current research on new schemes is for a large part oriented towards type-dependant schemes, such as flux-splitting schemes, modified flux techniques or flow oriented schemes. The flux-splitting schemes can be traced to early works of Warming, Beam, and Steger [1,2]. The idea is to use one-sided spatial differences for each split flux vector, even if the flow is locally subsonic, in order to model more closely wave propagation phenomena and avoid wiggles. Recently Harten introduced the notion of total variation diminishing schemes (TVD) and combined it with a modified flux approach chosen so that the scheme is second-order at regions of smoothness and first order at points of extrema. Although the scheme is an upwind scheme, it is written in a symmetric form; i.e. central difference plus an appropriate numerical dissipation term [3,4]. This approach has been applied and extended by Yee and Harten [5] and by Yee, Warming and Harten [6] to two space dimensions. The notion of flow oriented scheme can be attributed to Murman and Cole in the early 70's [7] with their mixed-scheme and the fully conservative four operator scheme of Murman [8]. The scheme uses centered differences at subsonic points and upwind differences at supersonic points. This scheme was studied in detail by Engquist and Osher [9,10] and lead to an entropy satisfying scheme constructed in a more mathematical framework. Other contributions along these lines are due to Van Leer [11] and Roe [12].

A more physical approach to flow oriented schemes is due to Moretti [13] and his followers [14-16], and consists in using one-sided differences according to the direction of propagation of the waves of the non-conservative system of equations. This approach suffers from the fact that it does not capture Rankine-Hugoniot shock waves and thus does not produce the correct shock speed.

The first mention of a box-scheme seems to be owed to Keller [17] in the context of numerical solution of viscous equations written as a system of first order differential equations and discretized in compact form with a two-point approximation, of second order accuracy. The two-point differencing idea appears to be a very attractive choice for numerically solving first order systems, such as the Euler equations and was used by Wornom [18]. Here we combine the idea of the box scheme with the more physical notion of direction of propagation as given by characteristic theory.

The paper is divided into three parts. In part one, we develop the basic features of the box-scheme in the scalar case with particular attention to the shock and expansion operators. In part two, the method is extended to the equations of gas dynamics. Test cases are used to exhibit the properties of the scheme for steady and unsteady flows with discontinuities. In the last part, preliminary results are presented in the case of two-dimensional supersonic flow using a small pertubation assumption for a shock reflection problem and a profile in a channel.

I. Box-scheme applied to a scalar conservation law.

Let us consider the inviscid Burgers equation:

$$\frac{\partial u}{\partial t} + \frac{\partial f(u)}{\partial X} = 0 \qquad \text{with } f(u) = \frac{u^2}{2}$$

A two-point conservative discretization of the flux in the box [i-1,i] yields

$$\left[\frac{\partial f(u)}{\partial X}\right]_{i-1,i} = \frac{f(u)_i - f(u_{i-1})}{\Delta X} = \frac{u_{i-1} + u_i}{2} \, \frac{u_i - u_{i-1}}{\Delta X}$$

Thus the propagation velocity associated with this discretization is

$$u_{i-1/2} = \frac{u_{i-1} + u_i}{2}$$

We consider first the case of a compression with a sonic point (fig.1).

Fig. 1

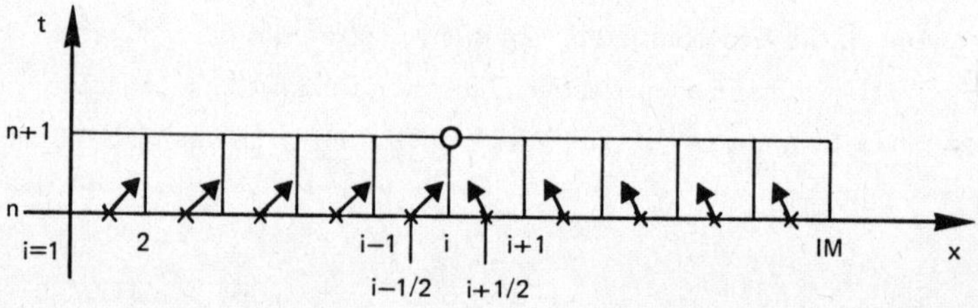

Clearly we have IM unknowns if IM is the maximum number of points and we have IM-1 "box equations" + 2 boundary conditions if we assume that the incoming flow is supersonic and the exhaust is subsonic.

We have 1 equation more than unknowns, not surprisingly since we can and we must attribute to the shock point defined by $u_{i-1/2} > 0$ and $u_{i+1/2} \leqslant 0$ the box [i−1,i] **and** [i,i+1]. If we neglect to introduce downstream influence at the shock then it will never move in the upstream direction. This problem has been encountered with the λ-scheme.

Since we must conserve fluxes in order to enforce the correct jump condition, the shock point must be discretized on the reunion of the two boxes, that is on the larger [i−1, i+1] box.

The shock point discretization is the following:

$$\frac{u_i^{n+1} - u_i^n}{\Delta t} + \frac{f(u_i^n) - f(u_{i-1}^n)}{\Delta X} + \frac{f(u_{i+1}^n) - f(u_i^n)}{\Delta X} = 0$$

We prove now that the shock velocity satisfies $\left(\frac{dx}{dt}\right)_s = \frac{u_l + u_r}{2}$ in the case of a single point structure shock, that is for low speed shocks.

Suppose that we have a constant state to the left $u_k = u_l > 0$ for $k = 1,2,\ldots,i-1$ and a constant state to the right $u_k = u_r \leqslant 0$ for $k = i+1,\ldots,IM$. Point i is the shock point introducing an artificial structure allowing the transition from supersonic to subsonic flow, fig.2

Fig. 2

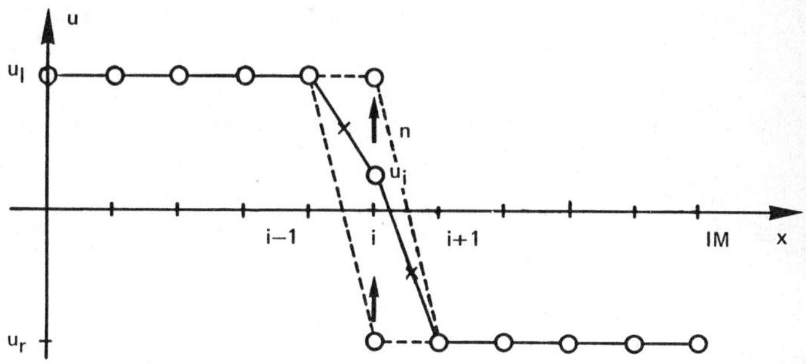

For a shock moving to the right, one says that the shock has moved from i to i+1 when u_i varies from u_r to u_l. But during Δt the shock moves $\Delta u_i = -\Delta t \frac{f(u_r) - f(u_l)}{\Delta x}$. Thus its velocity is

$$\left(\frac{dx}{dt}\right)_s = \frac{\Delta u_i}{u_l - u_r} \; \frac{\Delta x}{\Delta t} = \frac{f(u_l) - f(u_r)}{u_l - u_r} = \frac{u_l + u_r}{2}$$

It can be noted that steady shocks $\left(\frac{dx}{dt}\right)_s = 0$ will have a usual structure with **one** artificial intermediate value u_i, that is a structure on two boxes, and exceptionally no structure at all when $u_i = u_l$ or $u_i = u_r$. The residuals $\frac{\partial f(u)}{\partial x}$, about point i, will in general be non zero but will satisfy:

$$\left[\frac{\partial f(u)}{\partial X}\right]_{i-1,i} + \left[\frac{\partial f(u)}{\partial X}\right]_{i,i+1} = 0$$

The propagation of a shock wave is presented on fig.3 and compared with the analytic solution.

a) shock in the middle of a cell

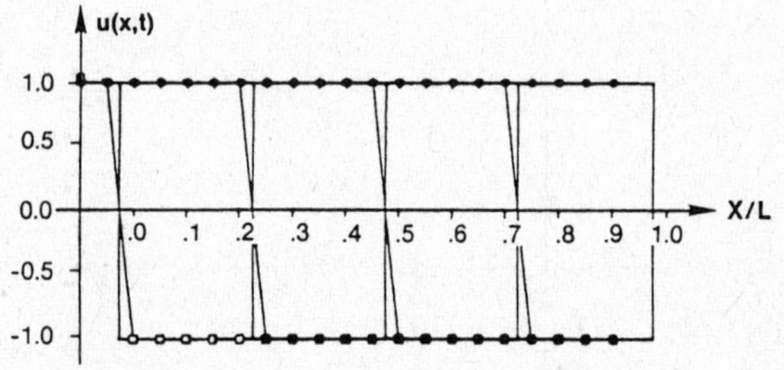

Fig. 3 $u_l = 1$ $u_r = -0{,}98$

b) shock at the interface of two cells

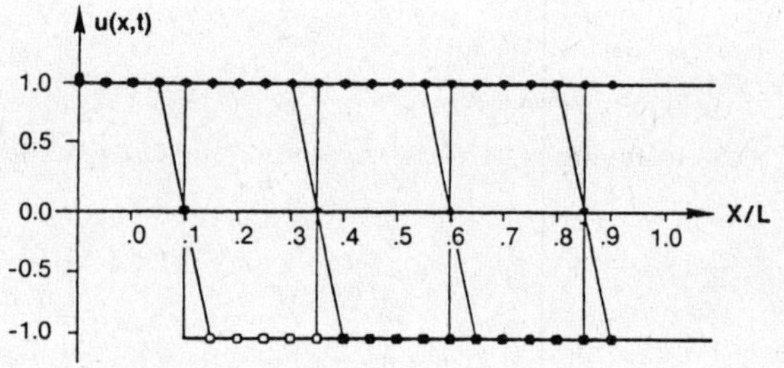

We consider next the case of an expansion with a sonic point, fig.4

Fig. 4

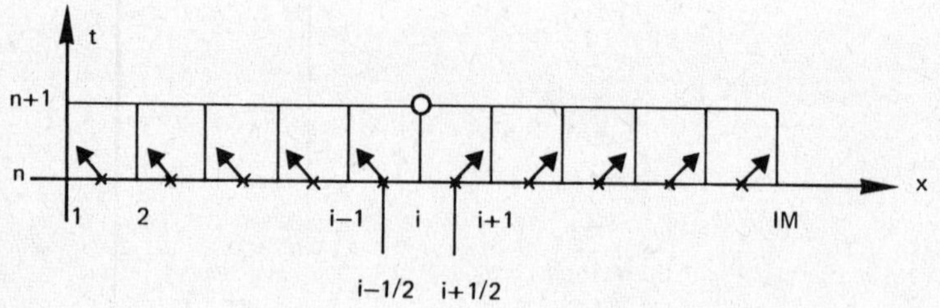

We have IM unknowns and IM−1 box equations. At the end points no boundary condition is given. We need an extra equation at the sonic point i since we cannot use the box [i−1,i] nor the box [i,i+1] because the propagation velocities $u_{i-1/2} \leqslant 0$ $u_{i+1/2} > 0$ are oriented away from point i, and the corresponding box schemes are unstable separately or combined. We use at the sonic point a non-conservative discretization of the equation in the smaller box bounded by point i and the sonic point. If $u_i > 0$ this is depicted fig.5.

Fig. 5

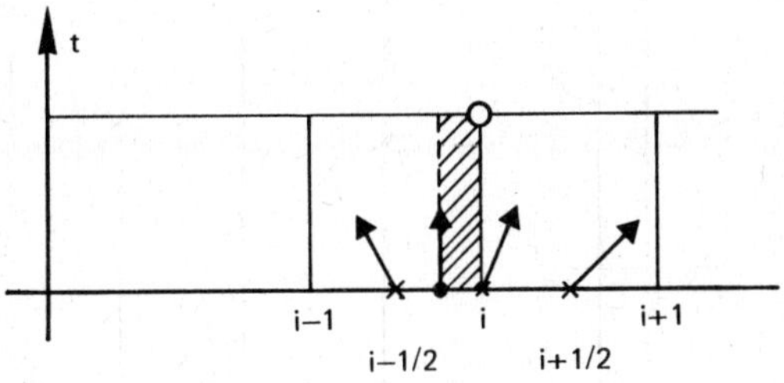

In this case the scheme reads:

$$\frac{u_i^{n+1} - u_i^n}{\Delta t} + \frac{u_i^n + u^*}{2} \; \frac{u_i^n - u_{i-1}^n}{\Delta X} = 0$$

In summary the box scheme applied to the inviscid Burgers equation is the following:

- $u_{i-1/2}^n > 0$ and $u_{i+1/2}^n > 0$ supersonic point

$$\frac{u_i^{n+1} - u_i^n}{\Delta t} + \frac{f(u_i^n) - f(u_{i-1}^n)}{\Delta X} = 0 \qquad \Delta t \leqslant \frac{\Delta X}{u_{i-1/2}^n}$$

- $u_{i-1/2}^n > 0$ and $u_{i+1/2}^n \leqslant 0$ shock point

$$\frac{u_i^{n+1} - u_i^n}{\Delta t} + \frac{f(u_i^n) - f(u_{i-1}^n)}{\Delta X} + \frac{f(u_{i+1}^n) - f(u_i^n)}{\Delta X} = 0 \qquad \Delta t \leqslant \frac{\Delta X}{u_{i-1/2}^n - u_{i+1/2}^n}$$

• $u^n_{i-1/2} \leqslant 0$ and $u^n_{i+1/2} > 0$ sonic point

$u^n_i > 0$

$$\frac{u_i^{n+1} - u_i^n}{\Delta t} + \frac{u_i^n}{2}\,\frac{u_i^n - u_{i-1}^n}{\Delta X} = 0 \qquad \Delta t \leqslant \frac{2\Delta X}{u_i^n}$$

$u^n_i \leqslant 0$

$$\frac{u_i^{n+1} - u_i^n}{\Delta t} + \frac{u_i^n}{2}\,\frac{u_{i+1}^n - u_i^n}{\Delta X} = 0 \qquad \Delta t \leqslant \frac{-2\Delta X}{u_i^n}$$

• $u^n_{i-1/2} \leqslant 0$ and $u^n_{i+1/2} \leqslant 0$ subsonic point

$$\frac{u_i^{n+1} - u_i^n}{\Delta t} + \frac{f(u_{i+1}^n) - f(u_i^n)}{\Delta X} = 0 \qquad \Delta t \leqslant -\frac{\Delta X}{u_{i+1/2}^n}$$

This scheme is very similar to Osher's and Roe's schemes except at the sonic point. All three schemes eliminate expansion shocks and insure a smooth transition from subsonic to supersonic flow, fig.6.

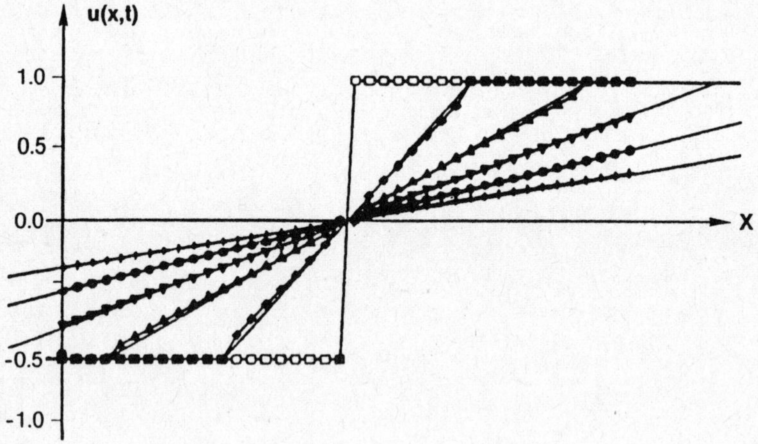

Fig. 6

A linear stability analysis indicates that the scheme is stable when a CFL condition, indicated above in each case, is satisfied.

The scheme is first oder accurate in time and space during the transient phase and yields second order accuracy at steady state, with a source term.

The extension to the quasi one-dimensional transonic small pertubation equation, modelling the isentropic near sonic flow in a slender nozzle, is straightforward:

$$\frac{\partial ug}{\partial t} - \frac{\partial\left[(1-u^2)g\right]}{\partial x} = 0 \qquad g = g(x) \text{ is the given nozzle area fig. 7}$$

Fig. 7

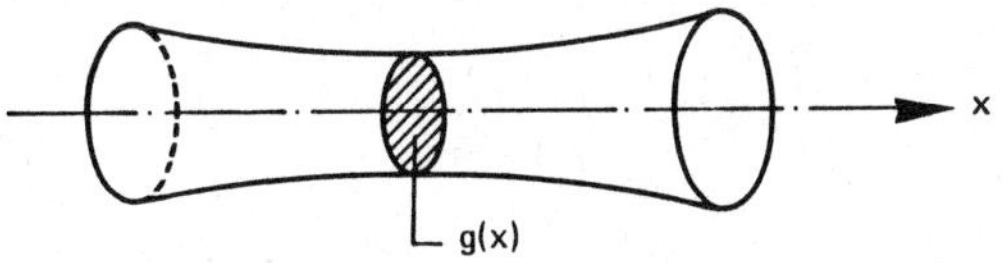

The equation can be developped and written as an inviscid Burgers equation plus a source term:

$$\frac{\partial u}{\partial t} + \frac{\partial u^2}{\partial X} = (1-u^2)\frac{g'}{g}$$

It can be verified numerically that, away from the sonic point which behaves non-linearily, the box-scheme is second order accurate in space at steady state when the source term is discretized as

$$\left[(1-u^2)\frac{g'}{g}\right]_{i-1,i} = \left[1-(u^n_{i-1/2})^2\right]\frac{g_i - g_{i-1}}{\Delta x g_{i-1/2}}$$

However the **exact solution** is obtained independently of Δx when the box discretization is applied directly to the flux $f(x,u) = (1-u^2)\, g(x)$. With this approach the discretization of the source term can be identified with:

$$\left[(1-u^2)\frac{g'}{g}\right]_{i-1,i} = \left[1-(u^n_{i-1})^2\right] \frac{g_i - g_{i-1}}{\Delta x\, g_i}$$

The second order accuracy is best depicted in the comparison with the first order accurate Engquist-Osher's scheme applied to the test case due to Salas, Abarbanel and Gottlieb [19] corresponding to the following equation (fig.8):

$$\frac{\partial u}{\partial t} + \frac{\partial}{\partial x}\left(\frac{u^2}{2}\right) = u(1-u)$$

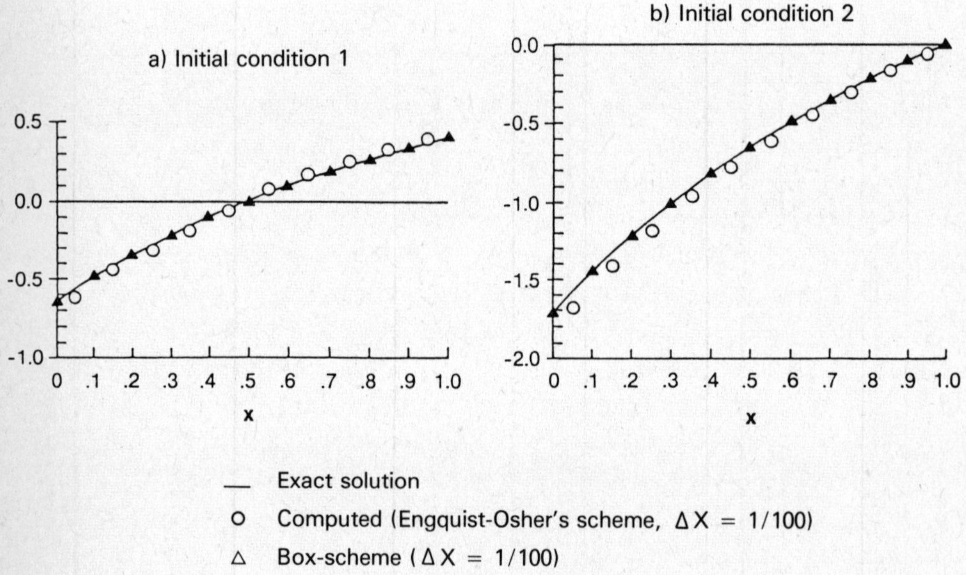

Fig. 8

II Box-scheme applied to the equations of gas dynamics.

The quasi-one dimensional equations of gas dynamics $\frac{\partial W}{\partial t} + \frac{\partial F(W)}{\partial X} = G(W)$ read:

$$\begin{cases} \frac{\partial \rho g}{\partial t} + \frac{\partial \rho u g}{\partial x} = 0 \\ \frac{\partial \rho u g}{\partial t} + \frac{\partial [(\rho u^2 + p) g]}{\partial x} - p g' = 0 \\ \frac{\partial \rho E g}{\partial t} + \frac{\partial \rho u H g}{\partial x} = 0 \,, \; E = \frac{p}{(\Upsilon - 1)\rho} + \frac{u^2}{2} \,, \; H = \frac{\Upsilon p}{(\Upsilon - 1)\rho} + \frac{u^2}{2} \end{cases}$$

Consider first the case of a compression (fig.9).

Fig. 9

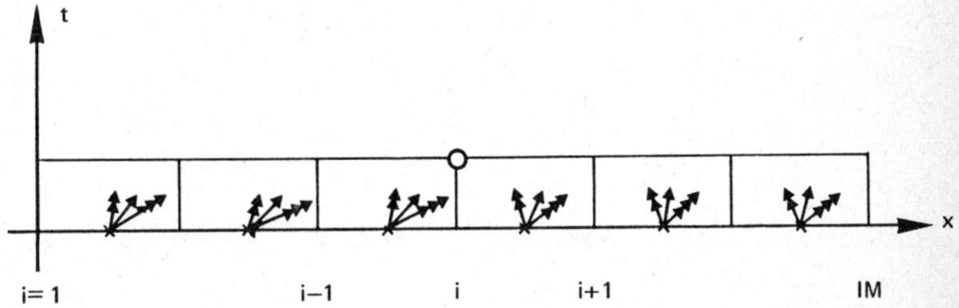

Assuming that the incoming flow is supersonic and the exhaust subsonic we have 3xIM unknowns and 3x(IM-1) box equations +3 upstream +1 downstream boundary conditions. Again we have one more equation than unknowns and this corresponds to the 4 characteristics converging towards the shock point i. Thus at a shock point the discretization is spread on two boxes:

$$\frac{W_i^{n+1} - W_i^n}{\Delta t} + \frac{F(W_i^n) - F(W_{i-1}^n)}{\Delta x} + \frac{F(W_{i+1}^n) - F(W_i^n)}{\Delta x} = G(W_i^n)$$

One can show that this scheme conserves fluxes and yields the correct shock velocity for upstream or downstream moving shocks.

At a supersonic point all the characteristics of the box [i−1,i] run into point i. The scheme is:

$$\frac{W_i^{n+1} - W_i^n}{\Delta t} + \frac{F(W_i^n) - F(W_{i-1}^n)}{\Delta x} = G(W_{i-1/2}^n)$$

At a subsonic point two characteristics corresponding to the eigenvalues $\lambda^{(1)} = u^n_{i-1/2}$ and $\lambda^{(3)} = (u+a)_{i-1/2}$ come from the left and one characteristic corresponding to $\lambda^{(2)} = (u-a)^n_{i+1/2}$ come from the right. The compatibility relations along the characteristics obtained by multiplying the equation by the $\alpha^{(l)}_k$ components of the left eigenvectors, are used, with:

$$\begin{cases} \alpha^{(l)}_1 = \left[\lambda^{(l)}\right]^2 - 3u\lambda^{(l)} + \dfrac{3+\gamma^2}{2}u^2 - \gamma(\gamma-1)E \\ \alpha^{(l)}_2 = \lambda^{(l)} - \gamma u \\ \alpha^{(l)}_3 = \gamma - 1 \end{cases}$$

Consider now the case of an expansion fig.10.

Fig. 10

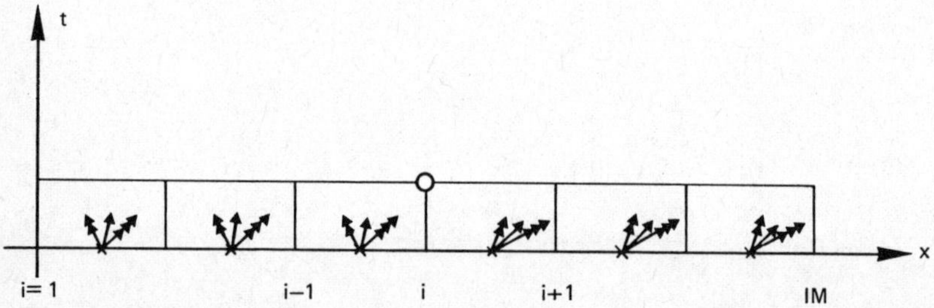

Assuming that the incoming flow is subsonic and the exhaust supersonic we have 3xIM unknowns and 3x(IM−1) box equations +2 upstream boundary conditions. No boundary condition can be specified downstream. We are short of one equation because only two characteristics run into the sonic point i. To compute the solution at this point we consider the 3 eigenvalues and eigenvectors at point i and use a non-conservative scheme to discretize the corresponding compatibility relations. This is summarized below (a is the local speed of sound):

- $(u-a)^n_{i-1/2} > 0$ and $(u-a)^n_{i+1/2} > 0$ supersonic point

$$\begin{pmatrix} 1 & 0 & 0 \\ 0 & 1 & 0 \\ 0 & 0 & 1 \end{pmatrix} \cdot \left[\left(\frac{\partial W}{\partial t}\right)_i + \left(\frac{\partial F(W^n)}{\partial X}\right)_{i-1/2} - \left(G(W^n)\right)_{i-1/2} \right] = 0$$

• $(u-a)^n_{i-1/2} > 0$ and $(u-a)^n_{i+1/2} \leqslant 0$ shock point

$$\begin{pmatrix} 1 & 0 & 0 \\ 0 & 1 & 0 \\ 0 & 0 & 1 \end{pmatrix} \bullet \left[\left(\frac{\partial W}{\partial t} \right)_i + \left(\frac{\partial F(W^n)}{\partial X} \right)_{i-1/2} - \left(G(W^n) \right)_{i-1/2} + \left(\frac{\partial F(W^n)}{\partial X} \right)_{i+1/2} - \left(G(W^n) \right)_{i+1/2} \right] = 0$$

• $(u-a)^n_{i-1/2} \leqslant 0$ and $(u-a)^n_{i+1/2} > 0$ sonic point

$(u-a)^n_i > 0$

$$\begin{pmatrix} 1 & 0 & 0 \\ 0 & 1 & 0 \\ 0 & 0 & 1 \end{pmatrix} \bullet \left[\left(\frac{\partial W}{\partial t} \right)_i + \left(\frac{\partial F(W^n)}{\partial W} \right)_i \bullet \left(\frac{\partial W^n}{\partial X} \right)_{i-1/2} - \left(G(W^n) \right)_{i-1/2} \right] = 0$$

$(u-a)^n_i \leqslant 0$

$$\begin{pmatrix} \alpha_1^{(1)} & \alpha_2^{(1)} & \alpha_3^{(1)} \\ 0 & 0 & 0 \\ \alpha_1^{(3)} & \alpha_2^{(3)} & \alpha_3^{(3)} \end{pmatrix} \bullet \left[\left(\frac{\partial W}{\partial t} \right)_i + \left(\frac{\partial F(W^n)}{\partial W} \right)_i \bullet \left(\frac{\partial W^n}{\partial X} \right)_{i-1/2} - \left(G(W^n) \right)_{i-1/2} \right] +$$

$$\begin{pmatrix} 0 & 0 & 0 \\ \alpha_1^{(2)} & \alpha_2^{(2)} & \alpha_3^{(2)} \\ 0 & 0 & 0 \end{pmatrix} \bullet \left[\left(\frac{\partial W}{\partial t} \right)_i + \left(\frac{\partial F(W^n)}{\partial W} \right)_i \bullet \left(\frac{\partial W^n}{\partial X} \right)_{i+1/2} - \left(G(W^n) \right)_{i+1/2} \right] = 0$$

• $(u-a)^n_{i-1/2} \leqslant 0$ and $(u-a)^n_{i+1/2} \leqslant 0$ subsonic point

$$\begin{pmatrix} \alpha_1^{(1)} & \alpha_2^{(1)} & \alpha_3^{(1)} \\ 0 & 0 & 0 \\ \alpha_1^{(3)} & \alpha_2^{(3)} & \alpha_3^{(3)} \end{pmatrix} \bullet \left[\left(\frac{\partial W}{\partial t}\right)_i + \left(\frac{\partial F(W^n)}{\partial X}\right)_{i-1/2} - \left(G(W^n)\right)_{i-1/2} \right] +$$

$$\begin{pmatrix} 0 & 0 & 0 \\ \alpha_1^{(2)} & \alpha_2^{(2)} & \alpha_3^{(2)} \\ 0 & 0 & 0 \end{pmatrix} \bullet \left[\left(\frac{\partial W}{\partial t}\right)_i + \left(\frac{\partial F(W^n)}{\partial X}\right)_{i+1/2} - \left(G(W^n)\right)_{i+1/2} \right] = 0$$

$\frac{\partial F(W^n)}{\partial W}$ is the Jacobian matrix. It is used in the non-conservative discretization at the sonic point and evaluated at point i. The coefficients β^l_k are the following:

$\beta^1_1 = 0$ $\beta^1_2 = 1$ $\beta^1_3 = 0$

$\beta^2_1 = -\frac{3-\gamma}{2} u^2$ $\beta^2_2 = (3-\gamma) u$ $\beta^2_3 = \gamma - 1$

$\beta^3_1 = (\gamma - 1) u^3 - \gamma u E$ $\beta^3_2 = \gamma E - \frac{3(\gamma-1)}{2} u^2$ $\beta^3_3 = \gamma u$

Note that the scheme requires a 3x3 matrix inversion at the sonic point when $(u-a)^n_i \leqslant 0$ and at the subsonic point.

A linear stability analysis requires the computation of the eigenvalues of the discrete Jacobian matrix which has not been carried out analytically.

On the basis of numerical tests it has been found that the scheme is stable when the CFL condition $\Delta t \leqslant \frac{\Delta X}{(u+a)^n_{i-1/2}}$ $i=2,\ldots,IM$ is satisfied over the domain.

The box scheme is first order accurate in the transient flow and second order accurate in space at steady state.

All the **boundary conditions are natural conditions.**

At a subsonic inlet, the two compatibility relations corresponding to the characteristics entering the domain are replaced by the linearized total enthalpy and entropy equations:

$$[(\gamma-1)u_1^2 - \gamma E_1]\delta(\rho_1 g_1) - (\gamma-1)u_1\,\delta(\rho_1 u_1 g_1) + \gamma\delta(\rho_1 E_1 g_1) = \rho_1 g_1 (H_0 - H_1)$$

$$[\tau_1 + \gamma\frac{(\gamma-1)}{2}\frac{u_1^2}{a_1^2} - \gamma]\delta(\rho_1 g_1) - \gamma(\gamma-1)\frac{u_1}{a_1^2}\delta(\rho_1 u_1 g_1) + \gamma(\gamma-1)\frac{1}{a_1^2}\delta(\rho_1 E_1 g_1) = \rho_1 g_1 (\tau_0 - \tau_1)$$

where $\delta(W_1) = W_1^{n+1} - W_1^n$, H_0 is the stagnation enthalpy and $\tau_0 = \log\frac{\gamma p_0}{\rho_0{}^\gamma}$ is the given entropy.

At a subsonic exhaust the pressure p_e is specified and the compatibility relation along the entering charasteristic is replaced by:

$$\frac{u_{IM}^2}{2}\delta(\rho_{IM} g_{IM}) - u_{IM}\,\delta(\rho_{IM} u_{IM} g_{IM}) + \delta(\rho_{IM} E_{IM} g_{IM}) = g_{IM}\left(\frac{p_e}{\gamma-1} + \rho_{IM}\frac{u_{IM}^2}{2} - \rho_{IM} E_{IM}\right)$$

The results presented concern first an upstream moving shock (fig.11). The numerical solution is compared with the analytic solution. The velocity of the shock is in good agreement with the exact solution. The shock structure is spread on at most three boxes.

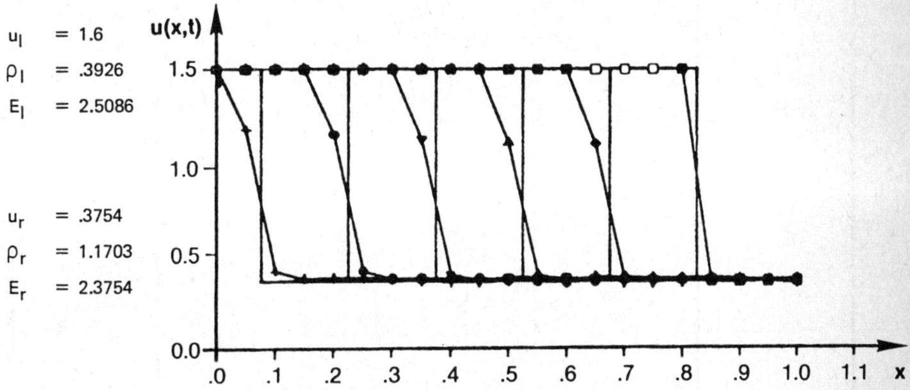

Fig. 11

The shock tube test case due to Sod is then examined. The box scheme results are compared with the analytic solution and with the Lagrangian Godunov's method used in ref [20], fig. 12. Clearly the box-scheme is less dissipative than Godunov's scheme in the expansion, and there is no overshoot in the energy plot. The shock is captured in 3 cells, but the contact discontinuity is spread in the usual way with an eulerian scheme.

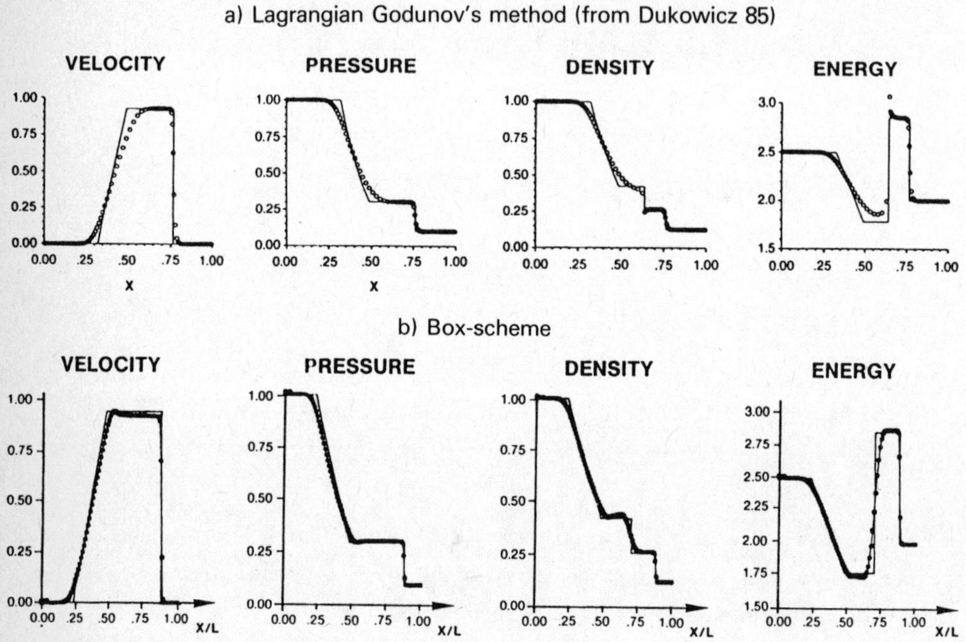

Fig. 12

Next we consider the steady shock in a diverging nozzle and compare with the exact solution and the results of various numerical experiments of Yee, Warming and Harten [6]. The box-scheme yields the exact solution regardless of Δx fig.13-14. The reason for this is not yet understood. The shock is captured in two cells and the intermediate point of artificial structure depends continuously on the data.

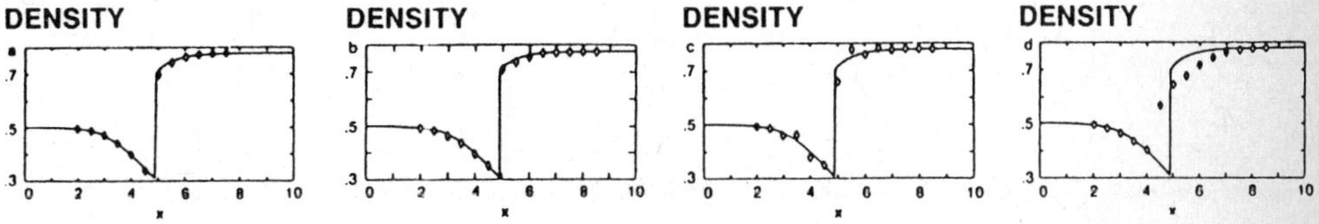

a) Explicit TVD

b) Implicit TVD

c) Conventional implicit

d) First order implicit flux vector splitting

(from Yee, Warming and Harten 85)

Fig. 13

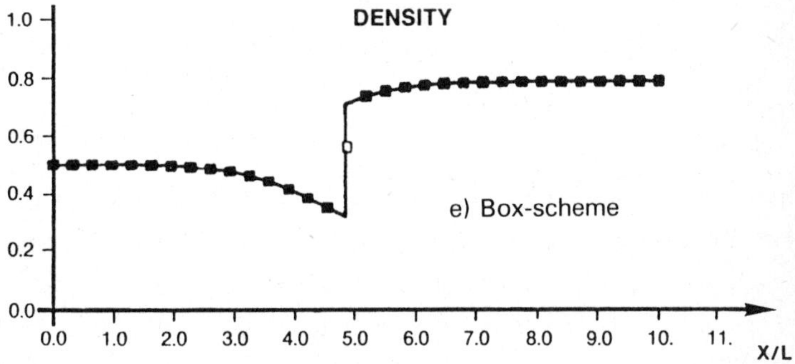

a) 16 point mesh

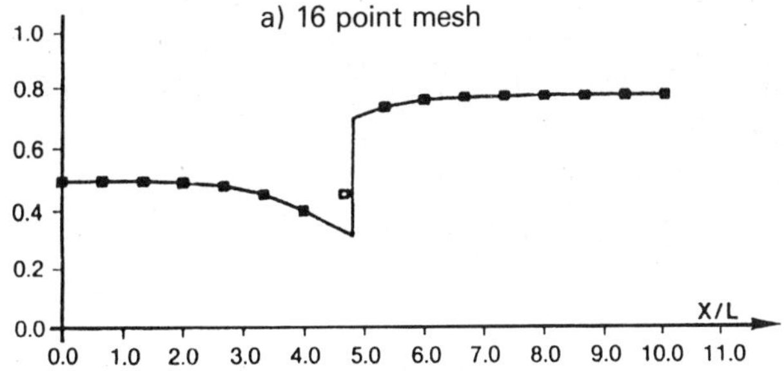

Fig. 14

b) 51 point mesh

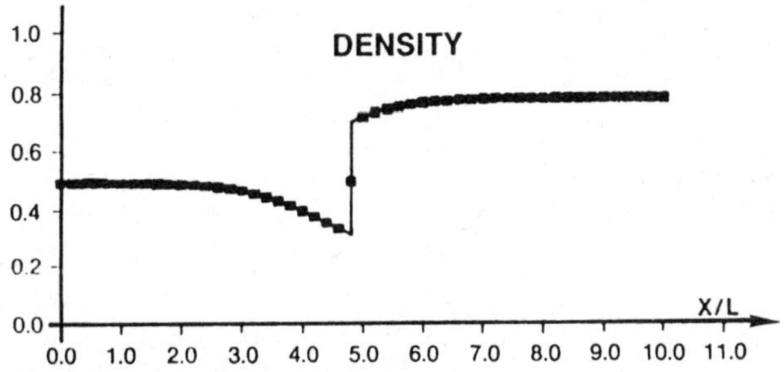

III Preliminary results in two-dimensions

There are not many test cases in two-dimensions for which an exact solution is known. In steady supersonic flow there is the shock reflection problem. A uniform horizontal supersonic stream is deviated down by an oblique shock wave which reflects on the x-axis and turns the flow back to its initial direction. The shock waves are sufficiently weak to maintain the flow supersonic everywhere and the solution can be found in shock tables.

In order to investigate the extension of the box-scheme to the Euler equations, in two dimensions we have chosen to solve a simpler set of equations obtained by perturbing of a uniform horizontal flow. The system reduces to two coupled first order partial differential equations which possess the main properties of the full system when weak shocks are computed:

$$\begin{cases} -M_0^2 \dfrac{\partial u}{\partial t} - u_0 \left\{ (M_0^2 - 1) \dfrac{\partial u}{\partial x} - \dfrac{\partial v}{\partial y} \right\} = 0 \\ \\ \dfrac{\partial v}{\partial t} + u_0 \left\{ \dfrac{\partial v}{\partial x} - \dfrac{\partial u}{\partial y} \right\} = 0 \end{cases}$$

The rectangular domain is discretized by a cartesian mesh with spacings Δx and Δy. One of the major advantages of the box scheme is that the control volumes used to write the flux balance coïncide with the mesh cells. This implies that at steady state a contour integral of fluxes following cell sides will vanish independently of the number of cells encircled.

The solution of the shock reflection problem is presented, fig. 15. The numerical solution coïncides with the exact solution for the 41x21 mesh with $M_0 = \sqrt{5}$.

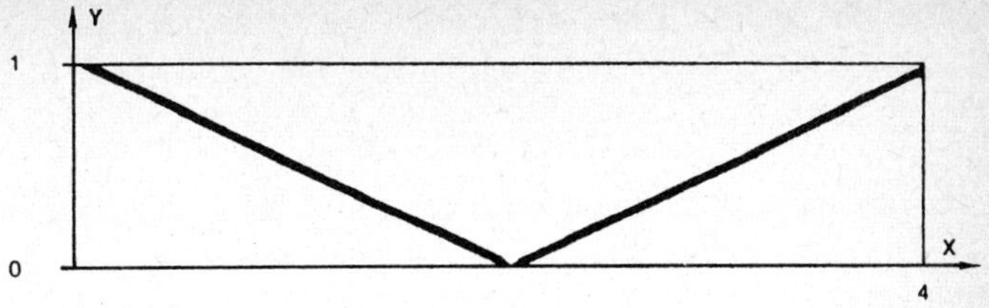

a) Iso mach lines

Fig. 15

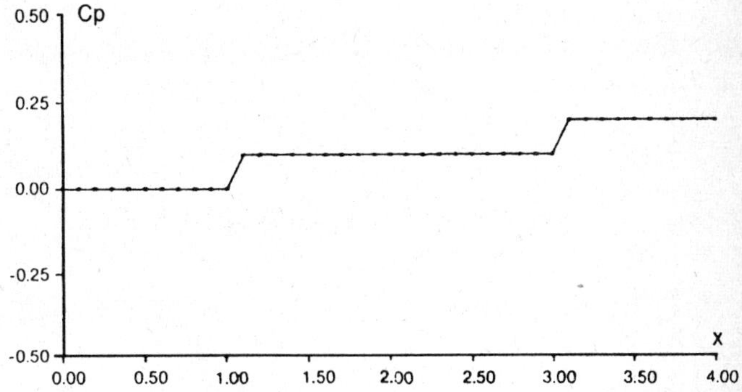

b) Pressure coefficient at y = 2

Another application is the supersonic flow past a biconvex profile in a channel. A small pertubation tangency condition is used on the profile and transfered to the x-axis. The profile is given by $y^+(x) = \varepsilon\,(1-x^2) \quad -1 \leqslant x \leqslant 1$.

The boundary condition on the slit is:

$$v_i^+ = -2\,\varepsilon x_i - \alpha \qquad (\,\alpha \text{ incidence in radian}).$$

The flow field is divided into two adjacent subdomains $y \geqslant 0$ and $y \leqslant 0$. The x-axis is a double line. The box scheme is easily matched along the double line in front and downstream of the profile. The result is presented fig. 16.

$$\varepsilon = 6\% \quad M_0 = 1.41 \qquad \alpha = 5^\circ$$

Fig.16

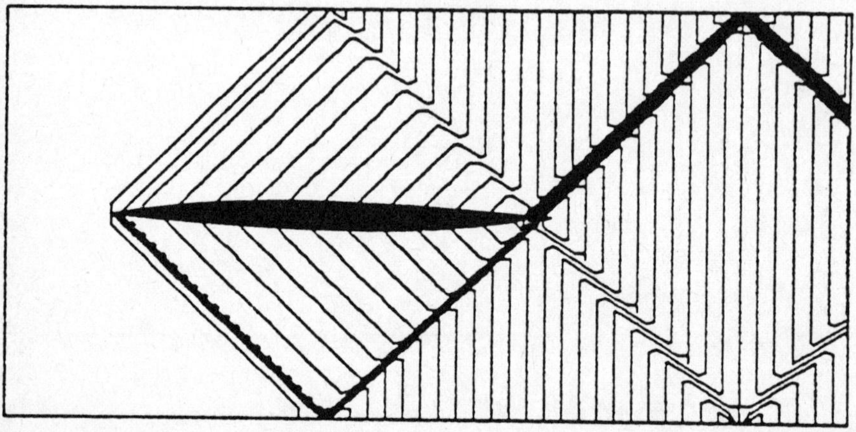

Conclusion

In all test cases performed to date, the box scheme exhibited good properties of convergence, stability and accuracy. The resolution of shock waves was found remarkably sharp.

For unsteady problems, the version presented is only first order accurate but it could be made easily second order within a two-step procedure. The extension to the full two-dimensional Euler system is in progress. From the view point of integration of first order partial differential equations with boundary conditions, the box scheme seems to be a natural scheme.

REFERENCES

[1] R.F. Warming and R.M. Beam, "upwind second-order difference schemes and applications in unsteady Aerodynamic flow", Proceedings of AIAA 2nd Comp . Fl. Dyn. Conf., Hartford, CT, june 19-20, 1975. AIAAJ, **14**, 1976.

[2] J.L. Steger and R.F. Warming, "Flux vector splitting of the inviscid gas dynamic equations with application to finite difference methods", J. Comp. Physics, **40**, 1981.

[3] A. Harten, "A high resolution scheme for the computation of weak solutions of hyperbolic conservation laws", J. Comp. Phys., vol.49,1983.

[4] A. Harten, "On a class of high resolution total-variation-stable finite difference schemes", SIAM J. Num. Anal., vol.21,1984.

[5] H.C. Yee and A. Harten, "implicit TVD schemes for hyperbolic conservation laws in curvilinear coordinates", AIAA 7th. CFD Conferences, Cincinnati, Ohio, July 15-17, 1985. AIAA Paper n° 85-1513.

[6] H.C. Yee, R.F. Warming and A. Harten, "Implicit total variation diminishing (TVD) schemes for steady-state calculation", J. Comp. Phys., Vol.57, 1985.

[7] E.M. Murman and J.D. Cole" Calculation of plane steady transonic flows", AIAA J., **9**, n° 1, 1971.

[8] E.M. Murman, "Analysis of embedded shock waves calculated by relaxation methods" AIAA J., **12**, 1974.

[9] B. Engquist and S. Osher, "Stable and entropy Condition satisfying approximations for transonic flow calculations", Math. Comp., **34**, 1980.

[10] B. Engquist and S. Osher, "one sided difference approximations for nonlinear conservation laws", Math. Comp., **36**, 1981.

[11] B. Van Leer, "Flux-Vector splitting for the Euler equations", ICASE Rept. n°82-30 sept. 1982.

[12] P.L. Roe, "Generalized formulation of TVD Lax-Wendroff schemes", ICASE Rept. n°84-53, oct. 1984.

[13] G. Moretti "the λ-scheme", Comp. and Fluids, **7**, 1979.

[14] G. Moretti and L. Zannetti, "a new improved computational technique for two-dimensional unsteady compressible flows", AIAA paper 82-0168, 1982.

[15] M. Pandolfi, "A contribution to the numerical prediction of unsteady flows", AIAA J, **22**, 1984.

[16] A Dadone and M. Napolitano, "Efficient Transonic flow solutions to the Euler equations", AIAA paper n° 83-0258, jan. 1983.

[17] H.B. Keller, "a new difference scheme for parabolic problems", in Numerical Solution of Partial Differential Equation, J. Bramble, ed., Vol.II, Academic Press, NY, 1970.

[18] S.F. Wornom, "Application of compact difference schemes to the conservative Euler equations for one-dimensional flows.", NASA TM 83262, may 1982.

[19] M.D. Salas, S. Abarbanel and D. Gottlieb, "Multiple steady states for characteristic initial value problems", ICASE Rept. n°84-57, november 1984.

[20] J.K. Dukowicz, "A general, non-iterative Riemann solver for Godunov's method", J. Comp. Phys., **61**, 1985.

NUMERICAL TECHNIQUES
IN ELASTOPLASTICITY

J.F. Colombeau and A.Y. Le Roux
UER de Mathématiques et Informatique
Université de Bordeaux I
33405 Talence - France

I - INTRODUCTION.

The equations of elastoplastic dynamics in two dimensional models, have the form of a seven equation hyperbolic system.

Four of them describe the conservation of mass, momentum in the two directions, and total energy, as in classical hydrodynamics. But here, the pressure p has to be replaced by the stress $\sigma = S - pI$ where I denotes the identity in $\mathbb{R}^2$, and S is the stress deviator ; here S is a symetric tensor. The pressure is connected to the density ρ and to the internal energy T by a state law $p = P(\rho,T)$. In many cases, this state law can be easily written as $T = \Phi(\rho,p)$. This is the case for a Mie-Gruneisen state law (recalled in Section VI).

The components $S_{11}, S_{12} = S_{21}, S_{22}$ of S are linked to the other parameters by three further equations involving partial derivatives but which do not have a conservative form. When the symmetries allow a plane-one dimensional model, we have the following system,

$$\rho_t + (\rho u)_x = 0 \ , \tag{1}$$

$$\rho(u_t + u\, u_x) + p_x = S_x \ , \tag{2}$$

$$p_t + u\, p_x + \rho\, c_o^2\, u_x = 0 \ , \tag{3}$$

$$S_t + u\, S_x - c^2 u_x = 0 \ , \tag{4}$$

where ρ denotes the density, u the velocity, $S = S_{11}$ the stress deviator and p the pressure. The parameter c_o is the sound speed, given by

$$c_o^2 = \frac{\partial p}{\partial \rho} + \frac{p}{\rho^2}\,\frac{\partial p}{\partial T} ,$$

or

$$\frac{\partial \Phi}{\partial p}\, c_o^2 = - \frac{\partial \Phi}{\partial \rho} + \frac{p}{\rho^2} .$$

The parameter c depends on the stress deviator :

$$c = \begin{cases} c_1 & \text{for } |S| < S_o , \\ 0 & \text{for } |S| \geqslant S_o , \end{cases}$$

where c_1 is the constant speed of the elastic waves, and S_o is the plasticity point.

Equation (3) may be replaced by the total energy conservation

$$E_t + [(E+p)u]_x = 0 , \quad \text{with } E = \rho(T + u^2/2). \qquad (5)$$

We remark that equation (4) has not a conservative form. This is also true for (3) which can however be written under the conservative form (5). Several questions may be asked for this model :

- <u>How to derive Rankine Hugoniot conditions for shocks</u> ?
- <u>How to build performant numerical schemes</u> ?
- <u>How to define a weak solution</u> ?

In this paper, we shall try to answer these questions, at first on the following simple model coupling the velocity and the stress

$$u_t + u\, u_x = \sigma_x , \qquad (6)$$

$$\sigma_t + u\, \sigma_t = k^2\, u_x , \qquad (7)$$

which allows the same questions. This model can be derived from (1), (2),(3),(4) when the density is a constant and in the elastic domain. Here $k^2 = c_o^2 + c^2$ is supposed to be a constant.

We can find this model in some other applications, such as a wave equation including inertia terms, for example the Rossby waves in an ocean simulation, with a flat bottom.

II - THE SIMPLE ELASTIC MODEL.

The (u,σ) model can also be written with a conservative form, by using the quantity $q = \sigma + u^2/2$. We get

$$u_t + (u^2 - q)_x = 0 , \tag{8}$$

$$q_t + (u^3/3 - k^2 u)_x = 0 . \tag{9}$$

Let us denote by [ ] the jump of a parameter for a shock wave. We put $\xi = [q]/[u]$. Thus ξ satisfies the equation

$$\xi^2 - (u_R + u_L)\xi + (u_L^2 + u_L u_R + u_R^2)/3 - k^2 = 0$$

where u_L , u_R are the values of u on the left and right sides of the discontinuity. This equation has real solutions for $12k^2 - [u]^2 \geq 0$, but for some flows, we can have $12k^2 - [u]^2 < 0$, forbidding the existence of ξ . Thus no Rankine Hugoniot condition may be derived from this conservative model. However, numerical experiments have shown that shocks waves really occur, even for $12k^2 - [u]^2 < 0$.

Now we come back to the system (6),(7), and propose a Lagrange-Euler numerical scheme to solve it. Let $h>0$ be the space meshsize and $\Delta t = rh$ the time increment. Here r is a constant parameter. The scheme is broken into two steps. The first one deals with the convection, that is the inertia terms, using the velocity field, and the second is devoted to the wave propagation. We set $t_n = n\Delta t$, $n \in \mathbb{N}$.

III - THE INERTIA STEP.

For this step, we introduce the space V_h of functions which are constant on any cell $I_i =]ih,(i+1)h[$, for $i \in \mathbb{Z}$. Let us suppose that for $t=t_n$, the values u_i^n and σ_i^n for the velocity and the stress are known in any cell I_i .

By starting from the middle of each cell I_i , we draw the characteristic, which cuts the line $t = t_{n+1}$ at

$$x = x_i^n = (i+1/2)h + u_i^n \Delta t \ . \tag{10}$$

We define now the set W_n of functions which are constant on any cell $J_i^n =]x_{i-1}^n, x_i^n[$. Here we suppose that $x_{i-1}^n \leqslant x_i^n$, that is $1+r(u_i^n-u_{i-1}^n) \geqslant 0$. The projection operators (associated to the usual L^2-norm) on V_h and W_n are respectively denoted by P_h and Q_n .

Let u_o and σ_o be the initial data, assumed to be bounded. We set

$$u^o = P_h u_o \quad , \qquad \sigma^o = P_h \sigma_o \ .$$

Then, at time $t=t_n$, we denote by u^n and σ^n the approximate solution belonging to V_h . First we compute u^{n+1} by

$$u^{n+1} = P_h \ Q_n \ G(\Delta t)u^n \tag{11}$$

where $G(\Delta t)$ denotes the semi group associated to the Burgers equation $u_t+uu_x = 0$. Now, if $T_n(t)$ is the semi group associated to the transport equation $\sigma_t+ G(t)u^n\sigma_x = 0$, we compute σ^{n+1} by

$$\sigma^{n+1} = P_h \ Q_n \ T_n(\Delta t)\sigma^n \ . \tag{12}$$

This gives a Lagrange Euler scheme which has a very simple formulation, since on J_i^n ,

$$Q_n \ G(\Delta t)u^n = \frac{1}{2} (u_i^n + u_{i-1}^n) \ , \tag{13}$$

$$Q_n \ T_n(\Delta t) \ \sigma^n = \frac{1}{2} (\sigma_i^n + \sigma_{i-1}^n) \ . \tag{14}$$

Moreover the projection is very easy to compute since only constant piecewise functions are concerned. An antidiffusion step would be useful to clear up the diffusion brought by the two successive projections. The technique described in [4] can be used for u and σ.

Note that from (13),(14) the velocity and the stress are convected exactly in the same way. This is the most important argument of this scheme, for it gives shocks on u and σ which are in phase. If not, we can see some singularities on the shock, or a wrong shock velocity.

IV - THE WAVE PROPAGATION.

This step corresponds to the approximation of the wave equation

$$u_t = \sigma_x \tag{15}$$

$$\sigma_t = k^2 u_x \quad . \tag{16}$$

We propose an extension of the Godunov scheme, which works without any stability condition. The two backwards characteristics of slope +k and -k are drawn from each point (ih, t_{n+1}), $i \in \mathbb{Z}$. We get the two sequences, at the intersections with the line $t = t_n$, for any $i \in \mathbb{Z}$,

$$y_i = ih - k\,\Delta t \qquad , \qquad z_i = ih + k\,\Delta t \ .$$

Since the functions $Y = \sigma - ku$ and $Z = \sigma + ku$ satisfy

$$Y_t + k\,Y_x = 0 \qquad , \qquad Z_t - k\,Z_x = 0 \ ,$$

we can easily compute the average of Y or Z on any cell I_i, at t_{n+1} from the values, at t_n , of Y on the cell (y_i, y_{i+1}) and of Z on the cell (z_i, z_{i+1}) for $i \in \mathbb{Z}$. We get, defining $j, m \in \mathbb{N}$ by

$$y_i \leqslant jh < y_{i+1} \qquad , \qquad z_i \leqslant mh < z_{i+1} \ ,$$

and

$$\alpha = i+1-rk-j \qquad , \qquad \beta = i+1+rk-m \ ,$$

$$Y_i^{n+1} = \alpha(\sigma_j^n - k\,u_j^n) + (1-\alpha)(\sigma_{j-1}^n - k\,u_{j-1}^n) \tag{17}$$

$$Z_i^{n+1} = \beta(\sigma_m^n + k\,u_m^n) + (1-\beta)(\sigma_{m-1}^n + k\,u_{m-1}^n) \ . \tag{18}$$

Since k is a constant, we have here α and β nondepending on i. Eventually we compute,

$$\sigma_i^{n+1} = \frac{1}{2}\,(Y_i^{n+1} + Z_i^{n+1}), \qquad u_i^{n+1} = \frac{1}{2k}\,(Z_i^{n+1} - Y_i^{n+1}). \tag{19}$$

For a CFL number less than one, this is exactly the Godunov scheme since $j=i$ and $m=i+1$. This scheme is stable in L^∞ norm and variation and the convergence towards a generalized solution can be proved, even for a CFL number larger than one. See [2] for a similar proof. The numerical results are similar to the results usually obtained by the Godunov method for conservation laws. They may be improved by an antidiffusion step adapted to the propagation step.

V - COMPARISON WITH A CONSERVATIVE FORM.

The conservation law

$$U_t + F(U)_x = 0 \qquad (20)$$

with $U=(u,q)$ and $F(U)=(u^2-q,u^3/3-k^2u)$ has been computed, using the two step Lax-Friedrichs scheme, with antidiffusion. This scheme is stable in L^∞ norm for many examples, under the Courant-Friedrichs Levy condition.

Denoting by W_h the set of functions whose translation of the rate $h/2$ belongs to V_h, this scheme may be written, for $U^n \in V_h^2$ being the approximate solution at $t=t_n$,

$$U^{n+1} = [A]\, P_h\, \Sigma(\Delta t/2)\, Q_h\, \Sigma(\Delta t/2) U^n \qquad (21)$$

where $\Sigma(t)$ is the semi group associated to (20), P_h and Q_h are respectively the projection operators on V_h^2 and W_h^2 and $[A]$ is an antidiffusion operator which may be the same as the one given in [4].

The computations by the scheme (21) and the previous scheme (11), (12),(17),(18),(19) have been performed with the same values of the parameters. We see on Figure 1 that similar results are obtained for $k=1.33$ and a jump $[u]\simeq 1.6$ after 100 time steps. This corresponds to the case (8),(9). The results are quite different if $12k^2-[u]^2<0$,

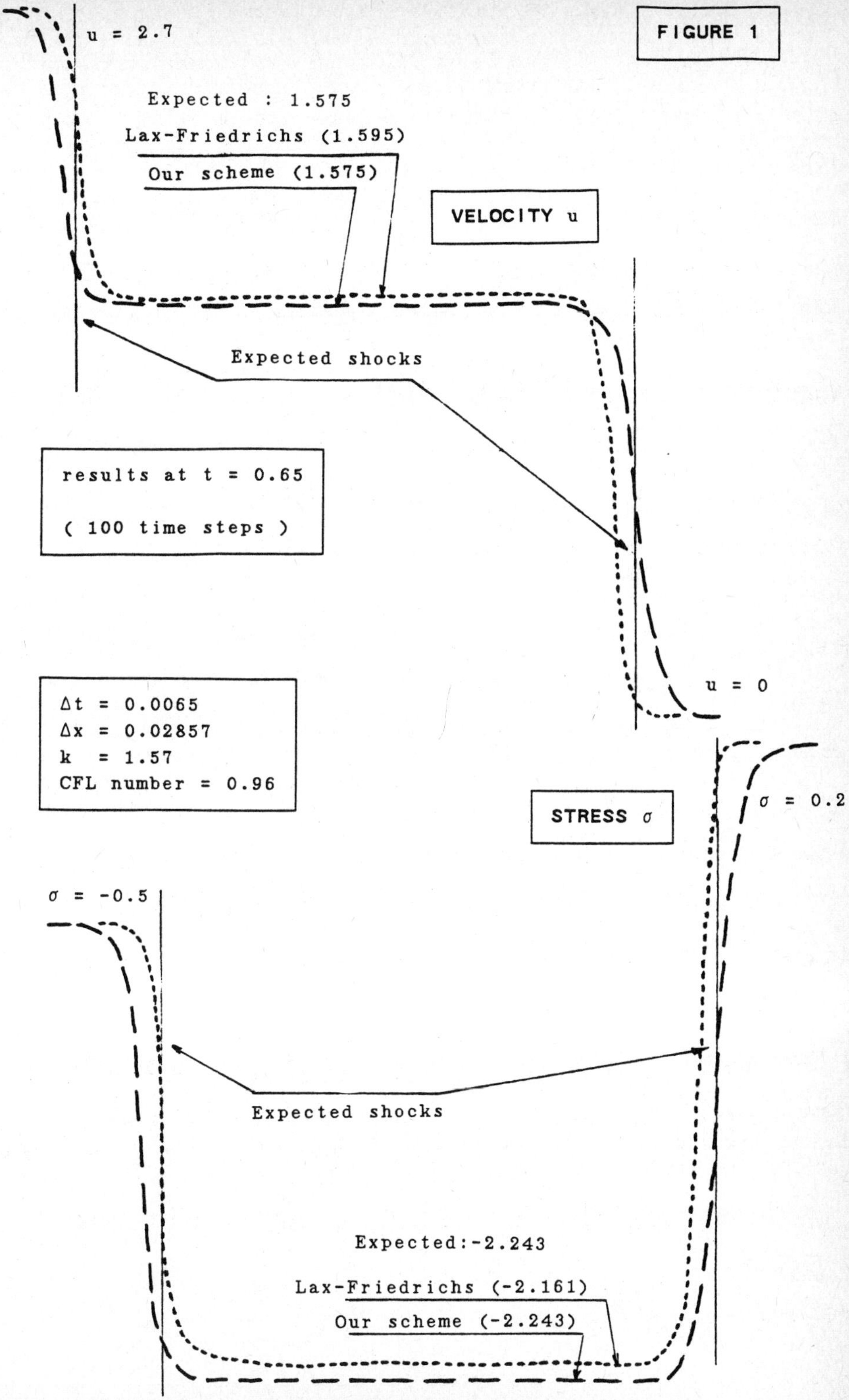
u = 2.7
FIGURE 1
Expected : 1.575
Lax-Friedrichs (1.595)
Our scheme (1.575)
VELOCITY u
Expected shocks
results at t = 0.65
(100 time steps)
u = 0
Δt = 0.0065
Δx = 0.02857
k = 1.57
CFL number = 0.96
STRESS σ
σ = 0.2
σ = -0.5
Expected shocks
Expected:-2.243
Lax-Friedrichs (-2.161)
Our scheme (-2.243)

even after a few time steps. As shown in Figure 2, the stress computed from the conservation law holds a top value at 16.57, for a stress lying in (0,1) by the non-conservative model. Other

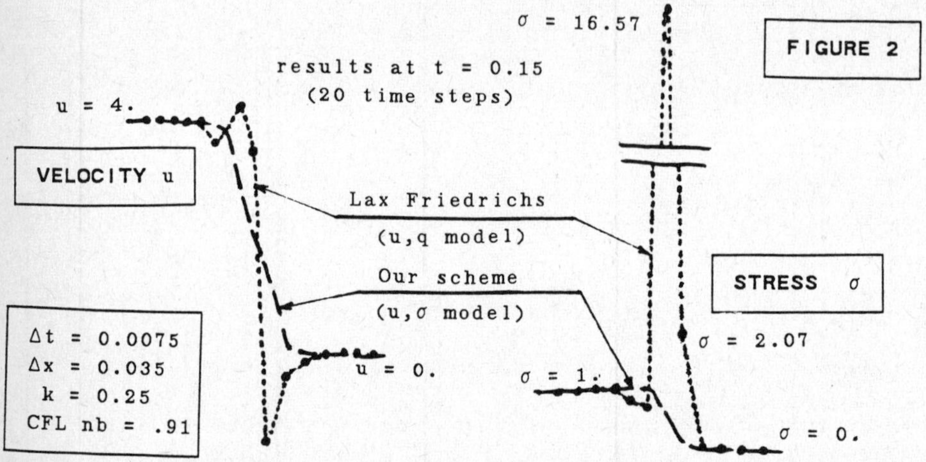

computations performed by the Godunov scheme give similar results as this last scheme, in agreement with the fact that a monotonous shape for the stress was expected by the engineers in practice. A smaller singularity also appears for the velocity computed from the conservation law, which is not seen in the nonconservative case.

VI - THE RANKINE HUGONIOT CONDITION.

To derive a Rankine Hugoniot condition from a conservative law is obvious. This is also easy to perform on a nonconservative system by assuming that all the shocks are shaped in the same fashion, that is in phase. This means here that the multiplication of the Heaviside function by the Dirac mass must be equal to the half of this Dirac mass : $H.\delta = \delta/2$. This is false when the functions are not in phase. For example, the following one dimension system of elastodynamics

$$v_t + u\, v_x - v\, u_x = 0 \tag{21}$$

$$u_t + u\, u_x + v(p_x - S_x) = 0 \tag{22}$$

$$p_t + u\, p_x + c^2/v\, u_x = 0 \tag{23}$$

$$S_t + u\, S_x - k^2\, u_x = 0 \tag{24}$$

gives exactly the same Rankine Hugoniot condition as (1),(2),(3), and (4), in the case of a Mie-Gruneisen state law, that is for

$$P(\rho,T) = \gamma\, \rho\, T + K(\rho) .$$

Here $v=1/\rho$ is the specific volume and $K(\rho)$ is a nonnegative function of ρ, which is given from experiments. This is no longer true by replacing (21) by (1) or (22) by (2). This shows that velocities and densities are never in phase, the momenta and specific volumes too. The formulation (21),(22),(23),(24) has many advantages. First it allows more general state laws for the stress (note that by replacing k^2 by $v\, k^2$ in (24), we can give a conservative form to this system with another Rankine Hugoniot condition). Then the computation of the pressure may be performed more easily, with the same scheme as for the stress deviator in the elastic domain.

VII - EXAMPLE : A NONCONSERVATIVE MODEL FOR HYDRODYNAMICS.

We consider for $1<\gamma<3$, the system

$$v_t + u\, v_x - v\, u_x = 0 ,$$

$$u_t + u\, u_x + v\, p_x = 0 ,$$

$$p_t + u\, p_x + \gamma\, p\, u_x = 0 ,$$

which is cut into two parts, the convection and the propagation of waves. This system corresponds to the usual hydrodynamics.

The convection corresponds to

$$u_t + u\, u_x = 0 , \quad v_t + u\, v_x = 0 , \quad p_t + u\, p_x = 0 .$$

and is discretized as follows, by the Godunov scheme, using the same notations as before in §3 and for a CFL-number less than 0.5.

For the <u>flux computation</u>, we set, for $j=i-1/2$,

$u_j^n = u_i^n$, $p_j^n = p_i^n$, $v_j^n = v_i^n$, if $u_i^n < 0$ and $u_i^n + u_{i-1}^n < 0$;

$u_j^n = u_{i-1}^n$, $p_j^n = p_{i-1}^n$, $v_j^n = v_{i-1}^n$,if $u_{i-1}^n > 0$, $u_i^n+u_{i-1}^n>0$;

$u_j^n = 0$, $p_j^n = [u_i^n p_{i-1}^n - u_{i-1}^n p_i^n] / [u_i^n - u_{i-1}^n]$

$v_j^n = [u_i^n v_{i-1}^n - u_{i-1}^n v_i^n] / [u_i^n - u_{i-1}^n]$

if $u_i^n > 0$ and $u_{i-1}^n < 0$.

For the <u>transport step</u>, we set

$$u_i^{n+1/2} = u_i^n + r \left((u_j^n)^2 - (u_{j+1}^n)^2 \right) / 2$$

$$p_i^{n+1/2} = p_i^n + r \left((u_i^n+u_j^n)(p_j^n-p_i^n) - (u_i^n+u_{j+1}^n)(p_{j+1}^n - p_i^n) \right) / 2$$

$$v_i^{n+1/2} = v_i^n + r \left((u_i^n+u_j^n)(v_j^n-v_i^n) - (u_i^n+u_{j+1}^n)(v_{j+1}^n - v_i^n) \right) / 2 .$$

The second part is the <u>wave propagation</u>: This may be performed by projecting the exact solution of a linearized Riemann problem, using the impedance $d=c/v$ where c is the sound speed. We note $m=n+1/2$ and $j=i+1/2$ in order to clarify the notations. We have on any cell

$$v_t - v_i^m u_x = 0,\ u_t + v_i^m p_x = 0,\ p_t + \gamma p_i^m u_x = 0 ,$$

with $v(x,t_n) = v_i^m$, $u(x,t_n) = u_i^m$, $p(x,t_n)= p_i^m$, for $x < jh$,

and

$$v_t - v_{i+1}^m u_x = 0 ,\ u_t + v_{i+1}^m p_x = 0,\quad p_t + \gamma p_{i+1}^m u_x = 0,$$

with $v(x,t_n) = v_{i+1}^m$, $u(x,t_n) = u_{i+1}^m$, $p(x,t_n) = p_{i+1}^m$ for $x > jh$.

By setting $d_i^m = [\gamma p_i^m / v_i^m]^{1/2}$ and $c_i^m = [\gamma p_i^m v_i^m]^{1/2}$, as the cell impedance and the cell sound speed respectively, we get

$$u_j^m = [\ u_i^m d_i^m + u_{i+1}^m d_{i+1}^m - p_{i+1}^m + p_i^m\ ] / (d_{i+1}^m + d_i^m) ,$$

$$p_j^m = p_i^m - d_i^m (\ u_j^m - u_i^m\ ) ,$$

$$v^m_{j,R} = v^m_{i+1} + (u^m_{i+1} - u^m_j) \,/\, d^m_{i+1} ,$$

$$v^m_{j,L} = v^m_i + (u^m_j - u^m_i) \,/\, d^m_i .$$

Then we compute, on each cell I_i,

$$u^{n+1}_i = r\, c^m_i \,(u^m_{j-1} + u^m_j) + (1 - 2\, r\, c^m_i)\, u^m_i ,$$

$$p^{n+1}_i = r\, c^m_i \,(p^m_{j-1} + p^m_j) + (1 - 2\, r\, c^m_i)\, p^m_i ,$$

$$v^{n+1}_i = r\, c^m_i \,(v^m_{j-1,R} + v^m_{j,L}) + (1 - 2\, r\, c^m_i)\, v^m_i .$$

This scheme may be extended to CFL numbers less than 1, by modifying the last step . An antidiffusion procedure may also be added after both steps. This scheme, without antidiffusion gives results which are similar to the usual results computed by the Godunov scheme for the conservative form. These results are strongly improved after the Antidiffusion step.

CONCLUDING REMARKS.

The knowledge of how to compute a non conservative law will be useful for unstationary elastoplastic models. The problems (see [3], [5]) occuring in the multiplication of distributions still influence the numerical results, and a scheme must be built in order to compute a well shaped solution, as expected by the engineers. This is true for the scheme given in §7. An extended version of this scheme is given in [1], for a general elastoplastic model. This work was supported by the Centre d'Etudes de Gramat.

REFERENCE.

[1] J.J. Cauret, *Thèse*, Université de Bordeaux (1986).

[2] J.J. Cauret, J.F. Colombeau, A.Y. Le Roux, *Comptes Rendus de l'Académie des Sciences* (1986) 302,I, p435-437.

[3] J.F. Colombeau, *New generalized functions and multiplication of*

Distributions. North-Holland (1984).

[4] A.Y. Le Roux, P. Quesseveur, SIAM J. Numer. Anal. vol.21, n°5, pp.985-994 (1984).

[5] L. Schwartz, Comptes Rendus de l'Académie des Sciences 239, Série A, pp.847-848 (1954).

Added in proof.

Derivation of jump conditions for shocks in nonconservative form is exposed in

J.F. Colombeau, A.Y. Le Roux : Multiplications of Distributions in Elasticity and Hydrodynamics.

A New Mathematical Analysis and Product of Distributions appearing in Nonlinear Equations of Physics.

Definition of a weak solution and a solution of the Cauchy problem for the system (6) (7) are exposed in

J.J. Cauret, J.F. Colombeau, A.Y. Le Roux : Discontinuous generalized solutions of nonlinear nonconservative hyperbolic equations.

Other numerical schemes originating from studies of "multiplications of distributions" will be published subsequently.

AN IMPLICIT CENTERED SCHEME WHICH GIVES NON-OSCILLATORY STEADY SHOCKS

Virginie DARU and Alain LERAT

Laboratoire de Simulation Numérique en Mécanique des Fluides
E.N.S.A.M.
151 Boulevard de l'Hôpital
75640 Paris cédex 13 - France

ABSTRACT :

We consider the calculation of steady weak solutions of an hyperbolic system in one-space dimension. By using an implicit centered scheme of second-order accuracy with an appropriate treatment of the boundary conditions, we obtain a quick convergence to a non-oscillatory steady solution.

1 - INTRODUCTION

This paper is devoted to the numerical solution of an hyperbolic system of conservation laws in one-space dimension using a centered 3-point scheme without artificial viscosity. Such a method is known to be non-TVD and to produce oscillatory shock profiles. However our purpose is to show that if the method is applied only to reach a steady-state and if it is implicit in time, then it can be an efficient and accurate approach, much simpler to implement than upwind methods.

The basic scheme we used is presented in Section 2. It is a block-linearly implicit sheme introduced in [1] from a study of a class of schemes of second-order accuracy (see also [2] and [3]). The question of its convergence rate to a steady solution is addressed in Section 3. Section 4 deals with the accuracy at steady-state. Using a large time step, we can obtain an accurate and non-oscillatory steady solution. Finally an application to a quasi-one dimensional flow in a nozzle is described in Section 5.

2 - THE IMPLICIT CENTERED SCHEME

We consider a system of m conservation laws in one-space dimension :

$$w_t + f(w)_x = 0 \qquad (2.1)$$

where $w = w(x, t) \in \Omega \subset \mathbb{R}^m$ and $f : \Omega \rightarrow \mathbb{R}^m$ is a smooth function. The system is hyperbolic, i.e. the jacobian matrix :

$$A(w) = \frac{df}{dw}(w)$$

has m real eigenvalues and can be diagonalized for every w in Ω.

We approximate system (2.1) by the implicit time-differencing :

$$\Delta w + \beta \frac{\Delta t^2}{2} [(A^n)^2 (\Delta w)_x]_x = - \Delta t\, f_x^n + \frac{\Delta t^2}{2} (A^n f_x^n)_x \qquad (2.2)$$

where w^n is the numerical solution at time $t_n = n\Delta t$, Δt is the time step,

$$f^n = f(w^n), \qquad A^n = A(w^n), \qquad \Delta w = w^{n+1} - w^n$$

and β is a parameter ($\beta \in \mathbb{R}$).

For $\beta = 0$, Eq. (2.2) reduces to the explicit time-differencing of the Lax-Wendroff scheme [4].

For $\beta = -1$, by using the relation

$$f_x^n = A^n w_x^n$$

Eq. (2.2) can be rewritten as

$$\Delta w = - \Delta t\, f_x^n + \frac{\Delta t^2}{2} [(A^n)^2 w_x^{n+1}]_x$$

which is also a kind of Lax-Wendroff approximation, but with an implicit treatment of the higher-order term.

The space-differencing is quite simple. It makes use of centered formulas involving only 3 points. To write down this space-differencing, let us define a first difference δ and an average μ, both over one mesh interval :

$$(\delta\Psi)_j = \Psi_{j+1/2} - \Psi_{j-1/2}$$

$$(\mu\Psi)_j = \frac{1}{2} (\Psi_{j+1/2} + \Psi_{j-1/2})$$

where the subscript j corresponds to the location $x_j = j\,\Delta x$ and the subscript $j + 1/2$ to $x_{j+1/2} = (j + 1/2)\,\Delta x$, Δx being the space increment.

Using the above operators and the step ratio

$$\sigma = \Delta t / \Delta x$$

the fully discretized scheme takes the form :

$$\overset{v}{\Delta w_j} = - \sigma\, \delta\, (\mu f^n)_j + \frac{\sigma^2}{2} \delta\, [(\mu A^n)\, \delta f^n]_j \qquad (2.3)$$

$$\Delta w_j + \beta \frac{\sigma^2}{2} \delta\, [(\mu A^n)^2\, \delta(\Delta w)]_j = \overset{v}{\Delta w_j} \qquad (2.4)$$

The scheme (2.3), (2.4) is in conservation form and second-order accurate in time and space. For $\beta \neq 0$, the scheme is implicit and, at each time step, it leads to the solution of an algebraic linear system $\mathcal{L}$ with a block-tridiagonal structure, i.e. :

$$(B_{-1})_j^n \Delta w_{j-1} + (B_0)_j^n \Delta w_j + (B_1)_j^n \Delta w_{j+1} = \overset{v}{\Delta w_j}$$

with

$$(B_{-1})_j^n = \beta \frac{\sigma^2}{2} (\mu A^n)^2_{j-1/2}$$

$$(B_1)_j^n = \beta \frac{\sigma^2}{2} (\mu A^n)^2_{j+1/2}$$

$$(B_0)_j^n = I - (B_{-1})_j^n - (B_1)_j^n$$

where I is the m x m identity matrix.

For the pure initial-value problem, we know [1] that if and only if we choose

$$\beta < -\frac{1}{2} \tag{2.5}$$

then all the following properties hold unconditionally, i.e. without restriction on the time step Δt :

- the scheme is linearly stable in L_2,
- the scheme is dissipative in the sense of Kreiss, whenever A^n is non-singular,
- $\mathcal{L}$ is strictly diagonally dominant in the scalar case (i.e. for m = 1)

3 - CONVERGENCE RATE TO A STEADY STATE

In many applications, implicit schemes are used only to reach a steady state by following some unsteady or pseudo-unsteady evolution. When using the scheme (2.3) - (2.4) for such a purpose, it is advantageous to select the value of β in the range (2.5) so as to obtain the quickest convergence to the steady state.

A linear analysis of the convergence rate has been performed in [5] for a class of schemes including the scheme (2.3) - (2.4), without boundary conditions. This analysis is based on the fact that, in the Fourier space, a linear scheme can be viewed as a purely algebraic iterative method in the form :

$$\hat{v}^{n+1}(\xi) = G(\xi, \sigma)\, \hat{v}^n(\xi)$$

where $\hat{v}^n$ is the Fourier space-transform of the numerical solution at time t_n, $G(\xi, \sigma)$ is the amplification matrix of the scheme which depends on the wave number k or more precisely on the reduced wave number

$$\xi = k\, \Delta x$$

and on the step ratio σ.

Using a classical result on iterative methods, it is clear that, for given ξ and σ, the convergence rate increases as the spectral radius of $G(\xi, \sigma)$ decreases between 1 and 0. So, our goal is to choose β in order to minimize this spectral radius.

For the present scheme used with large time steps (i.e. $\sigma \to \infty$), the optimal parameter has been found to be

$$\beta = -1 \tag{3.1}$$

independently of the reduced wave number. The corresponding amplification matrix is

$$G(\xi, \sigma) = [I + (1 - \cos \xi)(\sigma A)^2]^{-1} [I - \underline{i} \sin \xi (\sigma A)] \tag{3.2}$$

where the jacobian matrix A is supposed to be constant and $\underline{i}^2 = -1$.

The spectral radius of matrix (3.2) is

$$\rho[G(\xi, \sigma)] = \frac{[1 + \sin^2 \xi (\sigma\lambda_0)^2]^{1/2}}{1 + (1 - \cos \xi)(\sigma\lambda_0)^2} \tag{3.3}$$

with :

$$\lambda_0 = \min_q |\lambda_q|$$

where λ_q is an eigenvalue of A ($\lambda_q \in \mathbb{R}$) and $q = 1, 2, \ldots, m$.

Let us assume that the matrix A is non singular (i.e. $\lambda_0 \neq 0$). Then from expression (3.3), we obtain the additional result :

The convergence rate is a strictly increasing function :

- of the step ratio σ, for a given non-zero ξ,
- of the reduced wave number ξ (in the range $[0, \pi]$), for a given non-zero σ.

In other words, the analysis indicates that, for $\beta = -1$:

- The efficiency increases with the CFL number, where

$$CFL = \sigma\, \rho(A).$$

- Short waves converges more quickly than long ones.

4 - ACCURACY AT STEADY-STATE

A steady-state of the exact system (2.1) is obtained when $w_t = 0$ on the space domain. Steady weak solutions of (2.1) are thus defined by :

$$f(w) = \text{const.} \tag{4.1}$$

Taking into account the entropy condition and excluding the trivial solutions, a steady solution of (2.1) is nothing but a steady shock between two constant states.

For the numerical scheme (2.3), (2.4), a steady solution is obtained when $\Delta w_j = 0$, which yields :

$$\overset{v}{\Delta w_j} = 0$$

or

$$- \delta(\mu f)_j + \frac{\sigma}{2} \delta[(\mu A)\ \delta f]_j = 0 \qquad (4.2)$$

Obviously, the steady numerical solution depends on the step ratio σ and thus on the time step. But, this does not mean that the steady solution is necessarily unaccurate for a large σ. Let us consider the behaviour of the steady state in three different situations :

a) Δt small with respect to Δx ($\sigma \ll 1$)

Then, Eq. (4.2) behaves as :

$$\sigma(\mu f)_j = 0$$

or

$$f_{j+1} = f_{j-1} \qquad (4.3)$$

Even and odd mesh points are uncoupled. Steady shocks may be highly oscillatory.

b) Δt of the order of Δx ($\sigma \sim 1$)

Then, Eq. (4.2) corresponds to the usual Lax-Wendroff approximation, which is known to give oscillatory shock profiles.

c) Δt large with respect to Δx ($\sigma \gg 1$)

Then, Eq. (4.2) behaves as :

$$\delta\ [(\mu A)\ \delta f]_j = 0$$

or

$$(\mu A)_{j+1/2}\ (\delta f)_{j+1/2} = K \qquad (4.4)$$

where K is a constant vector.

By comparing this equation to the exact one (4.1), we see that the method can lead to a good approximation, provided that :

$$K = 0 \qquad (4.5)$$

To fulfill this requirement, we are going to modify slightly the explicit stage (2.3) of the scheme near a boundary. Let us first rewrite this explicit stage in the form :

$$\overset{v}{\Delta w}_j = - \sigma\ (F_{j+1/2} - F_{j-1/2}) \qquad (4.6)$$

for $j = 1, 2, \ldots, J$ with a numerical flux :

$$F_{j+1/2} = (\mu f^n)_{j+1/2} - \frac{\sigma}{2} (\mu_A^n)_{j+1/2} (\delta_f^n)_{j+1/2} \qquad (4.7)$$

If the second term of this numerical flux vanishes somewhere in the mesh, then it is clear that the relation (4.4) holds everywhere with a zero right-hand side. So, the idea is to modify the formula (4.7) at or near a boundary, that is to compute $F_{1/2}$ or $F_{J+1/2}$ dropping the term with coefficient σ. To be precise, we must distinguish between the finite-volume formulation and the finite-difference formulation.

In the finite-volume formulation, the first interior cell j = 1 will be calculated from (4.6) with a flux at the boundary side j = 1/2 defined by :

$$F_{1/2} = f^{n}_{1/2} \tag{4.8}$$

where the vector $f^{n}_{1/2}$ is deduced at time level n from the exact boundary conditions, completed if necessary by additional relations accounting for the inner influence. These relations must be independant of the ratio σ.

In the finite-difference, the first interior point j = 1 will be calculated from (4.6) with a special formula for the numerical flux between the boundary point j = 0 and the point j = 1, i.e.

$$F_{1/2} = (\mu f^{n})_{1/2} = \frac{1}{2}(f^{n}_{0} + f^{n}_{1}) \tag{4.9}$$

Concerning now the implicit stage (2.4) of the scheme, various treatments can be done near a boundary. Accuracy at steady-state is not involved in the choice of these treatments. The simplest choice is to cancel the implicit part of the numerical flux. At the left boundary, this comes to calculate the first interior cell (or point) j = 1 with the modified implicit stage :

$$\Delta w_{1} + \beta \frac{\sigma \eta^{2}}{2} [(\mu A^{n})^{2} \delta(\Delta w)]_{3/2} = \Delta w^{v}_{1} \tag{4.10}$$

Of course, this explicit treatment of the boundary conditions is not the optimal one for a fast convergence to the steady-state.

Now we return to the discussion of the accuracy of the steady-state. Due to the above modifications of the explicit stage (2.3), the behaviour of the steady solution for $\sigma \gg 1$ is given by :

$$(\mu A)_{j+1/2} (\delta f)_{j+1/2} = 0,$$

At every regular side j + 1/2 (i.e. a side for which the matrix μA has no zero eigenvalue), this yields :

$$(\delta f)_{j+1/2} = 0$$

or

$$f_{j+1} = f_{j} \tag{4.11}$$

which is an exact solution !

However this result is not quite satisfactory, because a steady shock satisfying Eq. (4.11) has no numerical structure. Let us assume that the steady shock has a well-defined location - this is not really the case for the homogeneous one-dimensional system (2.1) - and suppose that the mesh is not adapted to the exact location of the shock, then the steady numerical solution may not exist. For this reason, it is preferable to use a moderately large value of σ, which allows the first term of Eq. (4.2) to play some small part and thus the numerical shock structure to be spread over one or two mesh intervals.

5 - APPLICATION TO QUASI-ONE DIMENSIONAL FLOWS

In the purpose of applying the implicit scheme to a more interesting problem in one-space dimension, we extend the method to the inhomogeneous hyperbolic system :

$$w_t + f(w)_x = h(x, w) \qquad (5.1)$$

where the right-hand side is a smooth function of x and w. In addition to the matrix A(w), we introduce a new matrix :

$$D(x, w) = \frac{\partial h}{\partial w}(x, w)$$

Since the second time-derivative of w can be written as :

$$w_{tt} = - f(w)_{tx} + h(x, w)_t = -[A(w)\, w_t]_x + D(x, w)\, w_t = [A(w) r]_x - D(x, w)\, r$$

with :

$$r = f(w)_x - h(x, w),$$

the explicit time-differencing of Lax-Wendroff for system (5.1) reads :

$$\Delta w = - \Delta t\, r^n + \frac{\Delta t^2}{2}[(A^n r^n)_x - D^n r^n] \qquad (5.2)$$

where :

$$r^n = f_x^n - h^n, \quad f^n = f(w^n) \quad \text{and} \quad h^n = h(x, w^n).$$

Thus, the implicit time-differencing can be written as :

$$\Delta w + \beta \frac{\Delta t^2}{2}[(A^n)^2 (\Delta w)_x]_x = - \Delta t\, r^n + \frac{\Delta t^2}{2}[(A^n r^n)_x - D^n r^n] \qquad (5.3)$$

It is convenient to write Eq. (5.3) in several stages as :

$$\widetilde{\Delta w} = - \Delta t\, (f_x^n - h^n) \qquad (5.4\ a)$$

$$\tilde{f}^{n+1/2} = f^n + \frac{1}{2} A^n \widetilde{\Delta w} \qquad (5.4\ b)$$

$$\tilde{h}^{n+1/2} = h^n + \frac{1}{2} D^n \widetilde{\Delta w} \qquad (5.4\ c)$$

$$\overset{\vee}{\Delta w} = -\,\Delta t\,\left(\tilde{f}_x^{\,n+1/2} - \tilde{h}^{\,n+1/2}\right) \tag{5.4 d}$$

$$\Delta w + \beta \frac{\Delta t^2}{2}\left[(A^n)^2\,(\Delta w)_x\right]_x = \overset{\vee}{\Delta w} \tag{5.4 e}$$

$$w^{n+1} = w^n + \Delta w \tag{5.4 f}$$

Inviscid flows in a quasi-one dimensional nozzle are governed by system (5.1) with :

$$h(x, w) = \frac{ds}{dx}(x)\ \phi(w)$$

$$w = \begin{bmatrix} \rho\, s \\ \rho\, u\, s \\ \rho\, E\, s \end{bmatrix}, \qquad f(w) = \begin{bmatrix} \rho\, u\, s \\ (\rho\, u^2 + p)\, s \\ (\rho\, E + p)\, u\, s \end{bmatrix}, \qquad \phi(w) = \begin{bmatrix} 0 \\ p \\ 0 \end{bmatrix}$$

where $s = s(x)$ is the cross section of the nozzle, ρ is the density, p the pressure, u the fluid velocity and E the total specific energy. By assuming the gas to be perfect, with constant specific heats of ratio γ ($\gamma = 1.4$ in our calculations), we have :

$$E = \frac{1}{\gamma - 1}\ \frac{p}{\rho} + \frac{1}{2}\ u^2$$

and the matrices $A(w)$ and $D(x, w)$ are defined by :

$$A(w) = \begin{bmatrix} 0 & 1 & 0 \\ -\dfrac{3-\gamma}{2}\,u^2 & (3-\gamma)\,u & \gamma-1 \\ -\left(\dfrac{2-\gamma}{2}\,u^2 + \dfrac{a^2}{\gamma-1}\right)u & \dfrac{3-2\gamma}{2}\,u^2 + \dfrac{a^2}{\gamma-1} & \gamma u \end{bmatrix}$$

with the sound speed $a = (\gamma\, p\, /\, \rho)^{1/2}$, and

$$D(x, w) = \frac{ds}{dx}(x)\ \Pi(w)$$

with

$$\Pi(w) = (\gamma-1)\begin{bmatrix} 0 & 0 & 0 \\ \dfrac{1}{2}\,u^2 & -u & 1 \\ 0 & 0 & 0 \end{bmatrix}$$

We look for a steady flow in a nozzle defined by [6] :

$$s(x) = 1.398 + 0.347 \tanh(0.8x - 4), \qquad 0 < x < 10.$$

The fluid entering the nozzle is assumed to originate from an infinite reservoir where it is at rest with pressure p_0 and density ρ_0.

At the entrance section (x = 0), the flow is supersonic (Mach number 1.2639) and all the components of w are given. At the exit section (x = 10), the flow is subsonic and the pressure is fixed at the value :

$$p_{ex} = 0.7466 \; p_0$$

The space discretization of (5.4) is performed in the finite-volume formulation. The space domain is divided into J cells :

$$\Omega_j = [x_{j-1/2}, x_{j+1/2}], \qquad j = 1, 2, \ldots J$$

which are not necessary of equal length, with :

$$x_{1/2} = 0 \quad \text{and} \quad x_{J+1/2} = 10$$

For the solution of steady problems, we use a local time-step Δt such that the CFL number is the same in all the mesh cells during the whole computation, that is :

$$(\Delta t)^n_j = \sigma^n_j (x_{j+1/2} - x_{j-1/2})$$

with :

$$\sigma^n_j = \mathrm{CFL} \, / \, \rho(A^n_j) \tag{5.5}$$

where CFL is a given positive constant and :

$$\rho(A^n_j) = (|u| + a)^n_j$$

As a consequence, the method is consistent only at steady-state.

The fully-discretized method can be written as :

$$\widetilde{\Delta w}_{j+1/2} = -(\mu\sigma^n)_{j+1/2} \, [(\delta f^n) - (\delta s)(\mu\phi^n)]_{j+1/2} \tag{5.6 a}$$

$$\tilde{f}^{\,n+1/2}_{j+1/2} = [(\mu f^n) + \frac{1}{2}(\mu A^n)\,\widetilde{\Delta w}]_{j+1/2} \tag{5.6 b}$$

$$\tilde{\phi}^{\,n+1/2}_j = [\phi^n + \frac{1}{2}\Pi^n(\mu\,\widetilde{\Delta w})]_j \tag{5.6 c}$$

$$\overset{v}{\Delta w}_j = - \sigma_j^n \left[\!\left[\widetilde{\delta f}^{\,n+1/2} \right) - (\delta s)\, \tilde{\phi}^{\,n+1/2} \right]_j \tag{5.6 d}$$

$$\Delta w_j + \frac{\beta}{2} \sigma_j^n \,\delta[(\mu\sigma^n)\,(\mu A^n)^2\,\delta(\Delta w)]_j = \overset{v}{\Delta w}_j \tag{5.6 e}$$

$$w_j^{n+1} = w_j^n + \Delta w_j \tag{5.6 f}$$

Since we are not concerned here with the accuracy in time, the present calculations have been performed by reducing (5.6 c) to :

$$\tilde{\phi}_j^{\,n+1/2} = \phi_j^n \tag{5.7}$$

At steady state and for CFL $\gg 1$, the corresponding scheme behaves as :

$$(\mu A)_{j+1/2} \,[(\delta f) - (\delta s)(\mu\phi)]_{j+1/2} = \text{const.} \tag{5.8}$$

which is analogous to Eq. (4.4).

No artificial viscosity is added to the scheme.

Boundary conditions are treated explicitly with boundary fluxes such as (4.8). More precisely, at the entrance section ($j = \frac{1}{2}$), the flow is supersonic and the explicit part of the numerical flux is prescribed as :

$$\tilde{f}_{1/2}^{\,n+1/2} = f(w_{1/2}) \tag{5.9}$$

where $w_{1/2}$ is a given vector.

At the exit section ($j = J + \frac{1}{2}$), the explicit part of the numerical flux is calculated from the datum of the pressure and from two additional relations which express the conservation of mass and total enthalpy between the center of the last cell and the exit section, that is :

$$\tilde{p}_{J+1/2}^{\,n+1/2} = p_{ex} \tag{5.10}$$

$$(\tilde{f}_{J+1/2}^{\,n+1/2})^{(k)} = (f_J^n)^{(k)} \quad , \qquad \text{for } k = 1 \text{ and } 3 \tag{5.11}$$

where k denotes the number of the vector-component.

Finally, at the entrance and exit sections, the implicit part of the numerical flux is cancelled, i.e.

$$[(\mu\sigma^n)(\mu A^n)^2\,\delta(\Delta w)]_{j+1/2} = 0 \quad , \quad \text{for } j = 0 \text{ and } J \tag{5.12}$$

Let us now present the results obtained with the implicit method for $\beta = -1$ and various CFL numbers. A comparison is made on Fig. 1 of the steady numerical solutions (density versus space) for CFL = 1, 5 and 20 in a regular mesh, and for CFL = 50 in a mesh adapted to the location of the shock. Both meshes are composed of 100 cells. The exact solution is also represented on Fig. 1.

One notices that the numerical shock profile is oscillatory for CFL = 1 ; it becomes montonic for larger values of the CFL number. For CFL = 20, the numerical solution involves some local error in the vicinity of the shock, because the numerical shock structure tends to disappear and, in the mesh used, this does not allow the shock to be located at its exact position. By using an adapted mesh, one can obtain a very accurate solution for a large CFL number (see Fig. 1 d).

Fig. 2 illustrates the convergence histories relative to the results shown on Fig. 1. The residuals RES are defined with a discrete L_2 norm of the components of $\Delta w/\Delta t$. One can see that the convergence speed increases with the CFL number from CFL = 1 to CFL = 50. A high degree of convergence is obtained when the residual has reached the level 10^{-6}. For this convergence criterion, the number of iterations needed with the various CFL numbers is given in Table 1. Let us recall that these results have been obtained using an explicit treatment of the boundary fluxes. The use of implicit boundary conditions might improve the convergence rate.

CFL	1	5	20	50
Iterations	2 000	400	200	150

Table 1 : Number of iterations needed to reach a residual of 10^{-6}, using 100 cells

REFERENCES

[1] LERAT A. - Une classe de schémas aux différences implicites pour les systèmes hyperboliques de lois de conservation, Comptes Rendus Acad. Sc. Paris, Vol. 288A, pp. 1033-1036, 1979.

[2] LERAT A., SIDES J. and DARU V. - An Implicit Finite-Volume Method for Solving the Euler Equations, Lecture Notes in Physics, Vol. 170, pp. 343-349, 1982.

[3] LERAT A., Implicit Methods of Second-Order Accuracy for the Euler Equations, AIAA J., Vol. 23, pp. 33-40, 1985.

[4] LAX P.D. and WENDROFF B. - Systems of Conservation Laws, Comm. Pure Appl. Math., Vol. 13, pp. 217-237, 1960.

[5] DARU V. and LERAT A. - Analysis of an Implicit Euler Solver, in Numerical Methods for the Euler Equations of Fluid Dynamics, F. Angrand and al. Ed, SIAM Publ., pp. 246-280, 1985.

[6] SHUBIN G.R., STEPHENS A.B. and GLAZ H.M. - Steady Shock Tracking and Newton's Method Applied to One-Dimensional Duct Flow, J. Comput. Phys., Vol. 39, pp. 364-374, 1981.

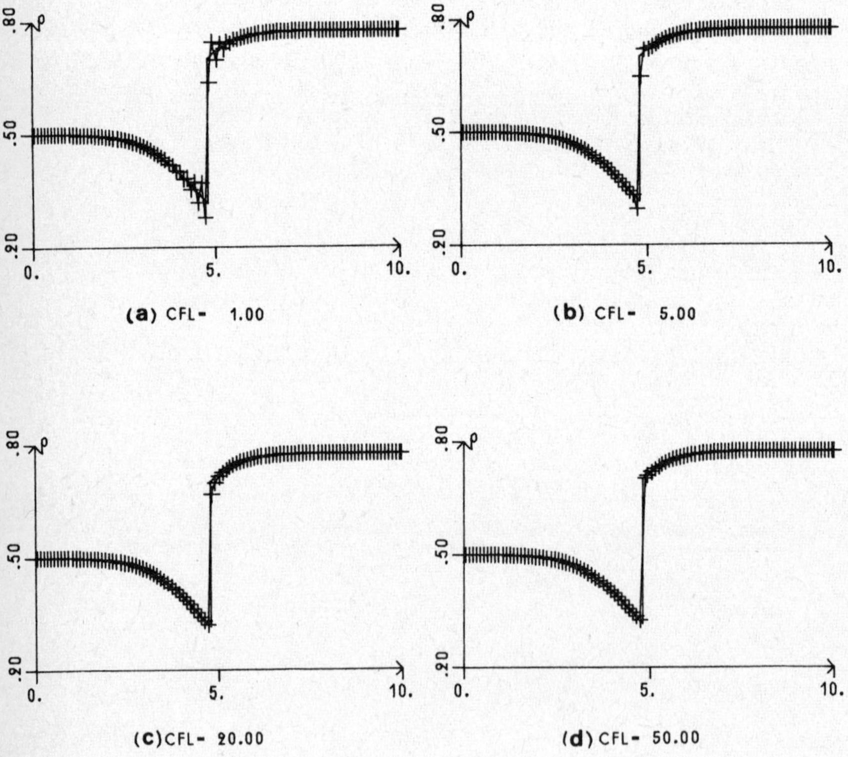

Figure 1 : Steady solution for various CFL numbers

(a) CFL = 1 (b) CFL = 5 (c) CFL = 20 (d) CFL = 50 (adapted mesh)

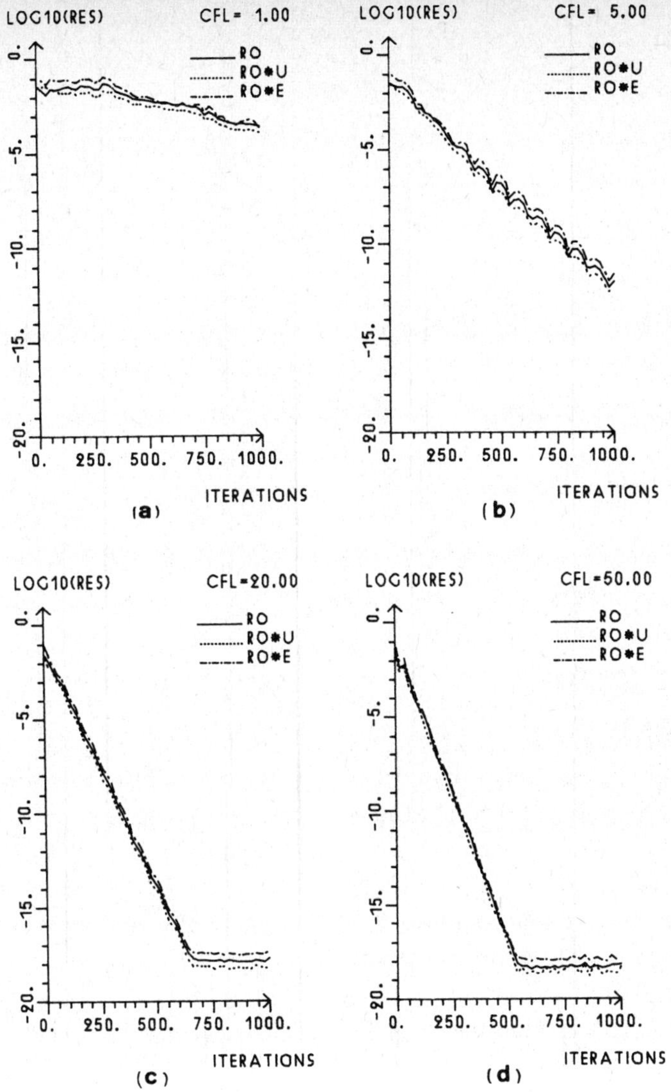

Figure 2 : Convergence history for various CFL numbers

(a) CFL = 1 (b) CFL = 5 (c) CFL = 20 (d) CFL = 50 (adapted mesh)

A COMPUTER-AIDED SYSTEM FOR INVESTIGATION AND CONSTRUCTION OF DIFFERENCE SCHEMES OF GAS DYNAMICS

Yu.I.Shokin
Computing Center
USSR Academy of Sciences
Siberian Division
660036 Krasnoyarsk 36
USSR

At present one of the main methods of solving problems of mathematical physics is the method of finite differences. The appearance and practical use of a great number of difference schemes have resulted in increasing the role of criteria of their selection and comparison.
An important technique of analysing properties of difference schemes, ways of their constructing and classification is the method of differential approximation [1 - 3] which has widely been used at present and whose effectivity has been supported by practice. The indicated method allows one to construct new difference schemes with predetermined properties, to make an analysis of properties of the existing and newly created difference schemes and also to classify them by a number of signs. Due to these possibilities the method of differential approximation (DA) was taken as a mathematical foundation for the computer-aided system of constructing and investigating difference schemes of gas dynamics. The developped system represents a program package with a problem-oriented language for the user and means for generation of a FORTRAN program.
The oversimplified structure of the computer-aided system can be presented as follows: INPUT of initial data - ANALYSER connected with the library of difference schemes - GENERATOR - COMPUTATION. Functioning of the system begins with the input of a multi-parametric family of difference schemes and a job list. Having analysed the initial information the system constructs an internal representation of the user's problem. Then the ANALYSER fulfils a detailed investigation of the difference scheme family by the method of differential approximation and selects a concrete difference scheme according to the user's requirement which is analysed as a computation algorithm. Then for the selected difference scheme the GENERATOR automatically constructs a

FORTRAN program whose operation is illustrated by the computation of a specially chosen system of tests, containing characteristic features and peculiarities of a wide class of gas dynamics problems.

The range of potential users of the system is supposed fairly wide: from a student wishing to familiarize himself with difference schemes of gas dynamics and their properties to engineers and scientific associates who have to solve gas dynamics problems and who can, to a different extent of detail, outline a class of difference schemes available for computing their problems. The user's level is defined by the completeness of a job list made by him.

At present the main attention is devoted to the development of the ANALYSER block where the investigation of difference schemes is fulfilled by the method of differential approximation. The following problems arise here: A - to outline a class of studied difference schemes; B - to give a representation (common form of writing) of the considered class of schemes; C - to construct an effective algorithm of computing difference approximations; D - to select suitable data structures for the computer and program language. The problems B,C,D are closely connected with one another: the representation of difference schemes and algorithm cause the choice of program means, in the first place the algorithmic language, and vice versa, the software available suggests what representation is the most convenient for formulating an algorithm of obtaining differential approximations and for its program realization.

Consider difference schemes of a straight through calculation for gas dynamics equations of the type

$$\sum_{i=0}^{S} A_i \frac{\partial}{\partial x_i} U_i = H \ . \qquad (1)$$

Here A_i are $(S+2)\times(S+2)$ size matrices, U , H are vectors, $x_0 = t$ is time, $x_1, \ldots, x_S$ are spatial variables (S = 1, 2 or 3). In common case matrices A_i and vectors U_i, H depend on gas dynamics functions $u_s (s=1, \ldots, S)$, ρ, p, ε (all notations are generally accepted), constants and independent variables. Depending on considered classes of problems and methods of their solution different forms of representation of gas dynamics equations are applied. We shall consider the system of equations (1) to be written in such a form that the elements of matrices A_i and vectors U_i , H are polynomials in variables entering them. Such restriction on the form of the gas dynamics system is not of principle. Only when constructing the Γ -form of differential approximation it allows one to avoid division into the Taylor series and to treat operations of addition, subtraction and multiplication of series. We shall consider the state equation $p = p(\varepsilon, \rho)$ to be an algebraic function of their arguments. Note that in the form (1) there can be written gas dynamics systems both in Eulerian coordinates and in Lagrangian ones, non-linear and linearized and so on. Introduce a number of notations to be used subsequently. By $a_1, \ldots, a_Q$ we denote constants (parameters and other constant values either present in the gasdynamic equations or arising when constructing a difference scheme). Besides, the function-parameters can enter a difference scheme, which are algebraic expressions of gasdynamic variables and concrete form of which is not necessary when constructing a differential approximation (for example, the coefficient of artificial viscosity $\mu = \mu(\rho)$). A concrete expression for a function-parameter is written only when it is necessary to obtain quite a definite scheme from a class of schemes. Denote the set of gasdynamic functions and functions-parameters by W. Consider further a non-fixed difference grid and R -layer star ($R \geqslant 2$). Let $\tau, h_1, \ldots, h_S$ be the steps of a difference grid by time and spatial coordinates. In accordance with the star form a set E^{S+1} of the shift vectors $\alpha = \{\alpha^0, \alpha^1, \ldots, \alpha^S\}$ so that for any function f from W the shift operator T_α is defined. Together with W and E^{S+1} consider the sets w and e^{S+1} consisting of the same elements, but not obligatory all, and probably with repetitions. For example, if $W = \{u, \rho, p, \varepsilon\}$ then w cannot be such: $w = \{u, u, \rho\}$

By the way of calculation of a differential approximation difference schemes can be divided into three classes: 1 - one-step schemes (two-layer or multi-layer); 2 - explicit τ -step schemes ($\tau \geqslant 2$); 3 - implicit τ -step schemes ($\tau \geqslant 2$). Dwell upon considering one-step schemes. Write the following difference scheme:

$$\sum_{k=1}^{K_i} \mathcal{P}_k^{(i)} \mathcal{M}_k^{(i)} = 0 \qquad (i = 1, \ldots, S+2),$$

$$\mathcal{M}_k^{(i)} = \prod_{\substack{\alpha \in e_k^{S+1\,(i)} \\ f \in w_k^{(i)}}} T_\alpha f ,$$

$$\mathcal{P}_k^{(i)} = \sum_{l=1}^{L_{ki}} \tau_{kl}^{(i)} \, \tau^{j_{kl}^{(i)}} \prod_{s=1}^{S} h_s^{n_{kls}^{(i)}} \prod_{q=1}^{Q_i} a_q^{m_{klq}^{(i)}}$$

Here $\mathcal{M}_K^{(i)}$ is the monomial of the power $M_K^{(i)}$ equal to the product of functions from $W_K^{(i)}$ taken at the star nodes. For example, consider a monomial of the 3 : ($M_K^{(i)} = 3$) $\rho_{j+1}^{n+1} \; u_{j+1}^{n+1} \; u_j^n$ power. It will be written like this:

$$T_{(1,1)}\rho \, T_{(1,1)} u \, T_{(1,0)} u = \prod_{\substack{f \in W_K \\ \alpha \in E_K^2}} T_\alpha f$$

where $W_K = \{\rho, u, u\}$, $E_K^2 = \{(1,1),(1,1),(1,1)\}$

The index K changes from 1 to $K^{(i)}$ where $K^{(i)}$ is the length of the scheme, that is the number of monomials in the polynomial; $\mathcal{P}_K^{(i)}$ is the polynomial of the length $L_K^{(i)}$ in parameters of the scheme and constants with real coefficients $\tau_{Kj}^{(i)}$. Indices of the powers $n_{Kls}^{(i)}$, $m_{Klq}^{(i)}$, $j_{Kl}^{(i)}$ ($K = 1,\dots,K_i$; $l = 1,\dots,L_{K_i}$; $s = 1,\dots,\mathcal{S}$; $q = 1,\dots,Q_i$) are integers including zeros.

A representation of a difference scheme of the form (2)-(4) will be called a T-form of the difference scheme. It is evident that one-step difference schemes of a straight-through calculation constructed by any of the ways given in [4], by opening parentheses and collecting like terms are reduced to the T-form. Note that if one considers the simplest transport equation $\frac{\partial u}{\partial t} = a \frac{\partial u}{\partial x}$, $a = Const$, then a two-layer one-step difference scheme of the T-form is usually written for it as follows (see [1]):

$$\sum_{\alpha=\alpha_0}^{\alpha_1} b_\alpha u(t+\tau, x+\alpha h) - \sum_{\beta=\beta_0}^{\beta_1} c_\beta u(t, x+\beta h) = 0,$$

where b_α and c_β depend on τ, h (τ and h entering the combinations $\varkappa = \tau / h$ and the constants a). Here all the monomials $\mathcal{M}_K$ have the first power. If a difference scheme includes τ steps then the reduction to the T-form is implemented by passing to the scheme in whole steps. To do this one should substitute auxiliary values computed at the first ($\tau - 1$) steps into the expression, by which the values on the ($n+1$)-th layer are determined. After exclusion a reduction to the T-form occurs by the same algorithm as for the above one-step scheme. Among multi-step implicit difference schemes we shall consider those, whose relationships on the first ($\tau - 1$) steps relative to sought-for functions are linear. All known to us implicit multi-step schemes including split schemes are just like those. On excluding fractional steps one has to deal with the reverse operator, as a rule, of the form $(E + \alpha\Lambda)^{-1}$, where α is the small parameter (in particular, a step of the grid). To bring the reduction of the considered scheme to the previous case, one has to represent the operator $(E+\alpha\Lambda)^{-1}$ as a segment of the series by the powers $\alpha\Lambda$ of a concrete length, connected a priori with the order of the calculated differential approximation.

The writing of a difference scheme in the T-form is very convenient for setting up algorithms and corresponding programs to calculate differential approximations in the languages of high level (PASCAL, FORTRAN). With such an approach to a difference scheme the main unit of analytical information is an algebraic polynomial

$$\sum_{i=1}^{N} \alpha_i u_1^{n_{1i}} u_2^{n_{2i}} \ldots u_N^{n_{Ni}}$$

where α_i are integers or rational fractions, and $u_1,\ldots,u_N$ are the names of scalar variables, which a polynomial depends on. Since one has to deal with truncated Taylor series, some scalars have the order of smallness which like the names of scalars can be changed during operation.

Thus, there arises a problem of describing two languages: the input language of the user, that is at most approximated to the "natural" writing of a difference scheme through the shift operator and their combination, and the language corresponding to the scheme representation of the T-form - it will be input for the ANALYSER block. A pretranslator, for whose creation the PASCAL language is used, should make the translation of the scheme from the user's language into the internal language of the computer-aided system.

The question put to the user, which of the existing schemes is to be chosen for a calculation of a concrete problem, has led to the necessity to classify schemes according to specific properties. A classification given in the monograph [3] gives essentially a comparative analysis of difference schemes and enables one to solve the problem of choosing a difference method, however, presentation of the results in the form of tables is always of a limited character, since an attempt to include a new scheme in the tables requires to repeat for it all the classification work done manually. At the same time a similar work can be computer-aided by creating a program package, oriented to a research work. In this case difference schemes and their properties will be stored in the scheme library, organized on a magnetic drum or a magnetic tape, that can be enriched by new results.

Another aspect of the problem is a choice from a multi-parametric family of schemes of such a scheme that would possess a set of properties necessary for a successful solution of a concrete gas dynamics problem. We shall give the name ANALYSER to a complex of programmes performing investigation and classification of difference schemes and fulfiling a choice of a concrete scheme. This complex of programmes performs the following work:

1. Analyses (identifies) a scheme, seeks it in the scheme library and if it is found it gives a complete information on the properties of the scheme.

2. If there is no scheme in the library, it is being investigated. To do this a computation is done of the Γ-form and Π-form of the differential approximation of the input difference scheme and also of its variants, approximating simplified (as a rule linear) models. Then an analysis is made on the basis of the obtained differential approximations.

3. After the investigation the scheme and its properties are placed in the library.

4. The ANALYSER has means to separate from the investigated class of such a scheme that would possess a set of properties necessary for the user. Some of the properties of difference schemes are known to be able to contradict one another. For example, the high order of approximation and monotony requirement are, as a rule, incompatible. Therefore, the problem for the analyser to separate the scheme is not always to be solved. If in the given family there are no required schemes, then the system suggests the user either to change the family or refuse from the number of properties not so important for the problem being solved.

It is convenient to write ANALYSER programs in the PAS AL language since it has approapriate data structures and dynamics.

The final stage of the work of the computer-aided system is generation of the program in FORTRAN, realizing the calculation algorithm according to the scheme separated by the ANALYSER.

After the ANALYSER finishes work and separates a scheme there begins functioning of the preparation block to generation. Its aim is to analyse a difference scheme as the algorithm of calculation of the initial-boundary problem. For this purpose the scheme is investigated as explicit-implicit, the number of points in the star and their interpositioning are determined (the number of extra boundary conditions depends on this). If a scheme is implicit then it is found out how many points are "tied" on the upper by the time layer (the form of the sweep depends on this). A multi-layer scheme (K -layer) as distinct from a two-layer one requires either to assign extra boundary conditions on the ($K-2$) layers by time or application at the beginning of the calculation of the two-layer scheme.

The work of the preparation block for generation comes to an end by forming an inquiry table that is passed to the generator. The operation of the generator is based on the library of macromoduli, each of them represents a parametrized "star" of this or another class of difference schemes. On the basis of inquiries there appears a necessary "star" and occurs its "filling" with a description of a concrete difference scheme. This results in creating a textual modulus, written in FORTRAN and corresponding to the difference scheme set at the input. After translating this modulus one can make calculations of concrete problems.

1. Shokin Yu.I. Metod differentsialnogo priblizhenia. - Novosibirsk: Nauka, 1979. - 221 s.
2. Shokin Yu.I. The method of differential approximation. - Springer-Verlag, 1983. - 296 p.
3. Shokin Yu.I., Yanenko N.N. Metod differentsialnogo priblizhenia. Primenenie k gazovoi dinamike. - Novosibirsk: Nauka, 1985. - 364 s.
4. Rozhdestvenskii B.L., Yanenko N.N. Sistemi kvazilineinykh uravnenii. - M.: Nauka, 1978. - 689 s.

LOIS DE CONSERVATION ET INTEGRALES D'ENERGIE DES EQUATIONS HYPERBOLIQUES

S.K.GODOUNOV
Novossibirsk

L'étude des propriétés des solutions discontinues des équations quasi linéaires est basée sur les lois de conservation qui sont des conséquences des équations. Toutes les équations n'admettent pas de telles conséquences. C'est pourquoi la théorie des solutions généralisées doit à mon avis se construire seulement pour les systèmes admettant les lois de conservation. Pour des classes différentes de tels systèmes, cette théorie peut avoir un contenu varié.

Conformément à ce point de vue, je me suis intéressé à la classification des formulations des équations concrètes de la physique mathématique. Je voulais trouver de telles formulations dont on déduirait facilement le type du système (par exemple son hyperbolicité) d'une part, et l'existence de lois de conservation d'autre part.

Dans mon rapport je ferai un examen des recherches sur cette question qui ont été accomplies par moi et mes collaborateurs. Je ferai un exposé sommaire d'une série de recherches analogues par d'autres auteurs. Dans la seconde partie du rapport je parlerai de l'étude des intégrales d'énergie pour les équations hyperboliques linéaires d'après Petrovski et sur l'usage d'une série d'intégrales nouvelles de l'équation classique des ondes qui peuvent être appliquées dans une théorie des ondes de choc.

En 1960 [1,2], faisant des efforts pour schématiser sous une forme commode pour les problèmes multidimensionnels les relations de la thermodynamique, j'ai dégagé la classe de systèmes :

$$(1) \qquad \frac{\partial L^0_{q_i}}{\partial t} + \frac{\partial L^j_{q_i}}{\partial x^j} = 0$$

dont chacun se définit par son vecteur générateur du "potentiel thermodynamique" avec les composantes $L^{\alpha}(q_1,q_2...)$, exprimées par les variables canoniques q_i. Dans la dynamique des gaz les fonctions de telles variables sont les paramètres de Plank $1/E_S$, $(E-SE_S+\rho E_\rho-1/2\ \Sigma u_i^2)/E_S, u_j/E_S$, et les fonctions des composantes du "potentiel thermodynamique" vectoriel prennent les valeurs :

$$L^0 = -\frac{\rho^2 E_\rho}{E_S}\ ,\ L^j = -\frac{u_j\rho^2 E_\rho}{E_S}$$

ou ce qui revient au même :

$$L^0 = -\frac{p}{T}\ ,\quad L^j = -\frac{u_j p}{T}$$

Tout le système d'équations de la dynamique des gaz se définit essentiellement par une fonction de deux variables $E = E(\rho,S)$. (Une équation d'état du gaz). Ici ρ est la densité, p la pression, T la température, S l'entropie, et les u_j sont les composantes de la vitesse).

Les équations du type (1) s'écrivent donc sous la forme suivante

$$(2)\qquad L^0_{q_i q_k}\frac{\partial q_k}{\partial t} + L^j_{q_i q_k}\frac{\partial q_k}{\partial x^j} = 0$$

D'où il est évident que (2) est un système hyperbolique symétrique si seulement L^0 est convexe. La conséquence des équations (1) est une loi de conservation complémentaire

$$(3)\qquad \frac{\partial(q_i L^0_{q_i} - L^0)}{\partial t} + \frac{\partial(q_i L^j_{q_i} - L^j)}{\partial x^j} = 0$$

laquelle dans le cas de la dynamique des gaz coïncide avec la loi de conservation d'entropie. La conséquence (3) ne peut être déduite de (1) que pour les solutions classiques, car sur les ondes de choc (les discontinuités de la solution) elle doit être remplacée sur une inégalité-la loi de croissance d'entropie.

Le travail [1] a été rapporté au mois de mai 1960 pendant le séminaire à l'Université de Moscou en présence de R. Courant et P. Lax.

En 1972 Lax et Friedrichs (sans référence sur [1,2]) ont publié le travail [3] où ils ont souligné que la présence dans le système à N-équations de N+1 lois de conservation entraine automatiquement la représentabilité de ces équations dans la forme (1).

J'ai analysé en détail le processus de réduction des équations de la dynamique des gaz à la forme (1), (2), (3) dans le quatrième chapitre de mon livre [4] où j'ai encore donné les transformations spéciales des fonctions génératrices L^j et des variables canoniques q_i, lesquelles permettent dans certains cas d'obtenir pour le même système des ensembles différents de lois de conservation. Dans la dynamique des gaz à une dimension la description de toutes les lois de conservation a été donnée par Rojdestvenski B.L. encore en 1957 [5].

Pour un système à N inconnues possédant N+1 lois de conservation indépendantes, nous exigeront l'éxécution de N d'entre elles, et remplaceront une combinaison linéaire de ces N+1 lois par une inégalité. Selon le choix ainsi fait, nous aurons plusieurs définitions distinctes de la solution faible. Ainsi, la théorie des ondes de choc a d'abord été traitée par Riemann [6], qui a mis à sa base la loi de conservation d'entropie tandis que pour les gaz réels les conditions de Hugoniot résultent d'une loi de conservation d'énergie. Au contraire l'entropie s'accroit quand le gaz traverse un front d'onde.

Il est intéressant de remarquer que certains problèmes unidimensionnels pour les équations de l'hydrodynamique magnétique ont été étudiés par Koulikovski A.G. [7] à partir de la réduction à la forme (1) ou bien à sa généralisation

$$(4) \qquad \frac{\partial L_{q_i}}{\partial t} + \frac{\partial L^1_{q_i}}{\partial x} = \frac{\partial}{\partial x}\left(b_{ik}\frac{\partial q_k}{\partial x}\right)$$

laquelle pour les solutions stationnaires $q_i = q_i(x-wt)$ se réduit à (1), (4), et au système des équations ordinaires :

$$\frac{\partial q_i}{\partial t} = \sigma_i, \quad \Lambda_{q_i} = \mathfrak{D}_{\sigma_i}$$

$$(5) \qquad \Lambda = L - \frac{1}{W} L^1 - \sum_i A_i q_i, \qquad \mathfrak{D} = \sum_{i,k} b_{ik}\, \sigma_i\, \sigma_k$$

Les équations ordinaires de la forme (5) ont été proposées en 1959 par Germain [8] pour la description d'une structure des ondes de choc dans l'hydrodynamique magnétique, mais son utilisation systématique a commencée depuis que dans [1] est apparue son interprétation géométrique. Bien que les problèmes unidimensionnels de l'hydrodynamique magnétique aient été étudiés avec succès à l'aide de la réduction des équations aux formes (4), (5) on n'avait pas encore réussi à écrire les équations tridimensionnelles de l'hydrodynamique magnétique sous la forme (1), ni sous aucune généralisation d'où résulteraient les lois de conservation utilisées dans la théorie des solutions discontinues. Je suis parvenu à en trouver une en 1972 [9]. Elle est de la forme :

$$(6)\qquad \frac{\partial L^0_{q_i}}{\partial t} + \frac{\partial (L^j_{q_i} + A^j \Phi_{q_i})}{\partial x^j} - \Phi_{q_i} \frac{\partial A^j}{\partial x^j} = 0$$

et se construit suivant les fonctions génératrices :

$$L^0 = L^0\ (q_1, q_2, \ldots),\ L^j = L^j(q_1, q_2, \ldots),\ A^j = A^j(q_1, q_2, \ldots)$$

$$\Phi = \Phi\ (q_1, q_2, \ldots)$$

où Φ doit être une fonction homogène du premier degré de ses arguments, c'est-à-dire doit satisfaire à la condition $\Phi - q_i\, \Phi_{q_i} = 0$. Avec cela on déduit de (6) la loi de conservation

$$(7)\qquad \frac{\partial (q_i L^0_{q_i} - L^0)}{\partial t} + \frac{\partial (q_i L^j_{q_i} - L^j)}{\partial x^j} = 0$$

qui a la même forme que (3). En fait, les équations (6), à la différence de (1), ne sont des lois de conservation que si

$$\frac{\partial A^j}{\partial x^j} = 0.$$

Dans le cas de l'hydrodynamique magnétique il faut choisir les variables canoniques $(E = E(\rho, S))$:

$$(8)\qquad q_0 = -\frac{1}{E_S}\ (E + \rho E_\rho - S E_S - \frac{1}{2} \sum_i u_i^2)$$

$$q_i = -\frac{u_i}{E_S},\quad q_{3+i} = -\frac{H_i}{4\Pi E_S}\ (i=1,2,3),\ q_7 = \frac{1}{E_S}$$

et poser

$$L^0 = -\frac{1}{E_S}\left[\rho^2 E_\rho + \frac{1}{8\Pi}\left(\sum_i H_i^2\right)\right], \quad L^j = u_j L^0, \quad A^j = H_j$$

(9) $$\Phi = \frac{1}{4\Pi E_S}\left(\sum_j u_j H_j\right) \equiv \frac{q_1q_4+q_2q_5+q_3q_6}{q_7}$$

Des équations (6) peut être déduite une conséquence (c'est une relation sur la caractéristique) :

$$\left(\frac{\partial}{\partial t} + u_j \frac{\partial}{\partial x^j}\right)\left[\frac{1}{\rho}\left(\frac{\partial H_1}{\partial x^1} + \frac{\partial H_2}{\partial x^2} + \frac{\partial H_3}{\partial x^3}\right)\right] = 0$$

qui garantit l'égalité $\frac{\partial H_j}{\partial x^j} \equiv \frac{\partial A^j}{\partial x^j} = 0$, si elle est réalisée par les données initiales. Ainsi la condition $\frac{\partial A^j}{\partial x^j} \equiv \mathrm{div}\, H = 0$ ne contredit pas les équations en question.

Cette équation dans l'hydrodynamique magnétique joue un rôle de loi de conservation complémentaire. A cause de cette condition les équations (6) se transforment aussi en lois de conservation.

Voici les équations, quand les fonctions génératrices sont décrites par les égalités (8), (9) :

$$\frac{\partial \rho}{\partial t} + \sum_j \frac{\partial \rho u_j}{\partial x^j} = 0,$$

$$\frac{\partial \rho u_k}{\partial t} + \sum_j \frac{\partial\left(\rho u_k u_j - \frac{1}{4\Pi} H_k H_j\right)}{\partial x^j} + \frac{\partial}{\partial x^k}\left(\rho^2 E_\rho + \frac{1}{8\Pi}\sum_i H_i^2\right) +$$

(10) $$+ \frac{H_k}{4\Pi}\left(\sum_j \frac{\partial H_j}{\partial x^j}\right) = 0$$

$$\frac{\partial H}{\partial t} + \mathrm{rot}\,[uH] + (\mathrm{div}\, H)u = 0$$

$$\frac{\partial}{\partial t}\left[\left(\rho \frac{\sum u_i^2}{2} + \rho E\right) + \frac{1}{8\Pi}\left(\sum H_i^2\right)\right] + \frac{(u,H)}{4\Pi}\ \mathrm{div}\, H +$$

$$+\frac{\partial}{\partial x^j}\left[\rho u_j\left(\frac{\sum u_i^2}{2}+E+\rho E_\rho\right)+\frac{u_j}{4\Pi}\left(\sum H_j^2\right)-\frac{H_j}{4\Pi}(u,H)\right]=0$$

Elles ne diffèrent de la forme acceptée que par la présence des termes proportionnels à div H et nuls pour les solutions ayant un sens physique.

Les équations (6) dans le cas général se réécrivent encore comme suit :

$$\frac{\partial L^0{}_{q_i}}{\partial t}+\frac{\partial L^j{}_{q_i}}{\partial x^j}+A^j\,\frac{\partial \Phi_{q_i}}{\partial x^j}=0 \tag{11}$$

d'où il résulte que si $L^0(q_1,q_2,\dots)$ est convexe, alors on a la forme hyperbolique symétrique de ces équations :

$$L^0_{q_iq_k}\,\frac{\partial q_k}{\partial t}+\left(L^j_{q_iq_k}+A^j\,\Phi_{q_iq_k}\right)\frac{\partial q_k}{\partial x^j}=0 \tag{12}$$

Les équations de la magnétohydrodynamique sous la forme habituelle ne peuvent pas être réduites au type (1) puisqu'en déduisant la loi de conservation d'entropie il est nécessaire d'utiliser la condition complémentaire div H = 0 qui est la conséquence des données initiales spéciales. Pour surmonter cette difficulté on doit réécrire les équations initiales sous la forme (10) en y introduisant des termes complémentaires proportionaux à div H = 0, nuls pour les solutions en question.

A la fin de 1972 j'ai envoyé mon travail [9] à Friedrichs et à Lax en demandant de le publier dans Comm. Pure and Appl. Math. mais j'ai reçu en réponse une lettre de refus de Friedrichs contenant l'affirmation qu'il savait symétriser l'hydrodynamique magnétique. Friedrichs faisait référence au rapport fait au Colloque de magnétohydrodynamique à Lille. En 1972 dans un recueil consacré au quatre-vingtième anniversaire de l'académicien Mouskhelichvili N.I., Friedrichs a publié un article [32] sur les équations de la physique mathématique écrites sous la forme (1) à la fin duquel il y a déjà une référence à mon travail [2] de 1961. Dans cet article Friedrichs a donné sous la forme usuelle le système d'équations de l'hydrodynamique magnétique qui est différent de (10) car dans ce système sont absents les termes nuls proportionnels à div H. Il y affirme qu'une certaine combinaison linéaire des équations en question donne la loi de conservation d'entropie et c'est pourquoi le système admet l'écriture sous forme (1). Le calcul détaillé et les valeurs des coefficients dans les combinaisons linéaires ne sont pas donnés. J'ai déjà dit que par cette voie, on ne réussit pas à faire une symétrisation des équations de la magnétohydrodynamique.

En 1974, Friedrichs a publié un travail [35] où est décrite la réduction au type symétrique hyperbolique des formes différentes des équations de l'hydrodynamique électromagnétique. L'idée de ce travail au fond coïncide avec l'utilisation de la schématisation (11), (12). Le travail de Friedrichs [11] est aussi consacré à l'exposé de la technique développée dans [35].

En 1978 apparut encore un travail [12] où indépendamment de [9], [10], [11], était proposé le schéma analogique de symétrisation des équations admettant, comme la magnétohydrodynamique, des lois de conservation complémentaires.

Continuant le travail [9] et étudiant son applicabilité aux autres classes d'équations de la physique mathématique, en particulier aux équations de théorie de l'élasticité, Romenski E.I. a généralisé la classe d'équations (11), introduisant au lieu d'une seule fonction homogène du premier degré $\Phi(q_1, q_2, \ldots)$ plusieurs Φ^1, $\Phi^2, \ldots$. Il a montré que les équations du milieu élastique peuvent être mises sous la forme :

$$\frac{\partial L^0_{q_i}}{\partial t} + \frac{\partial (L^j_{q_i} + A^j_m \Phi^m_{q_i})}{\partial x^j} - \sum_{j,m} \Phi^m_{q_i} \frac{\partial A^j_m}{\partial x^j} = 0 \tag{13}$$

$$\Big(\sum_i q_i \Phi^m_{q_i} - \Phi^m = 0\Big)$$

assurant ainsi les lois de conservation (3), (7) et la forme de système hyperbolique symétrique :

$$L^0_{q_i q_k} \frac{\partial q_k}{\partial t} + (L^j_{q_i q_k} + A^j_m \Phi^m_{q_i q_k}) \frac{\partial q_k}{\partial x^j} = 0 \tag{14}$$

Pour celà il utilise la nullité des divergences $\frac{\partial A^j_m}{\partial x^j} = 0$. Cette condition de compatibilité des déformations est analogue à l'égalité div H=0 en magnétohydrodynamique. Si cette condition est remplie dans un moment initial du temps alors elle sera remplie par la suite comme une conséquence des équations en question. Cette circonstance est à la base des travaux [9] et [10]

Blokhin A.M. [14] utilisant le même schéma que dans [13] a réduit au type hyperbolique symétrique les équations de l'hydrodynamique superfluide (hélium II).

Un travail de Soultangasin et de l'auteur [33] est consacré à la mise sous la forme (1) des équations de Boltzmann de la cinétique des gaz.

Un exposé des recherches suivantes sur ce problème se trouve dans le livre de Soutangasin [34].

Une écriture des équations sous forme de système hyperbolique symétrique quasi linéaire permet d'utiliser la théorie relativement simple et examinée en détail, qui affirme que pour les solutions régulières de ces équations, les théorèmes d'existence et d'unicité sont vrais. Dans le cas des solutions discontinues, même si ces discontinuités sont isolées et forment des surfaces régulières, il y a des difficultés. Même dans la théorie linéarisée des ondes de choc, pendant longtemps on n'a pas réussi à trouver les intégrales d'énergie qui sur l'onde de choc seraient dissipatives et qu'on pourrait utiliser pour prouver que le problème est correctement posé. Le fait est que malheureusement, nous trouvons trop peu de lois de conservation avec le schéma (2) - (3). Blokhin A.M. a réussi à explorer une autre voie au récit de laquelle est consacrée la fin du rapport actuel. A l'aide d'intégrales nouvelles Blokhin A.M. a construit une théorie de l'onde de choc selon une surface régulière (les théorèmes d'existence et d'unicité), non seulement dans le cas linéarisé mais dans le cas quasi linéaire général. Notons qu'à l'aide de la technique des opérateurs pseudo-différentiels des affirmations analogues un peu plus tard ont été argumentées dans [15], [16]. Le travail de Blokhin A.M. est basé sur l'utilisation des idées qui ont été développées dans le cycle des recherches scientifiques faites à Novossibirsk et consacrées à la possibilité d e la réduction de l'équation hyperbolique à la forme symétrique. Dans ce cycle des nouveaux faits intéressants ont même été obtenus pour un objet aussi classique que l'équation des ondes. Les travaux de Gordienko V.M., Malychev A.N., Martchouk N.G. ont été consacrés à cette équation. Ils se faisaient en même temps avec les recherches de Blokhin A.M. et tous ces travaux influaient essentiellement l'un sur l'autre. Je vais tâcher d'en dégager les idées générales.

Les travaux concernants l'équation des ondes ont été commencés pendant l'examen de l'article de Miyatake S. [16] aux idées duquel on a réussi à donner un autre sens. Gordienko V.M. [17], [18] étudiait l'équation vectorielle des ondes

$$(15) \qquad u_{tt} - u_{xx} - u_{yy} = 0, \qquad u = \begin{pmatrix} u_1 \\ u_2 \\ \cdot \\ \cdot \\ u_n \end{pmatrix}$$

avec la condition au bord pour x = 0 :

$$(16) \qquad \left(P\frac{\partial}{\partial t} + Q\frac{\partial}{\partial x} + R\frac{\partial}{\partial y} \right) u = 0$$

et des données initiales Cauchy pour t=0. Un problème mixte est notoirement mal posé s'il admet une suite de solutions particulières (k=1,2,3,..) de la forme (Re $\tau>0$, Re $\xi<0$) ;

$$u^{(k)} = e^{k(\tau t+\xi x\pm iy)-\sqrt{k}} \cdot \tilde{u}$$

Cette suite est analogue à la suite étudiée dans un exemple classique d'Hadamard du problème Cauchy mal posé pour l'équation de Laplace.

On peut exiger pour de tels exemples qu'on ne puisse pas en construire, non seulement pour le problème (15), (16), mais pour toutes les petites perturbations de ce problème. Dans ce cas on dit que le problème satisfait à la condition uniforme de Lopatinski. Dans ce cas, comme cela est montré par Gordienko V.M., il faut et il suffit que det $(P+R)\neq 0$ et que la matrice $2n\times 2n$ Λ, définie par

$$\Lambda = \begin{pmatrix} P+R & 0 \\ 0 & P+R \end{pmatrix}^{-1} \begin{pmatrix} 0 & P+R \\ -P+R & 2Q \end{pmatrix}$$

ait toutes ses racines caractéristiques dans le demi-plan de gauche $\{\mathrm{Re}\, z < 0\}$. Dans ce cas il s'avère qu'on peut réduire l'équation vectorielle des ondes au système hyperbolique symétrique, les intégrales d'énergie correspondantes étant dissipatives à cause de la condition au bord.

La construction des intégrales d'énergie est basée sur une identité suivante entre opérateurs. Nous donnant les matrices hermitiennes $n\times n$: K,L,M,N arbitraires, nous composons les matrices hermitiennes $3n\times 3n$ suivantes

$$A = \begin{pmatrix} K & L & M \\ L & K & iN \\ M & -iN & K \end{pmatrix}, \qquad B = \begin{pmatrix} L & K & iN \\ -K & L & M \\ -iN & M & -L \end{pmatrix}$$

$$C = \begin{pmatrix} M & -iN & K \\ iN & -M & L \\ K & L & M \end{pmatrix}.$$

et nous nous convainquons que

$$(\tau A+\xi B+\eta C)\begin{pmatrix}\tau I_n\\ \xi I_n\\ \eta I_n\end{pmatrix} = (\tau^2 - \xi^2 - \eta^2)\begin{pmatrix}K\\ L\\ M\end{pmatrix}$$

En remplaçant $\tau \to \frac{\partial}{\partial t}$, $\xi \to \frac{\partial}{\partial x}$, $\eta \to \frac{\partial}{\partial y}$ et en appliquant cette identité au vecteur u satisfaisant à l'équation des ondes (15) nous voyons que les dérivées de ce vecteur sont liées par le système symétrique

$$(17) \qquad (A\frac{\partial}{\partial t} + B\frac{\partial}{\partial x} + C\frac{\partial}{\partial y}) \begin{matrix}u_t\\ u_x\\ u_y\end{matrix} = 0$$

Les solutions de (17) satisfont la forme différentielle de l'identité des intégrales d'énergie

$$(18) \qquad \frac{\partial}{\partial t}(AU,U) + \frac{\partial}{\partial x}(BU,U) + \frac{\partial}{\partial y}(CU,U) = 0 \;;U=\begin{pmatrix}u_t\\ u_x\\ u_y\end{pmatrix}$$

Pour achever ces constructions il est nécessaire de choisir K,L,M,N de manière que A soit définie positive et que la condition au bord (16) soit dissipative pour l'intégrale (18). On peut satisfaire à toutes ces exigences en se donnant une matrice hermitienne définie positive 2n×2n arbitraire G pour laquelle l'équation de Liapounov

$$(19) \qquad \Lambda^*\begin{pmatrix}K-M & -L-iN\\ -L+iN & K+M\end{pmatrix} + \begin{pmatrix}K-M & -L-iN\\ -L+iN & K+M\end{pmatrix}\Lambda = -G$$

définit K,L,M,N. On peut montrer qu'elles satisfont à toutes les conditions posées. (L'équation de liapounov (19) est résoluble puisque toutes les racines caractéristiques Λ se trouvent dans le demi-plan de gauche complexe).

Martchouk N.G. [19] a démontré l'équivalence du système ainsi construit [17] avec l'équation des ondes initiales (15).

Pour une dimension arbitraire cette construction a été généralisée par Malychev A.N. [20], mais seulement pour l'équation des ondes scalaire.

Attardons-nous brièvement sur l'idée posée par Blokhin A.M. à la base de l'application des intégrales d'énergie de l'équation vectorielle des ondes à l'étude du problème de l'onde de choc régulières. Il a montré que ce problème est bien posé. En supposant que le gaz coule avec une vitesse supersonique (M>1) sur le front faiblement courbé de l'onde de choc, situé près du plan x=0, alors derrière le front (pour x>0) la vitesse est subsonique (M<1). Pour la description des petites perturbations on peut se servir des équations de l'acoustique qui sous forme adimensionnelle sont de la forme :

$$\left\{\begin{pmatrix}1&0&0&0\\0&1&0&0\\0&0&1&0\\0&0&0&1\end{pmatrix}\frac{\partial}{\partial t}+\begin{pmatrix}M&1&0&0\\1&M&0&0\\0&0&M&0\\0&0&0&M\end{pmatrix}\frac{\partial}{\partial x}+\begin{pmatrix}0&0&1&0\\0&0&0&0\\1&0&0&0\\0&0&0&0\end{pmatrix}\frac{\partial}{\partial y}\right\}\begin{pmatrix}P\\u\\v\\s\end{pmatrix}=\begin{pmatrix}0\\0\\0\\0\end{pmatrix}$$

Ici M est un nombre de Mach derrière le front ($0<M<1$). Les conditions au bord dans le front $x = 0$ sont de la forme

(21) $$u+\beta p=0; \quad v+\sigma F_y=0, \quad \mu p+F_t = 0, \quad s+\gamma p = 0$$

$F(t,y)$ est la déviation du front perturbé de $x=0$.

De (20) et (21) on déduit l'équation des ondes pour la perturbation de la pression p :

(22) $$p_{tt} + 2Mp_{tx} - (1-M^2)p_{xx} - p_{yy} = 0$$

et la condition au bord ($\lambda=\sigma\mu$) :

$$(\beta+M)p_{tt} - (1-M^2)p_{tx} + \lambda M p_{yy} = 0$$

Les premières dérivées p_t, p_y satisfont, toutes les deux, à la même équations des ondes que p elle-même, et aux conditions au bord suivantes

$$\left[(\beta+M)\frac{\partial}{\partial t} - (1-M^2)\frac{\partial}{\partial x}\right]p_t + \lambda M\frac{\partial}{\partial y}p_y = 0$$

(23) $$-\frac{\partial}{\partial y}p_t + \frac{\partial}{\partial t}p_y = 0$$

Cette construction amène naturellement au problème mixte pour l'équation vectorielle des ondes avec des conditions au bord liant les premières dérivées des fonctions inconnues. Par une transformation un peu compliquée du système de coordonnées, l'équation des ondes peut être réduite au type standard

$$p_{t't'} - p_{x'x'} - p_{y'y'} = 0$$

Développant et élargissant en détail l'idée tracée ici, Blokhin A.M. a su construire des intégrales d'énergie qui sont dissipatives dans tous les cas lorsque l'équation d'état du gaz assure que le problème est bien posé, c'est-à-dire que la condition uniforme de Lopatinski est satisfaite. A cela ont été consacré les travaux [24] - [25]. Pour conclure l'exposé des recherches sur les lois de conservation pour les équations quasilinéaires de la physique mathématique, je ferai un court exposé sur le problèmede la réduction de l'équation strictement hyperbolique linéaire, d'ordre homogène N

$$P_n\left(\frac{\partial}{\partial t}, \frac{\partial}{\partial x_1}, \frac{\partial}{\partial x_2}, \ldots\right)u=0$$

à la forme symétrique hyperbolique en la fonction vectorielle composée de toutes les dérivées d'ordre N-1 de la fonction cherchée u. Dans les travaux [26] de Godounov et Kostin et [27] Tychtchenko, on donne le schéma de telle réduction et à l'aide d'une théorie des représentations du groupe linéaire complet on calcule le nombre de paramètres arbitraires dont dépendent les matrices des coefficients des systèmes symétriques obtenus. Dans [26] on a montré que parmi des symétrisations formellement admissibles des équations t-hyperboliques à deux variables d'espace, il en existe une dont la matrice des coefficients est strictement définie positive. Dans les travaux de Kostin et Mikhaïlova [28], [29], il est montré que pour un nombre quelconque de variables, toute équation proche d'une équation hyperbolique invariante par les rotations admet aussi la symétrisation hyperbolique. Un examen de tous ces travaux avec un exposé court des idées de démonstration est donné dans [30]. Dans le même temps, Ivanov V.V. [31] a construit un exemple d'équation strictement hyperbolique du quatrième ordre à quatre variables d'espace, qui ne peut pas être réduite au système hyperbolique symétrique. Cette équation $Lu = 0$ a l'opérateur différentiel suivant :

$$L = \left(\frac{\partial^2}{\partial t^2} - \frac{\partial^2}{\partial x_1^2} - \frac{\partial^2}{\partial x_2^2} - \frac{\partial^2}{\partial x_3^2} - \frac{\partial^2}{\partial x_4^2}\right)^2$$

$$- \frac{1}{1+2\varepsilon}\left(\frac{\partial^4}{\partial x_1^2\, \partial x_2^2} + \frac{\partial^4}{\partial x_1^2\, \partial_3^2} + \frac{\partial^4}{\partial x_2^2\, \partial x_3^2} + \frac{\partial^4}{\partial x_4^4} - 4\,\frac{\partial^4}{\partial x_1 \partial x_2 \partial x_3 \partial x_4}\right.$$

$$- \frac{\varepsilon}{1+2\varepsilon}\left(\frac{\partial^2}{\partial x_1^2} + \frac{\partial^2}{\partial x_2^2} + \frac{\partial^2}{\partial x_3^2} + \frac{\partial^2}{\partial x_4^2}\right)^2$$

dans lequel le paramètre ε est n'importe quel nombre positif inférieur à 1/4. Il est démontré que dans le cas de quatre variables spatiales l'ensemble des équations non symétrisables forme un ensemble ouvert.

REFERENCES

1. Godounov S.K. Sur la notion de solution généralisée. DAN 134, 6(1960), p. 1279 - 1282.

2. Godounov S.K. Classe intéressante de systèmes quasi linéaires. DAN 139, 3 (1961), p. 521 - 523.

3. Friedrichs K.O., Lax P.D. Systems of conservation equations with a convex extension. Proc. Nat. Acad. Sci. Usa 68, 8(1971), p. 1686 - 1688.

4. Godounov S.K. Eléments de mécanique des milieux continus. Moscou, "Nauka", 1978. - 304 p.

5. Rojdestvenski B.L. Sur les systèmes d'équation quasi linéaires. DAN, 115 (1957), p. 454-457.

6. Riemann B. Sur la diffusion des ondes planes d'amplitude finie. Compositions. Moscou. Gostekhizdat, 1948, p. 376-395.

7. Koulikovski A.G. Sur la structure des ondes de choc magnétohydrodynamiques sous une loi arbitraire de dissipation. PMM 26, ed. 2(1962), p. 273.

8. Germain P. Contribution à la théorie des ondes de choc en magnétodynamique des fluides. Office Nat. d'Etudes et de Rech. Aeronaut. Publ. No. 97, Paris, 1959.

9. Godounov S.K. Forme symétrique des équations de la magnétohydrodynamique. Méthodes numériques de la mécanique des milieux continus, v. 3, No. 1, 1972. Novossibirsk, p. 26-34.

10. Friedrichs K.O. A limiting process leading to the equations of relativistic and nonrelativistic fluid dynamics. Colloque Magnétohydrodynamique, Lille, 1969, p. 177-188.

11. Friedrichs K.O. Conservation equations and the Laws of Motion in Classical Physics. Comm. Pure and Appl. Math., XXXI (1978), p. 123-131.

12. Boillat G. Symétrisation des systèmes d'équations aux dérivées partielles avec densité d'énergie convexe et contraintes. Comptes Rendus. Série 1, 1982, t. 295, No. 9, p. 551-554.

13. Romenski E.I. Lois de conservation et forme symétrique des équations de la théorie de l'élasticité non linéaire. Dans : Problèmes aux limites pour les équations aux dérivées partielles (Trudy du séminaire de Sobolev S.L.), Novossibirsk, 1984, No. 1, p. 132 - 143.

14. Blokhin A.M. Sur la symétrisation des équations de Landau dans la théorie de superfluidité de l'helium II. Dynamique des milieux polyphasés (Dynamique du milieu continu 68), Novossibirsk, 1984, ed. 68, p. 13 - 34.

15. Majda A. The stability of multi-dimensional shock fronts. Memoir of the American Mathematical Society, 1983, vol. 41, No. 273, 95 p.

16. Miyatake S. Mixed problem for hyperbolic equations of second order with first order complex boundary operators. -Japanese J. Math. 1975, v. 1, No. 1, p. 111-158.

17. Gordienko V.M. Un problème mixte pour l'équation vectorielle des ondes : Cas de dissipation de l'énergie ; Cas mal posés. - C.r. Acad. Sci. 1979, t. 288, No. 10, Série A, p. 547-550.

18. Gordienko V.M. Symétrisation du problème mixte pour l'équation hyperbolique du second ordre à deux variables spatiales. -Sib. Math. J., 1981, v. 22, No. 2, p. 84 - 104.

19. Martchouk N.G. Sur l'existence des équations du problème mixte pour l'équation vectorielle des ondes. Dokl. AN SSSR, 1980, 252, No. 3, p. 546 - 550.

20. Malychev A.N. Problème mixte pour l'équation hyperbolique du second ordre avec une condition complexe au bord du premier ordre. Sib. Math. J., 1983, v. 24, No. 6, p. 102 - 212.

21. Blokhin A.M. Problème mixte pour le système des équations de l' acoustique avec conditions au bord sur l'onde de choc. Izv. SO AN SSSR, Série de sciences techniques, 1979, No. 13, ed. 3, p. 25 - 33.

22. Blokhin A.M. Estimation de l'intégrale d'énergie dans le problème mixte pour les équations de la dynamique des gaz avec conditions au bord sur l'onde de choc. -Sib. Math. J., 1981, v. 22, No. 4, p. 23 - 51.

23. Blokhin A.M. Unicité de la solution classique du problème mixte de la dynamique des gaz avec conditions au bord sur l'onde de choc. -Sib. Math. J., 1982, v. 23 No. 5, p. 17 - 30.

24. Godounov S.K., Blokhin A.M. Energy Intégrals in the Theory of Shock Wave Stability. Nonlinear Deformation Waves, IU'TAM Symposium, Tallin, 1982, Springer-Verlag, 1983, p. 18- 29

25. Blokhin A.M. Intégrales d'énergie dans le problème sur la stabilité de l'onde de choc. Novossibirsk, IM SO AN SSSR, 1982, - 176 p.

26. Godounov S.K., Kostin V.I. Réduction de l'équation hyperbolique au système hyperbolique symétrique dans le cas de deux variables spatiales. -Sib. Math. J., 1980, v. 21, No. 6, p. 3 - 20.

27. Tychtchenko A.V. Sur la base des solutions de l'identité homogène de Hörmänder. - Sib. Math. J., v. XXVI, No. 1 (1985), p. 150 - 158.

28. Kostin V.I. Sur la symétrisation des opérateurs hyperboliques. -Dans : Problèmes aux limites correctement posés pour les équations non classiques de la physique mathématique. Novossibirsk, 1980, p. 112 - 116.

29. Mikhailova T. Yu. Symétrisation des équations hyperboliques invariantes. DAN, 1983, v. 270, No. 3, p. 246-250.

30. Godounov S.K. Intégrales d'énergie des équations hyperboliques d'après Petrovski. Commentationes mathematicae universitatis carolinae, Praga 26.1, 1985, p. 41-74.

31. Ivanov V.V. Polynômes strictement hyperboliques n'admettant pas la symétrisation hyperbolique. Préprint No. 77 de l'Institut de Mathématiques SO AN SSSR, Novossibirsk, 1984, p. 1 - 16.

32. Friedrichs K.O. Symmetric hyperbolicity and conservation laws. Dans : Mécanique du milieu continu et problèmes affines de l'analyse. "Nauka", Moscou, 1972, p. 575 - 580.

33. Godounov S.K., Soultangazin Ou. M. Sur les modèles discrets de l'équation cinétique de Boltzmann. Uspekhi Math. nauk, 1971, v . 36, No. 3 (159), p. 1 - 15.

34. Soultangazin Ou. M. Modèles discrets non linéaires de l'équation de Boltzmann. "Nauka" Kaz. SSSR, Alma-Ata, 1985.-190 p.

35. Friedrichs K.O. On the laws of Relastivistic Electro-Magneto-Fluid Dynamics. Comm. Pure Appl. Math., vol . XXVII (1974), p. 749 - 808.

ON SYMMETRIZING HYPERBOLIC DIFFERENTIAL EQUATIONS

By Peter D. Lax

In his interesting paper [2], published over 25 years ago, Godunov has shown that a number of hyperbolic equations of mathematical physics, those of compressible gas dynamics par excellence, can be written in a form that is both symmetric and in conservation form. He also showed that in this case there is an additional conservation law satisfied by all smooth solutions of the original system. In 1972 he showed how to write the MHD equation in a form that is both symmetric and in conservation form.

It has been known for a long time that the equations of gas dynamics can be put in symmetric form, but by transformations that spoil the conservation form of the equations. In 1969 Friedrichs has shown how to put the MHD equations in symmetric form, again at the expense of the conservation form.

In 1971, [1], Friedrichs and I gave a partial converse of Godunov's result: if a system of n conservation laws implies a new conservation law where the conserved quantity is a convex function of the original n conserved quantities, then the original system can be put in symmetric form. We did not claim, nor did we show, that the symmetric version is in conservation form; this step was accomplished by M. Mock (Sever). All this is explained carefully in Section 1 of Harten and Lax, [3].

The main point of [3] is to describe approximate solutions of Riemann problems that suffice for Glimm's modification of Godunov's method for numerically solving initial value problems for systems of conservation laws.

In [5], I have shown that discontinuous solutions of systems of conservation laws that are limits of solutions of dissipative equations do <u>not</u> satisfy the additional conservation law described above, but an inequality.

[4] contains an example of a 3 x 3 <u>linear</u> system of first order equations in two space variables that cannot be put in symmetric form.

REFERENCES

[1] Friedrichs, K. O., and Lax, P.D., "Systems of Conservation Equations with a Convex Extension",Proc. Nat. Acad. Sci. U.S.A., 68, 8, pp.1686-88, (1971).

[2] Godunov, S., "An Interesting Class of Quasilinear Systems",Dokl. Acad. Nauk,SSSR, 139 (1961), pp. 521-523.

[3] Harten, A., and Lax, P.D., "Random Choice Finite, Difference Schemes", SIAM J. Num. Anal., 18, (1981), pp. 289-315.

[4]Lax, P.D., "Differential Equations, Difference Equations and Matrix Theory", Comm. Pure Appl. Math.,XI (1958), pp. 175-194.

[5] Lax, P.D., "A Concept of Entropy", Contributions to Nonlinear Functional Analysis, Ed. Zarantonello,(1971), pp. 603-634, Academic Press.

A SURVEY OF NONSTRICTLY HYPERBOLIC CONSERVATION LAWS.

Barbara Lee Keyfitz*
University of Houston - University Park
Houston, Texas 77004/USA

In this work I would like to describe recent results on Riemann problems and admissibility conditions for some systems of conservation laws which are not of the standard, strictly hyperbolic type. Such systems have been found as model equations in many fields which appear to have little else in common: in elasticity, in phase transitions, in multiphase flow in petroleum reservoirs. These potential applications motivate a study of the mathematical structure of the solutions; in particular, questions about well-posedness of the hyperbolic (and nonhyperbolic) problems and about the proper limiting procedures, from either a mathematical or a physical viewpoint, are just beginning to surface. A theme for the subject might be that some systems of conservation laws that look very peculiar have solutions that are almost classical, while other systems that appear almost classical have only very peculiar solutions (if they have any at all).

I. Introduction.

A system of conservation laws in one space variable can be written in the form

$$\frac{\partial U}{\partial t} + \frac{\partial}{\partial x} F(U) \equiv U_t + A(U)U_x = 0. \tag{1}$$

Here $U = (u_1,\dots,u_n)$ and A is the Jacobian matrix $\partial F/\partial U$. The system is hyperbolic if the eigenvalues $\lambda_1,\dots,\lambda_n$ of A are real; it is strictly hyperbolic if they are real and distinct. We shall use the term "classical theory" somewhat loosely to refer to the considerable collection of results on weak solutions for the Cauchy problem for strictly hyperbolic equations of type (1): to begin with, it is well known that smooth solutions of (1) exist only for a finite time, beyond which generalized (weak) solutions must be introduced. Solutions in (1) break down because their first derivatives blow up, and typically weak solutions are bounded but contain jump discontinuities. A starting point for the analysis is the study of Riemann problems: initial-value problems of the form

$$U(x,0) = \begin{cases} U_L & x < 0 \\ U_R & x > 0 \end{cases} \tag{2}$$

where the constant states are separated by a jump. Riemann problems are important

* Supported in part by the National Science Foundation under grant DMS-8504031.

in proving general existence theorems following the work of Glimm [7]; they are also important in numerical algorithms, such as the Glimm scheme and the Godunov scheme. For both of these applications one would like (1), (2) to have, for arbitrary U_L and U_R, a unique solution which depends smoothly on U_L and U_R. It is also well known that weak solutions are not uniquely defined by the standard weak formulation of (1), and there is an extensive literature on admissibility conditions for weak solutions. The most useful admissibility conditions are those which can be stated in purely mathematical terms, such as the Lax entropy condition, which stipulates that n+1 characteristics enter the shock and n-1 leave it. The shock is then associated with the family for which characteristics enter the shock from both front and back. This is basically a local condition, for rather small shocks; Liu introduced an alternative criterion which examines only the relation between the shock and characteristic speeds of its own family, but does so along an entire curve (the Hugoniot locus) of states between the front and back of the shock. But all such criteria are justified by appealing to limiting arguments which take into account the physical setting from which (1) was derived. For the classical examples of model problems in gas dynamics or elasticity, the relation between mathematical criteria, such as the Lax or Liu entropy conditions, and conditions motivated by physical considerations, such as entropy or viscosity, is reasonably well understood; it has been nicely described by Dafermos in [4]. The situation for nonstrictly hyperbolic conservation laws, and especially for conservation laws of mixed type, is going to be much more complicated, as we shall indicate here. Ultimately, well-posedness for mixed type systems can be determined only by a satisfactory resolution of these open questions.

Even for strictly hyperbolic systems the study of Riemann problems (1), (2) is not complete, but existence of a unique admissible centered solution consisting of shocks, rarefaction waves and contact discontinuities is known if $|U_L-U_R|$ is not too large and each characteristic family is either genuinely nonlinear or linearly degenerate or when genuine nonlinearity fails in a generic manner [24,25]. Global existence for all states U_L, U_R requires some additional conditions (for example, see [19]) to specify that the Hugoniot locus, defined as

$$H(U_0) = \{U \mid s(U-U_0) = F(U)-F(U_0) \text{ for some } s\}, \tag{3}$$

have globally the same structure as it has locally: 2n distinct branches which leave any neighborhood of U_0. Some examples are known where the failure of global existence appears to be related to the fact that the characteristic speeds, while everywhere distinct, overlap in the sense that $\lambda_1(U) > \lambda_2(V)$ for some states U and V, where λ_1 is the slower speed locally [22]. This situation, which occurs in models for anisotropic wave propagation [23], suggests that strict hyperbolicity may not be the precise hypothesis required for well-posedness: in this example, studied in [22], the only candidates for solutions are far from classical; on the other hand, the examples in the next section show that many equations without strict hyperbolicity have straightforward solutions.

II. <u>Systems that are hyperbolic but not strictly hyperbolic.</u> When one looks at the simplest situation where strict hyperbolicity can fail, the case of a pair of conservation laws where $\lambda_1(U) = \lambda_2(U)$ for some state U, one observes that this corresponds to a single condition, $(\mathrm{tr}\, A)^2 = \det A$, on the matrix A. On the other hand, A will be diagonalizable there only if a second constraint holds and, furthermore, if A is not diagonalizable and if the dependence of A on U is sufficiently smooth (for example, analytic), there will be states arbitrarily close to U at which A has complex eigenvalues. Thus, systems which are hyperbolic everywhere but for which strict hyperbolicity fails somewhere arise by virtue of some special structure - for example, the symmetry of A or the coupling of a linear and nonlinear family which cannot merge, or a degeneracy in the system.

One system of this type arises if one studies propagation of waves in an infinitely long elastic string in the plane. The governing equations (exact for an ideal string with no thickness and no resistance to bending) are

$$\begin{aligned} u_{tt} + (\phi u_x)_x &= 0 \\ v_{tt} + (\phi v_x)_x &= 0 \end{aligned} \tag{4}$$

where (u,v) are Cartesian coordinates in the plane, x is a Lagrangian coordinate representing arclength along the string in a reference configuration and $\phi = T/(1+\varepsilon)$ where $\varepsilon = \sqrt{u_x^2+u_x^2} - 1$ is the strain and $T = T(\varepsilon)$ the stress. The eigenvalues $\pm\sqrt{\phi}$ and $\pm\sqrt{T'}$ characterize speeds of propagation of transverse and longitudinal displacements respectively. Under assumptions with obvious physical significance they are always real, but coincide in pairs for states $U = (u_x, u_t, v_x, v_t)$ at which $\phi = 0$, a generic situation for nonlinear constitutive relations. In [20], we indicated how to solve the Riemann problem for this system; further details were worked out by Shearer [36]; Rascle and Serre have also studied this [30,33]. One notes in this problem that the transverse mode of propagation is linearly degenerate, while longitudinal waves behave locally like pressure waves in gas dynamics: in [20] we assumed genuine nonlinearity, though a more realistic assumption permits genuine nonlinearity to fail on curves transversal to the rarefaction curves [36]. Note also that strict hyperbolicity fails on a three-dimensional manifold in this sytem.

A simpler system for which the solution to the Riemann problem exhibits the same structure as (3) is given by

$$\left.\begin{aligned} u_t + (u\phi)_x &= 0 \\ v_t + (v\phi)_x &= 0 \end{aligned}\right\} \tag{5}$$

where we may generalize slightly to let $\phi = \phi(u,v) = \phi(r,\theta)$. The use of polar coordinates serves to emphasize that $\phi = \phi(r)$ corresponds to the actual structure of the elastic string constitutive relation. Again $\lambda_1 = \phi$ is linear, $\lambda_2 = \partial(r\phi)/\partial r$ is nonlinear, and $\lambda_1 = \lambda_2$ when $\partial\phi/\partial r = 0$; if there is no θ dependence in ϕ then $\lambda_1 = \lambda_2$ on a circle in the (u,v) plane; in fact the curves ϕ = constant, corresponding to contact discontinuities in the linear family, are concentric circles, while the shock and rarefaction curves of the other family are radial lines. (This system has the property that the shock and rarefaction curves coincide, even in the case that ϕ depends on θ; such systems have been characterized by Temple [44]. Here is a clear example of a structure, in this case geometric, which results in real eigenvalues for all (u,v) whether the matrix A is diagonalizable - as it is when ϕ is independent of θ - or not.) The speed of a discontinuity joining U to U_0 is $s = (r\phi - r_0\phi_0)/(r - r_0)$, and clearly $s = \phi = \lambda_1$ at a point where $\phi = \phi_0$. Under the assumption of genuine nonlinearity of λ_2, s is a monotone function along the shock curve, and hence Liu's entropy condition is satisfied. The Lax entropy condition is also satisfied (if inequalities are replaced by weak inequalities where $\phi = \phi_0$) with an interesting wrinkle: the shock speed is "fast" or "slow" according as the states U and U_0 are on the same side or on opposite sides of the curve $\phi = \phi_0$: the λ_2 characteristics always enter the shock, and the λ_1 characteristics always cross it, but they may cross either from right to left or from left to right. In the case $\phi = \phi(r)$, the Reimann problem for an arbitrary pair of states (U_L, U_R) has a solution with a single intermediate state U_1, which is either on the circle $\phi = \phi_L$ or on the circle $\phi = \phi_R$ depending on whether the shock is fast or slow.

The case where $\phi = \phi(r,\theta)$ is similar and is also described in [20]; it is interesting that a system with exactly the structure of (5) in which the dependence of ϕ on θ is nontrivial occurs in models for miscible flow in a petroleum reservoir; ignoring capillary pressure leads to a hyperbolic conservation law, the Buckley-Leverett equation; when a solvent is carried passively with the injected water, one obtains a system of two equations which approximate the dynamics of the flow; the existence of a curve along which $\lambda_1 = \lambda_2$ is a consequence of the physics of the model - that the purpose of the solvent is to alter the viscosity. This model, its applications and other comments on the Riemann and Cauchy problems can be found in [12]. We mention that the fact that A is not diagonalizable in this case does not seem to be important.

The admissibility of the solutions found - by physical entropy criteria, by the construction of viscous shock profiles using an artificial viscosity, and by numerical simulation using a shock-capturing method - seems to be consistent with the admissibility criteria of the mathematical conditions. By contrast with other problems we will describe, these solutions are very robust. Even here there is an example of mild ill-posedness: if one considers a one-parameter family of Riemann problems with, say, U_L fixed and $U_R = U_R(\zeta)$, then as U_R crosses the curve where $\phi = \phi_L$, the shock switches from fast to slow and the intermediate state, U_1, from the curve $\phi = \phi_L$ to $\phi = \phi_R$. At this value of ζ, of course, $s = \lambda_1$, and so speeds of the shock and the contact discontinuity are equal, and the intermediate state is not observed: one sees instead a sort of composite shock/contact which breaks up when perturbed. If one measures the solution in L^p, for $p<\infty$, the dependence on ζ is smooth; however it is the total variation norm which is important in solving the Cauchy problem by Glimm's method, and here one must introduce some weighting of the norm that is singular near the critical curve. This was done by Isaacson and Temple [13,43]; on the other hand Liu and Wang [26] were able to use a straightforward approach in the case $\phi = \phi(r)$, which can be reduced to a single conservation law.

Strict hyperbolicty can also fail because of a degeneracy. The system

$$\left.\begin{aligned} u_t &= u_x \\ u_t &= (\tfrac{1}{3}\, u^3)_x \end{aligned}\right\} \qquad (6)$$

is an example of a class of problems studied in [21]. Here the characteristic speeds are $\pm u$; both families are genuinely nonlinear; $\lambda_1 = \lambda_2$ on the line $u = 0$. The Riemann problem was solved for all states U_L and U_R in [21], and the shock waves are all admissible under a number of criteria. A small perturbation of the flux function to $u^3/3 + \varepsilon u$ results in a strictly hyperbolic system if $\varepsilon > 0$, and the classical solution to this system is close to the solution of (5) constructed in [21] when ε is small [18]. If $\varepsilon < 0$, a system of mixed type is obtained; it will be discussed in the next section.

An interesting class of systems which are hyperbolic but not strictly hyperbolic consists of systems in the plane that are strictly hyperbolic except at a single point, say $u = v = 0$. These systems are classified and analysed in [14], [32]. (Such equations are conjectured to play a role in some multiphase flow problems; the "inviscid Burgers' system," $z_t + (\bar{z}^2)_x = 0$, where $z = u + iv$ [3], is also an example of this type.) Shearer, Schaeffer, et al. [38], have considered homogeneous quadratic systems in order to understand the qualitative behavior of solutions. It appears that not all Riemann problems in the neighborhood of the origin have classical solutions; on the other hand certain shocks, called "undercompressive," which violate the Lax entropy condition, turn out to have viscous profiles and therefore to satisfy some admissibility criterion in this case. These systems also exhibit a three-fold symmetry (three branches of the Hugoniot locus, three rarefaction curves) at the origin, and a breakdown of genuine nonlinearity in an unusual way along rarefaction curves. The connection with physical examples is rather tenuous, and more work remains to be done on the problems; however one can certainly point them out as examples which are very close to satisfying the classical hypotheses but whose solutions are very far from classical.

III. Conservation Laws of Mixed Type. If the eigenvalues of the matrix A(U) in (1) are complex for certain values of U, then the system is nonhyperbolic or elliptic there. Since system (1) is an evolution equation, this differs from the situation in steady transonic flow, where one studies a boundary value problem for an equation of mixed type. Indeed, for a linear evolution equation, the Hadamard counterexample shows that the Cauchy problem is classically ill-posed: high-frequency perturbations of the data make large-amplitude changes in the solution. One effect of the nonlinearity is to limit the amplitude of perturbations; and there is possibly a sense in which (1) is not ill-posed for a system of mixed type. In this section, I shall present some examples which have led to mixed-type systems, and describe their solutions. Just as in the nonstrictly hyperbolic case, new types of shocks appear and admissibility conditions must be generalized. Again, the eventual justification will involve both physical and mathematical considerations, as in the less controversial cases already discussed. It is too early to predict what sort of theory will eventually emerge, but it is fair to say that just as the classical, strictly hyperbolic system (1) is not exactly well-posed (existence of classical solutions fails; uniqueness of weak solutions does not hold without admissibility conditions) so may the system of mixed type not be as ill-posed as might appear at first glance.

One easily described example that leads to a system of mixed type arises in the study of phase changes in elastic bars [15,16]. If thermal and viscous effects are ignored, the dynamic equations for the deformation, u, and velocity, v, of the point on an elastic bar with position x in a reference configuration are

$$\begin{aligned} &\frac{\partial u}{\partial t} - \frac{\partial v}{\partial x} = 0 \\ &\frac{\partial v}{\partial t} - \frac{\partial}{\partial x}\,\sigma(u) = 0 \end{aligned} \tag{7}$$

where $\sigma = \sigma(u) = \frac{\partial W}{\partial u}$ is the stress; W is the strain energy function. For a single-phase material, W is a convex function of u and σ is monotone; system (6) is strictly hyperbolic. A material for which W has an interval $\alpha < u < \beta$ on which it is concave can be described as exhibiting two-phase behavior: deformations $u < \alpha$ correspond to the "α-phase" and deformations $u > \beta$ are in the "β-phase"; intermediate states are "unstable" by elementary principles of minimization of energy. If equation (7) is used to describe the dynamic behavior of a rod that allows two phases, then the characteristic speeds,

$$\lambda = \pm\sqrt{\sigma'(u)}$$

are complex in the unstable interval $\alpha < u < \beta$, which corresponds to a strip in phase space (u,v). A mathematical example of such a system is the perturbation of equation (6), $\sigma(u) = u^3/3 + \varepsilon u$, with $\varepsilon < 0$.

When one examines the Hugoniot locus for (7), there appear to be four types of discontinuities:

1. Classical, single-phase shocks whose admissibility can be justified in the

classical way, with W playing the role of "entropy" and decreasing strictly when shocks are present.

2. Almost-classical rapidly-moving phase boundaries. These are solutions joining a state with $u_0 < \alpha$ to a state with $u_1 > \beta$ which satisfy the Lax and Liu entropy conditions on shock speeds. They have an analogue among solutions to nonstrictly hyperbolic systems of type (6) constructed in [21], which were also limits of viscosity solutions and could be computed approximately by finite-difference numerical techniques.

3. A new kind of shock, joining a point in the α- or β-phase to a point in the unstable phase. These cannot satisfy classical conditions on characteristic speeds, since those are complex on one side. Some experience suggests that they be admitted if they satisfy the Lax or Liu condition on the real part of the speed or, equivalently, if they are on the continuation of an admissible branch of the Hugoniot locus into the nonhyperbolic region. They will not have monotone viscosity profiles, but there may be a sense in which they are limits of solutions of a higher-order system of viscoelastic equations. They could also be compared to super- to subsonic shocks in steady transonic flow.

4. Slow-moving or stationary phase boundaries. This term applies to discontinuities involving pairs of states on the Hugoniot locus which definitely violate the Lax entropy condition on shock and characteristic speeds. These are reminiscent of the "undercompressive" shocks described in the last section. Some of these shocks are admissible under some criteria; for example, stationary phase boundaries (and no others) have discontinuous viscosity profiles if the second equation in (7) is modified by adding a "viscosity" term, εv_{xx}, to the right hand side [29]. Different "undercompressive" solutions are admissible if a combination of second- and third-order terms is used there. (The motivation for this in elasticity theory is unclear; however there is a related and important group of problems arising from phase transitions in fluid dynamics where system (7) has an analogue in which u is specific volume, v velocity and $p = -\sigma(u)$ is an equation of state of van der Waals type. The third-order terms represent capillary forces and influence the admissibility of certain phase transitions. In the thermodynamic model, the so-called Maxwell line and the equal area rule offer physical criteria for determining admissibility of phase transitions; these have been related to viscosity-capillarity criteria by Slemrod and Roytburd [39],[40], and to other criteria by Hattori [8],[9],[10]).

Shocks from all four of these categories were used by Shearer [34], to solve the Riemann problem for (7): Shearer's solution involves shocks from the first three groups, whose admissibility is widely accepted, and all stationary shocks from the fourth. A unique solution to the Riemann problem is obtained. To some extent, the work of Pego [29] on viscosity profiles for (7) provides justification of this, as does a study of a related problem [17] whose solution is qualitatively similar to Shearer's.

In this connection, work of Marchesin, Medeiros and Paes-Leme describes another application in which solutions corresponding to phase boundaries appear [27]. Here also there is a qualitative resemblance; on the other hand, as [35],[37] show, quite different definitions of admissibility will also give "qualitatively similar" solutions. Among other mathematical results on this problem we should mention Andrews' and Ball's existence theorem for the viscoelastic equations [1], and also results of Ebin and Saxton [5],[6]. Both of these give existence for the parabolic-type system with viscous dissipation, results which are valid when $\sigma(u)$ is not monotonic, but neither theorem allows one to pass to the limiting system (7). The work of Roytburd and Slemrod [31],[41], which answers similar questions for viscosity-capillarity perturbations in the context of van der Waals type phase transitions, should be compared with this. To decide between competing theorems may take more detailed comparisons with experimental results. It is also possible that no single mathematical criterion for admissibility will emerge: this would be rather a negative result for well-posedness of the mixed problem.

I'd like to conclude this list of examples with an intriguing model problem that arises when one considers multiphase flow in a reservoir - so-called porous medium flow - for example, in the context of secondary or enhanced oil recovery. The full system of equations for such flow couples an elliptic equation for the pressure with parabolic equations for the transport of the components, but if one looks at a model in one space dimension, and ignores interfacial tension (i.e. assumes no capillary forces) then the pressure equation can be solved, and there results a system of hyperbolic conservation laws (or, as it turns out, conservation laws of mixed type) for the phase saturations. In three-phase flow (oil, water, gas) one gets a pair of equations:

$$\begin{aligned} \frac{\partial}{\partial t} s_w + \frac{\partial}{\partial x} f_w(s_w, s_g) &= 0 \\ \frac{\partial}{\partial t} s_g + \frac{\partial}{\partial x} f_g(s_w, s_g) &= 0 \end{aligned} \tag{8}$$

for the water and gas saturations. This system is derived by using Darcy's law, which relates the phase velocity to the relative permeability and phase viscosity in the medium, and to the known pressure gradient. The relative permeabilities are assumed to be functions of the phase saturations: for two-phase flows, these functions can be measured experimentally and fit to empirical curves. For three-phase flow, the precise functional dependence is unknown and must be modelled. Stone's model [28],[42] has recently been studied by Bell, Trangenstein and Shubin [2], but other models of multiphase flow may lead to this phenomenon also: if one analyses the characteristics in (8) inside the "saturation triangle," the physically relevant region given by

$$0 \leqslant s_w \leqslant 1, \quad 0 \leqslant s_g \leqslant 1, \quad s_w + s_g \leqslant 1,$$

then typically there is a small tear-drop-shaped region in the interior where the characteristics are complex. Bell, Trangenstein and Shubin [2] have done a numerical study of the Riemann problem for (8), with a choice of parameters that makes the nonhyperbolic region quite large, using an upwind finite difference scheme (the characteristics have positive real part, even when they are complex). Their computations show that high-frequency oscillations are, in fact, damped out, and that the solutions have a wave structure consistent with the construction of Shearer [34], and also reminiscent of hyperbolic but not strictly hyperbolic problems such as the Buckley-Leverett system [12]. They do not find undercompressive shocks, and I conjecture that the compactness of the nonhyperbolic region, along with other differences between the structures of systems (7) and (8), provides a mechanism whereby one can construct a unique solution of the Riemann problem, consisting of rarefactions and shocks that everywhere satisfy Liu's entropy condition. Specifically, there may be points U_0 whose Hugoniot locus includes a point U on the boundary of the nonhyperbolic region at which the shock speed $s = \lambda_1 = \lambda_2$. These points will figure prominently in solving the Riemann problem; they have no analogue in the mixed-type problems originating from elasticity or the van der Waals model, where the elliptic region is a strip. It is, however, possible to construct similar systems by perturbing the systems studied by Shearer et al. [38], and described at the end of section II of this paper; one can in fact construct them with quadratic flux functions. Very recently, Holden [11] has analyzed a problem of this type: however, his solutions to the Riemann problem differ appreciably from the numerical results of Bell, Trangenstein, and Shubin [2]. At the moment, a comparison of the mixed-type model equation (8) with the elliptic-parabolic system including the pressure equation appears formidable, but, in the long run, this must be the test for the well-posedness of this problem, and its usefulness for applications in multiphase flow.

IV. References

[1] G. Andrews and J.M. Ball, "Asymptotic behavior and changes of phase in one-dimensional nonlinear viscoelasticity", Jour. Diff. Eqns., 44 306-341.

[2] J.B. Bell, J.A. Trangenstein and G.R. Shubin, "Conservation laws of mixed type describing three-phase flow in porous media", to appear, SIAM Jour. Appl. Math.

[3] J.M. Burgers, "A mathematical model illustrating the theory of turbulence", Advances in Appl. Mech., 1(1948) 171-199.

[4] C.M. Dafermos, "Hyperbolic systems of conservation laws" in Systems of Nonlinear Partial Differential Equations, ed. J.M. Ball, Reidel, Dordrecht, 1983 25-70.

[5] D.G. Ebin and R.A. Saxton, "The initial value problem for elastodynamics

of incompressible bodies", to appear in Arch. Rat. Mech. Anal.

[6] D.G. Ebin and R.A. Saxton, "The equations of incompressible elasticity", to appear in Nonstrictly Hyperbolic Conservation Laws, ed. B. Keyfitz, Amer. Math. Soc., Providence.

[7] J. Glimm, "Solutions in the large for nonlinear hyperbolic systems of equations", Comm. Pure Appl. Math. 18 (1965) 697-715.

[8] H. Hattori, "The Riemann problem for a van der Waals fluid with entropy rate admissibility criterion - isothermal case", to appear in Arch. Rat. Mech. Anal.

[9] H. Hattori, "The Riemann problem for a van der Waals fluid with entropy rate admissibility criterion - nonisothermal case", to appear in Jour. Diff. Eqns.

[10] H. Hattori, "The entropy rate admissibility criterion and the double phase boundary problem", to appear in Nonstrictly Hyperbolic Conservation Laws, ed. B.L.Keyfitz, Amer. Math. Soc., Providence.

[11] H. Holden, "On the Riemann problem for a prototype of a mixed type conservation law."

[12] E. Isaacson, "Global solution of a Riemann problem for a non-strictly hyperbolic system of conservation laws arising in enhanced oil recovery", to appear J. Comp. Phys.

[13] E. Isaacson and J.B. Temple, "Analysis of a singular hyperbolic system of conservation laws", to appear.

[14] E. Isaacson, D. Marchesin, B. Plohr, and J.B. Temple, "Classification of quadratic Riemann problems I", in preparation.

[15] R.D. James, "Co-existent phases in the one-dimensional static theory of elastic bars", Arch. Rat. Mech. Anal., 72 (1979) 99-140.

[16] R.D. James, "The propagation of phase boundaries in elastic bars", Arch. Rat. Mech. Anal. 73 (1980) 125-158.

[17] B.L. Keyfitz, "The Riemann problem for nonmonotone stress-strain functions: a 'hysteresis' approach", in Nonlinear Systems of PDE in Applied Mathematics, ed. Basil Nicolaenko, Lectures in Appl. Math. 23 part 1 (1986) 379-395.

[18] B.L. Keyfitz, "Some elementary connections among nonstrictly hyperbolic conservation laws", to appear in Nonstrictly Hyperbolic Conservation Laws, ed. B.L. Keyfitz, Amer. Math. Soc., Providence.

[19] B.L. Keyfitz and H.C. Kranzer, "Existence and uniqueness of entropy solutions to the Riemann problem for hyperbolic systems of two nonlinear conservation laws," Jour. Diff. Eqns., 27 (1978) 444-476.

[20] B.L. Keyfitz and H.C. Kranzer, "A system of hyperbolic conservation laws arising in elasticity theory", Arch. Rat. Mech. Anal., 72 (1980) 219-241.

[21] B.L. Keyfitz and H.C. Kranzer, "The Riemann problem for a class of hyperbolic conservation laws exhibiting a parabolic degeneracy", Jour. Diff. Eqns. 47 (1983) 35-65.

[22] B.L. Keyfitz and H.C. Kranzer, "A system of conservation laws with no classical Riemann solution."

[23] A.Y. LeRoux and J.F. Colombeau, "Techniques numeriques en elasticité dynamique", this volume.

[24] T.P. Liu, "The Riemann problem for general 2x2 conservation laws", Trans. Amer. Math. Soc., 199 (1974) 89-112.

[25] T.P. Liu, "Admissible solutions of hyperbolic conservation laws", Amer. Math. Soc. Memoirs, #240, Providence, 1981.

[26] T.P. Liu and C.H. Wang, "On a hyperbolic system of conservation laws which is not strictly hyperbolic", Jour. Diff. Eqns., to appear.

[27] D. Marchesin, H.B. Medeiros, and P.J. Paes-Leme, "A model for two-phase flow with hysteresis", to appear in Nonstrictly Hyperbolic Conservation Laws, ed. B.L. Keyfitz, Amer. Math. Soc., Providence.

[28] D.W. Peaceman, "Fundamentals of Numerical Reservoir Simulation", 1977, Elsevier, New York.

[29] R.L. Pego, "Phase transitions: stability and admissibility in one dimensional nonlinear viscoelasticity", IMA preprint, Minneapolis, 1985.

[30] M. Rascle, "Sur la convergence de la méthode de viscosité pour un problème modèle: le système de l'elasticité," Thesis.

[31] V. Roytburd and M. Slemrod, "Positively invariant regions for a problem in phase transitions", to appear Arch. Rat. Mech. Anal.

[32] D.G. Schaeffer and M. Shearer, "The classification of 2x2 systems of nonstrictly hyperbolic conservation laws, with application to oil recovery".

[33] D. Serre, "Bounded variation solutions for some hyperbolic systems of conservation laws", Jour. Diff. Eqns., to appear.

[36] M. Shearer, "The Riemann problem for the planar motion of an elastic string", preprint.

[37] M. Shearer, "Nonuniqueness of admissible solutions of Riemann initial value problems for a system of conservation laws of mixed type", Arch. Rat. Mech. Anal., to appear.

[38] M. Shearer, D.G. Schaeffer, D. Marchesin, and P. Paes-Leme, "Solution of the Riemann problem for a prototype 2x2 system of non-strictly hyperbolic conservation laws", to appear.

[39] M. Slemrod, "Admissibility criteria for propagating phase boundaries in a van der Waals fluid", Arch. Rat. Mech. Anal. 81 (1983) 301-315.

[40] M. Slemrod, "Dynamic phase transitions in a van der Waals fluid", Jour. Diff. Eqns., 52 (1984) 1-23.

[41] M. Slemrod and V. Roytburd, "Measure-valued solutions to a problem in dynamic phase transitions", to appear in Nonstrictly Hyperbolic Conservation Laws, ed. B.L. Keyfitz, Amer. Math. Soc., Providence.

[42] H.L. Stone, "Estimation of three-phase relative permeability and residual oil data," J. Can. Pet. Tech., 12 (1973) 53-61.

[43] J.B. Temple, "Global solution of the Cauchy problem for a class of 2x2 non-strictly hyperbolic conservation laws", Adv. Appl. Math., 3 (1982) 335-375.

[44] J.B. Temple, "Systems of conservation laws with coinciding shock and rarefaction curves", in Nonlinear Partial Differential Equations, ed. J.A. Smoller, Contemporary Mathematics, Vol. 17 (1983), 143-151.

ADMISSIBILITY CRITERIA FOR PHASE BOUNDARIES

M. Slemrod#
Department of Mathematical Sciences
Rensselaer Polytechnic Institute
Troy, N. Y. 12180

The purpose of this note is to review some ideas on admissibility criteria for phase boundaries in materials. Specifically we are concerned with one dimensional motions of elastic fluids or elastic solids which possess a constitutive relation of the form shown in Figure 1.

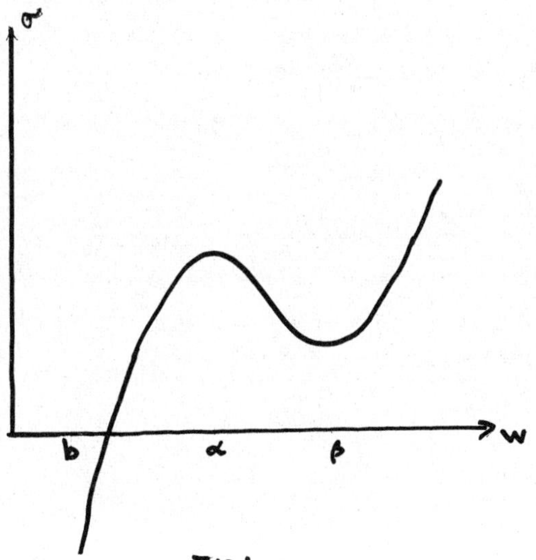

Fig.1

#Research supported by the Air Force Office of Scientific Research, Air Force Systems Command, USAF, under Contract/Grant No. AFOSR-85-0239. The United States Government is authorized to reproduce and distribute reprints for Government purposes not withstanding any copyright herein.

Here σ is Piola-Kirchoff stress and w is the specific volume (= $(\text{density})^{-1}$ for an elastic fluid (the deformation gradient for an elastic solid.)

We let $u(x,t)$ denote the velocity of the fluid (solid) at Lagrangian coordinate x and time t. The balance of mass and momentum can be expressed by the 2×2 system of conservation laws

$$\begin{aligned} u_t &= \sigma(w)_x , \\ w_t &= u_x , \quad t > 0 , -\infty < x < \infty . \end{aligned} \tag{1}$$

Of course (1) should be coupled with initial conditions for the motion

$$u(x,0) = u_0(x) , \ w(x,0) = w_0(x) . \tag{2}$$

Due to the fact that $\sigma' < 0$ in (α,β) and $\sigma' > 0$ elsewhere, (1) is a mixed hyperbolic-elliptic system.

An ambitious program would be to investigate solvability of (1) , (2). A less ambitious task is to study solvability for a simpler test problem namely the Riemann problem where

$$\begin{aligned} u_0(x) &= u_r , \quad w_0(x) = w_r , \quad x > 0 , \\ u_0(x) &= u_\ell , \quad w_0(x) = w_\ell , \quad x < 0 , \end{aligned} \tag{3}$$

u_r, u_ℓ, w_r, w_ℓ constants. We may then try to piece together a solution of the Riemann problem in terms of elementary waves. As usual the waves of interest are shock waves whose speed of propagation s satisfies the Rankine-Hugoniot jump condition

$$\begin{aligned} - s[u] &= [\sigma] , \\ - s[w] &= [u] , \end{aligned} \tag{4}$$

rarefaction waves, and a contact discontinuity for which $[\sigma] = 0$. Here $[u] = u_+ - u_-$, etc. where + , - denotes the limits of u from the right and left of the shock. When w_+ , w_- lies (b,α) or (β,∞) or vice-verse the shock is said to be a phase boundary.

This concept of phase boundary just reflects the usual elementary notions of a model like that in Figure 1. That is $w\varepsilon(b,\alpha)$ and $w\varepsilon(\beta,\infty)$ are supposed to denote different phases of the same material, e.g. (b,α) = liquid phase, (β,∞) = vapor phase in a van der Waals fluid.

Of course as is well known even for strictly hyperbolic problems we cannot expect a unique weak solution for the Cauchy initial value problem. We need some admissibility criteria for choosing preferred solutions.

In this note we consider four admissibility criteria: the viscosity criterion, the entropy criterion, the viscosity-capillarity criterion, and the entropy rate criterion. To keep matters simple we shall consider the simplest Riemann problem exhibiting a phase transition i.e. the case where $w_\ell \in (b,\alpha)$, $w_r \in (\beta,\infty)$ and u_ℓ, u_r, w_ℓ, w_r are constants consistent with the Rankine-Hugoniot jump condition for some s. In this case the shock wave solution

$$\begin{aligned} w &= w_\ell \ , \ x < st \ ; \quad w = w_r \ , \ x > st \ , \\ u &= u_\ell \qquad\qquad\qquad u = u_r \end{aligned} \tag{5}$$

is a <u>phase boundary</u>.

(I) <u>The viscosity criterion</u>.

Our phase boundary (5) is admissible according to the viscosity criterion if the wave is a limit as $\mu \to 0+$ of traveling wave solutions

$$u = \hat{u}\left(\frac{x-st}{\mu}\right) \ , \ w = \hat{w}\left(\frac{x-st}{\mu}\right) \tag{6}$$

to the viscous problem

$$\begin{aligned} u_t &= \sigma(w)_x + \mu u_{xx}, \\ w_t &= u_x \ . \end{aligned} \tag{7}$$

Traveling wave solutions $\hat{u}$, $\hat{w}$ satisfy

$$- s\hat{u}' = \sigma' + \hat{u}'' \ ,$$

$$- s\hat{w}' = \hat{u}' \ ,$$

where $' = \dfrac{d}{d\xi}$, $\xi = \dfrac{x-st}{\mu}$.

Since we wish $\hat{u} \to u_\ell$, $\hat{w} \to w_\ell$, $x < st$;

$\hat{u} \to u_r$, $\hat{w} \to w_r$, $x > st$; as $\mu \to 0_+$

it is natural to impose the boundary conditions

$$\hat{u}(-\infty) = u_\ell, \ \hat{u}(+\infty) = u_r, \ \hat{w}(-\infty) = w_\ell, \ \hat{w}(+\infty) = w_r. \tag{9}$$

Integration of (8) coupled with (9) yields

$$- s(\hat{u}-u_\ell) = \sigma(\hat{w}) - \sigma(w_\ell) + \hat{u}' \ ,$$

$$- s(\hat{w}-w_\ell) = \hat{u}-u_\ell \ ,$$

or simply

$$s\hat{w}' + s^2(\hat{w}-w_\ell) - \sigma(\hat{w}) + \sigma(w_\ell) = 0 \ . \qquad (10)$$

The equilibrium points of (10) are w_ℓ, w_r, and possibly an intermediate value w where the chord connecting $(w_\ell, \sigma(w_\ell))$ and $(w_r, \sigma(w_r))$ cuts the graph of σ. For (10) to have a continuous solution satisfying (9) it is impossible to have such a middle equilibrium point. For example a solution $s = 0$, $u_\ell = u_r = 0$, $\sigma(w_\ell) = \sigma(w_r)$ as shown in Fig. 2 would not be admissible from this point of view.

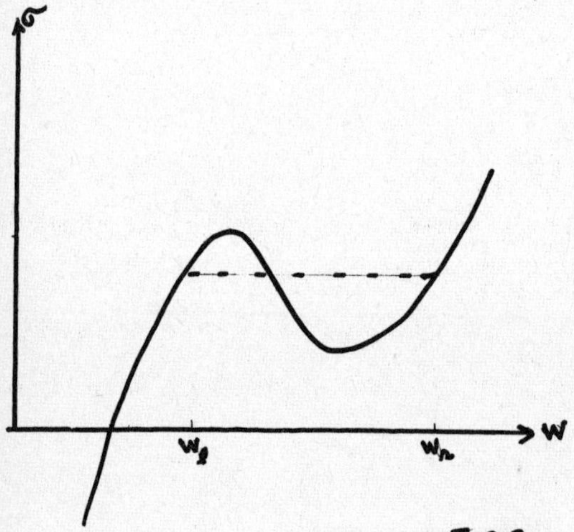

Fig.2

On the other if one allows discontinuous traveling waves then $\hat{w} = w_\ell$ $\xi < 0$; $\hat{w} = w_r$, $\xi > 0$; is a solution of (10).

If one allows such discontinuous traveling waves Shearer [1] has proven existence of solutions to the general Riemann problem (1), (3) whose solutions are admissible according to the viscosity criterion.

(II) Entropy criterion.

The entropy criterion postulates that there is a non-trivial function H(u,w) which satisfies an additional conservation law

$$H_t + Q_x = 0 \qquad (11)$$

for smooth solutions (u,w) of (1) but for which $H_t + Q_x$ has a preferred sign for non-smooth solutions. For example in our problem the natural "entropy" is the total mechanical energy

$$H(u,w) = \frac{1}{2} u^2 + \int^w \sigma(\xi)d\xi .$$

It is easy to check that (11) is satisfied for smooth solutions of (1) with $Q(u,w) = - u\sigma(w)$. For non-smooth solutions the entropy criterion asserts

$$H_t + Q_x \leq 0 . \tag{12}$$

Mechanistically (12) reflects that the fact that an isothermal non-conductor of heat with no heat sources will dissipate mechanical energy. More simply, shock formation does mechanical work.

For our simple solution (12) implies

$$- s \{(\frac{\sigma(w_r) + \sigma(w_\ell)}{2})(w_\ell - w_r) + \int_{w_\ell}^{w_r} \sigma(\xi)\theta\xi\} \geq 0 . \tag{13}$$

This inequality also has a geometric interpretation: The phase boundary joining the state (u_ℓ, w_ℓ) to (u_r, w_r) must have area > area B for $s > 0$, area A < area B for $s < 0$. For $s = 0$ all equilibria satisfying $\sigma(w_\ell) = \sigma(w_r)$ are admissible as was the case in the viscosity criterion (I).

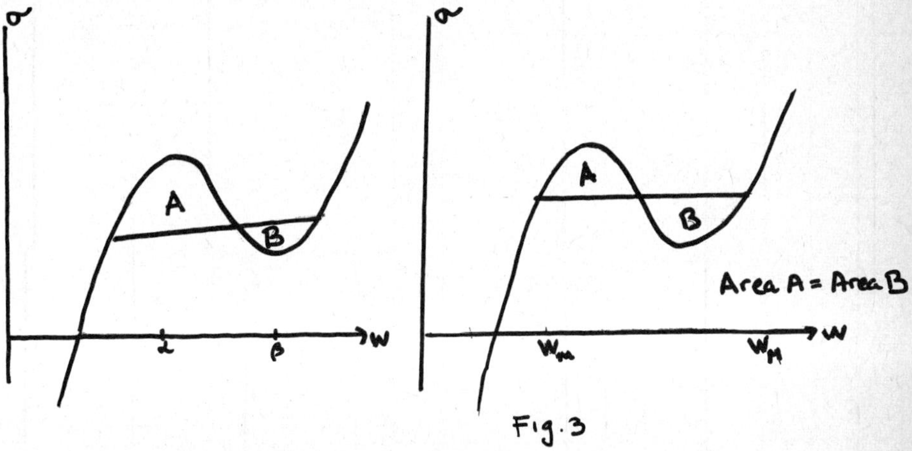

Fig.3

(III) <u>Viscosity-capillarity</u> <u>criterion</u>.

A third approach to the phase boundary problem admissibility criterion was proposed in [2]. In that paper it was suggested that perhaps viscosity and capillarity should play a role in studying shock structure. The idea of including capillarity in studying interfaces in phase transitions can be traced to the work of van der Waals [3] and has been reconsidered by many others since then, e.g. Cahn and Hilliard [4], Aifantis and Serrin [5].

Simply put in the framework of our problem we amend the balance law (1) by including the effects on viscosity and capillarity:

$$u_t = \sigma(w)_x + \mu u_{xx} - \mu^2 A w_{xxx} \,, \qquad w_t = u_x \,. \tag{14}$$

As in (I) we wish to approximate the discontinuous phase boundary solution by traveling wave solution $u = \hat{u}\left(\frac{x-st}{\mu}\right)$, $w = \hat{w}\left(\frac{x-st}{\mu}\right)$ as $\mu \to 0_+$. Here A is a positive constant.

A simple argument shows

$$A\hat{w}'' + s\hat{w}' + s^2(\hat{w}-w_\ell) - \sigma(\hat{w}) + \sigma(w_\ell) = 0 \tag{15}$$

$$\hat{w}(-\infty) = w_\ell \,, \quad \hat{w}(+\infty) = w_r \,,$$

$$\hat{w}'(-\infty) = 0 \,, \quad \hat{w}'(+\infty) = 0 \,.$$

For fixed (u_ℓ, w_ℓ) in (b,α) there is always some state to which it can be connected by a phase boundary. This result and several others may be found in the papers of Slemrod [2], Hagan and Slemrod [6], Shearer [7,8]. It is easy to see, however, that where s = 0 w must satisfy

$$A\hat{w}'' - \sigma(\hat{w}) + \sigma(w_\ell) = 0. \tag{16}$$

Multiply (16) by w' and integrate from $\xi = -\infty$ to $\xi = +\infty$. This shows

$$\int_{-\infty}^{\infty} (\sigma(w(\xi)) - \sigma(w_\ell))\, \hat{w}'(\xi)d\xi = 0$$

or

$$\int_{w_\ell}^{w_r} (\sigma(w) - \sigma(w_-))dw = 0 \tag{17}$$

i.e. the only equilibrium states consistent with a stagnant phase boundary must have area A = area B. (The Maxwell equal area rule.)

Notice the difference between this highly restricted stagnant phase boundary condition and the continuum of possibilities in (I) and (II). We also note that phase boundaries satisfying the viscosity-capilarity criterion satisfy the entropy criterion (see [2]).

In a remarkable paper [8] Shearer has shown existence of a solution to the general Riemann problem for w_ℓ, w_r close to w_m, w_M all of whose shocks satisfy the viscosity-capillarity criterion.

(IV) Entropy rate criterion.

The entropy rate criterion was proposed by Dafermos [9] for the study of admissible solutions to hyperbolic conservation laws. It has been extended by Hattori [10],[11], to the case of van der Waals like materials.

The idea behind the entropy rate criterion is as follows. The total mechanical energy on any interval [a,b] satisfies at time τ

$$\frac{D^+}{D_t} \int_a^b H(u,w)dx = \sum_{\substack{\text{jump}\\ \text{discontinuities}}} \sigma(\tau)\, A(w_-,w_+) \qquad (18)$$

for any piecewise smooth solutions u,w of (1) which possesses a finite number of shock waves. Here $\sigma(\tau)$ is the speed of the jump discontinuity and

$$A(w_-,w_+) = \frac{1}{2}\,(\sigma(w_-)+\sigma(w_+)(w_+-w_-)) - \int_{w_-}^{w_+} \sigma(\xi)d\xi.$$

Thus (18) computes the rate of energy dissipation. The entropy rate criterion says that among all solutions which agree up to time τ the preferable one is the one that maximizes the rate of energy dissipation at time τ, i.e. we choose the process which renders

$$\sum_{\substack{\text{jump}\\ \text{discontinuities}}} \sigma(\tau)A(w_-,w_+)$$

a minimum. Philosophically the criteria asks that nature should choose a solution which renders the already decaying mechanical energy decay as rapidly as allowed by the balance laws and constitutive equations.

The difficulty in applying the entropy rate criterion is the need to check a candidate for an admissible solution against all other solution competitors at each time τ. To check the admissibility of shocks Dafermos [9] has suggested a modified version of the entropy rate criterion. In the modified version the shock is admissible when compared against solutions of the Riemann problem (defined by the shock) made up of the usual fan of shocks, rarefaction waves, contact discontinuities. Hattori [10] has applied this modification to check admissiblity of phase boundaries for (1).

As an illustrative example consider once again the equilibrium Riemann problem $u_\ell = u_r = 0$, $\sigma(w_\ell) = \sigma(w_r)$. Hattori has shown that if $\sigma(w_\ell)$, $\sigma(w_r)$ are not on the Maxwell line (given by the equal area rule) there is another solution of the Riemann problem which dissipates energy more rapidly at $\tau = 0+$. Hence such a solution will <u>not</u> be admissible according to the entropy rate criterion. Moreover if $\sigma(w_\ell)$, $\sigma(w_r)$ are on the Maxwell line i.e. $w_\ell = w_m, w_r = w_M$. Hattori has shown that when compared against competitive solutions made up of shocks, rarefraction waves, and phase boundaries (in a manner motivated by Dafermos's entropy rate shock criterion) this Maxwell solution dissipates energy most rapidly. These results are similar to those given by the viscosity-capillarity criterion (III).

In conclusion we see the viscosity criterion and entropy criterion play no role in distinguishing stagnant phase boundaries. On the other hand the viscosity-capillarity criterion and entropy rate criterion do. In fluids the classical theory of phase transitions (both theoretically and experimentally) gives a preferred equilibrium with co-existing phases. Hence for fluid problems one might think either the viscosity-capillarity criterion or entropy rate criterion will be appropriate. For solids where viscous forces may dominate Pego [12] has argued for the viscosity criterion. Pending further results (theoretical, numerical, and experimental) I would be hesitant to say there is any "correct," universal, admissibility criteria for all materials modeled by (1).

References

1. M. Shearer, The Riemann problem for a class of conservation laws of mixed type, J. Differential Equations 46(1982), 426-443.

2. M. Slemrod, Admissibility criteria for propagating phase boundaries in a van der Waals fluid, Arch. Rational Mechanics and Analysis 81(1983), 301-315.

3. J. D. van der Waals, Translation of J. D. van der Waals, "The Thermodynamic theory of capillarity under the hypothesis of the continuous variation of density" by S. Rowlinson, J. Statistical Physics 20(1979), 197-244.

4. J. W. Cahn and J. E. Hilliard, Free energy of a nonuniform system, J. Chemical Physics 28(1958), 258-267.

5. E. C. Aifantis and J. B. Serrin, The mechanical theory of fluid interfaces and Maxwell's rule, J. Colloid and Interface Science 96(1983), 517-529.

6. R. Hagan and M. Slemrod, The viscosity-capillarity criterion for shocks and phase transitions, Archive for Rational Mechanics and Analysis 83(1984), 333-361.

7. M. Shearer, Admissibility criteria for shock wave solutions of a system of conservation laws of mixed type, Proc. Royal Soc. Edinburgh 93A(1983), 233-244.

8. M. Shearer, Nonuniqueness of admissible solutions of Riemann initial value problems for a system of conservation laws of mixed type, to appear J. Differential Equations.

9. C. Dafermos, Hyperbolic systems of conservation laws, in Systems of Nonlinear Partial Differential Equations, ed. J. M. Ball, Reidel (1984).

10. H. Hattori, The Riemann problem for a van der Waals fluid with entropy rate admissibility criterion, isothermal case, to appear Archive for Rational Mechanics and Analysis.

11. H. Hattori, The Riemann problem for a van der Waals fluid with entropy rate admissibility criterion, nonisothermal case, to appear J. Differential Equations.

12. R. Pego, Phase transitions: stability and admissibility in one dimensional nonlinear viscoelasticity, Institute for Mathematics and its Applications, Univ. of Minnesota, Preprint No. 180 (1985).

THE TRANSFORMATION FROM EULERIAN TO LAGRANGIAN COORDINATES FOR SOLUTIONS WITH DISCONTINUITIES

David H. Wagner
Department of Mathematics
University of Houston -- University Park
Houston, Texas, 77004, USA

Abstract

We demonstrate the equivalence of Eulerian and Lagrangian coordinates for weak, discontinuous solutions in one space dimension. This transformation also induces a one to one correspondence between the convex extensions, or "entropy functions" of the system of conservation laws in either coordinate system. Such entropy functions are of interest in the theory of compensated compactness, and our results imply the equivalence of some of the elements of that theory in either coordinate system.

As an application, we translate a large-data existence result of DiPerna for the Euler equations of isentropic gas dynamics into a similar theorem for the Lagrangian equations.

More detailed proofs of these results will appear elsewhere [11].

1. INTRODUCTION

The conservation laws of continuum mechanics in one space dimension may be formulated in two different coordinate systems, namely Eulerian coordinates and Lagrangian coordinates. We will demonstrate the equivalence of these two coordinate systems for weak, discontinuous solutions, using the equations of inviscid gas dynamics as a concrete example.

The Euler equations result naturally from choosing, as independent variables, a linear coordinate x on physical space, and time, t. We obtain:

(1.1)
(a) $\rho_t + (\rho u)_x = 0$
(b) $(\rho u)_t + [\rho u^2 + p(\rho, S)]_x = 0$
(c) $[\rho e(\rho, S) + \rho u^2/2]_t + [(\rho e(\rho, S) + \rho u^2/2 + p(\rho, S))u]_x = 0$
(d) $(\rho S)_t + (\rho u S)_x \geq 0.$

These equations describe the conservation of mass, momentum, and energy, and the increase of entropy across shock waves, respectively [2]. Here ρ, u, and S are the mass density, velocity, and entropy, respectively, and e and p are internal energy per unit mass and the pressure, expressed as functions of ρ and S. If we choose, instead of x, a material coordinate:

$$y = \int_{x(t)}^{x} \rho(s,t)\, ds, \tag{1.2}$$

where $x(t)$ is a well defined particle path satisfying $x'(t) = u(x(t), t)$, then we obtain the Lagrangian equations [2]:

$$(1.3) \qquad \begin{aligned} &(a) \quad \tau_t - u_y = 0 \\ &(b) \quad u_t + \tilde{p}(\tau, S)_y = 0 \\ &(c) \quad (\tilde{e}(\tau, S) + u^2/2)_t + (u\tilde{p}(\tau, S))_y = 0 \\ &(d) \quad S_t \geq 0 , \end{aligned}$$

where $\tau = 1/\rho$ is the specific volume, $\tilde{p}(\tau,S) = p(1/\tau, S)$, and $\tilde{e}(\tau, S) = e(1/\tau, S)$.

A tedious calculation using the chain rule and product rule shows that (1.1) and (1.3) are equivalent for classical solutions [2]. However, solutions of these equations are known to develop discontinuities, which represent shock waves. Consequently one must consider weak solutions. Even though we may define the weak derivatives D_t or F_x for any of the conserved densities D or any flux F in (1.1) or (1.3), the product rule and chain rule do not hold in any sense that permits us to say that (1.1) and (1.3) are equivalent for weak solutions; but see [10]. One may check that the Rankine-Hugoniot conditions for shock wave solutions of (1.1) and (1.3) are equivalent [2], however, this is not sufficient to prove mathematically that the Cauchy problems are equivalent.

In this paper we give a simple and elegant proof that (1.1) and (1.3) are equivalent for weak bounded measurable solutions on $\mathbb{R}\times\mathbb{R}^+$. To be precise, we prove the following theorem.

THEOREM 1. *The change of variables (1.2) induces a one to one correspondence between L^∞ weak solutions of (1.1) satisfying $\|S\|_\infty < \infty$, $\|u\|_\infty < \infty$, $0 < \delta \leq \rho(x,t) \leq M < \infty$ a.e. for some δ and M, and L^∞ weak solutions of (1.3) satisfying $0 < \varepsilon \leq \tau(y, t) \leq N < \infty$ a.e. for some ε and N. In addition, if $p(0, S) = 0$ and $e(0, S)$ is finite for all finite S, then there is a one to one correspondence between equivalence classes of bounded measurable solutions of (1.1) for which*

$$(1.4) \qquad \int_0^\infty \rho(x,t)\, dx = \int_{-\infty}^0 \rho(x,t)\, dx = \infty ,$$

and equivalence classes of weak solutions of (1.3), in which τ is a Radon measure on $\mathbb{R}\times\mathbb{R}^+$ that dominates two dimensional Lebesgue measure m_2, and in which u and S are bounded.

The equivalence classes mentioned above pertain to the following equivalence relations: In Eulerian coordinates, two solutions are equivalent if the mass densities are equal a.e. with respect to m_2, and u and S are equal a.e. with respect to ρ. In Lagrangian coordinates, two solutions are equivalent if the specific volumes τ are equal as measures and u and S are equal a.e. with respect to m_2.

Our methods make extensive use of change of variable formulae from geometric measure theory [5, 10]. The measure theoretic details are in [11]. In this paper we will use the formulae without comment.

In section 3 we will show that in the presence of vacuum states, the definition of weak solution in Lagrangian coordinates must be strengthened in order for equivalence to hold. One must eliminate certain non-physical weak solutions by requiring the definition of weak solution to hold with test functions whose distributional gradient is a measure which is absolutely continuous with respect to the specific volume, τ. In other words, we must admit test functions which are discontinuous at the vacuum.

Theorem 1 is a special case of the following general theorem which applies to many other important systems of conservation laws, including the isentropic gas dynamics equations, and the shallow water equations.

Theorem 2. *Let*

(1.5)
$$U_t + F(U)_x = 0$$
$$U(x, t) = (u_1, \dots, u_n)(x, t) \in \mathbb{R}^n$$
$$F(U) = (f_1, \dots, f_n)(U) \in \mathbb{R}^n$$

be a system of conservation laws. For any bounded measurable solution of (1.5), with $u_1(x, t) > 0$, *let* $y(x, t)$ *satisfy*

$$\frac{\partial y}{\partial x} = u_1(x, t), \quad \frac{\partial y}{\partial t} = -f_1(U(x, t)) \ ,$$

in the sense of distributions. Then $T:(x, t) \to (y(x, t), t)$ *is a Lipschitz continuous transformation, which induces a one to one correspondence between* L^∞ *weak solutions of (1.5) on* $\mathbb{R}\times\mathbb{R}^+$ *satisfying* $0 < \varepsilon < u_1(x, t) < M < \infty$ *for some* ε *and* M, *and* L^∞ *weak solutions of*

(1.7)
$$(1/u_1)_t - (f_1(U)/u_1)_y = 0 \ ,$$
$$[(u_2, \dots, u_n)/u_1]_t + [(f_2, \dots, f_n)(U) - f_1(U)(u_2, \dots, u_n)/u_1]_y = 0$$

on $\mathbb{R}\times\mathbb{R}+$ *satisfying* $\varepsilon < u_1(y, t) < M$. *In addition, if* $F(U)/u_1$ *is bounded for* $u_1 > 0$, *then there is a one to one correspondence between equivalence classes of bounded measurable solutions of (1.5) for which* u_j/u_1 *is bounded for* $j = 2, \dots, n$ *and*

(1.8)
$$\int_0^\infty u_1(x, t)\, dx = \int_{-\infty}^0 u_1(x, t)\, dx = \infty \ ,$$

and equivalence classes of weak solutions of (1.7) for which $v_1 = 1/u_1$ *is a Radon measure which dominates Lebesgue measure, and* $v_j = u_j/u_1$ *is bounded for* $j = 2, \dots, n$. *If* $\eta(U)$ *is any convex extension of (1.5), i.e., there is a flux* $q(U)$ *such that* $D\eta DF = Dq$, *so that* $\eta_t + q_x = 0$ *for classical solutions, then any solution of (1.5) satisfying*

(1.9)
$$\eta(U)_t + q(U)_x \le 0$$

corresponds to a solution of (1.7) satisfying

(1.10) $$\tilde{\eta}(V)_t + \tilde{q}(V)_y \leq 0 ,$$

where $V = (v_1, \ldots, v_n)$, $\tilde{\eta}(V) = \eta(U)/u_1$, *and* $\tilde{q}(V) = q(U) - f_1(U)\tilde{\eta}(V)$. *Furthermore* η *is convex if and only if* $\tilde{\eta}$ *is convex as a function of* V. *Thus Lax's generalized entropy condition [7] holds for a solution of (1.5) if and only if it holds for the corresponding solution of (1.7).*

As in Theorem 1, in Theorem 2 the definition of weak solution for (1.7) must be strengthened, in case u_1 equals zero on a set of positive measure, to admit discontinuous test functions. The equivalence relation referred to is similar to that used in Theorem 1.

As a consequence of these theorems, theorems on the existence, uniqueness, and behavior of solutions for one system may be carefully translated into theorems on solutions of the other. For example, in [3], DiPerna used compensated compactness methods to prove the existence of global solutions, via the limit of vanishing diffusion, to the Eulerian isentropic gas dynamics equations

(1.11)
$$\begin{aligned} \rho_t + (\rho u)_x &= 0 \\ (\rho u)_t + (\rho u^2 + \rho^{1+2-n})_x &= 0 \\ \rho(x, 0) &= \rho_0(x) \\ u(x, 0) &= u_0(x) \end{aligned}$$

for any $n \varepsilon \mathbb{N}$, with large initial data satisfying $0 < \varepsilon \leq \rho_0(x) \leq M < \infty$, $\|u_0\|_\infty < \infty$, and $(\rho_0 - \bar{\rho}, u_0 - \bar{u}) \varepsilon H^2(\mathbb{R}) \cap C^2(\mathbb{R})$, for some constants $\bar{\rho} > 0$ and $\bar{u}$. As an application of our results, we may translate the existence part of this result as follows:

Theorem 3. *Let* $\tau_0(y)$ *and* $u_0(y)$ *satisfy* $(\tau_0 - \bar{\tau}, u_0 - \bar{u} \varepsilon H^2(\mathbb{R}) \cap C^2(\mathbb{R})$, *and* $0 < \varepsilon \leq \tau_0(y) \leq M < \infty$ *for all* $y \varepsilon \mathbb{R}$, *and some constants* $\bar{\tau} > 0$ *and* $\bar{u}$. *Then there is a global weak solution to the Cauchy problem*

(1.12)
$$\begin{aligned} \tau_t - u_y &= 0 \\ u_t + (\tau^{-1-2-n})_y &= 0 \\ \tau(y, 0) &= \tau_0(y) \\ u(y, 0) &= u_0(y) , \end{aligned}$$

which satisfies:

$$\eta(\tau, u)_t + q(\tau, u)_y \leq 0$$

for all convex functions η *which satisfy* $D\eta D(-u, \tau^{-1-2-n}) = Dq$.

Remarks. Of course, the central theme of this paper is that the coice of coordinate system is not important, provided that proper physical laws are upheld.

A related paper, [1], has appeared, in which it is shown that the Principle of Virtual Work, i.e., the definition of weak solutions to systems of conservation laws in several space dimensions, is equivalent to the integral laws of motion, using some similar techniques.

2. THE TRANSFORMATION

The proofs of Theorems 1 and 2 are almost identical and we concentrate on Theorem 1.

The transformation from (1.1) to (1.3) is effected by a change of space coordinate, (1.2). Actually this is simply a classical formula for a solution to the gradient system

$$\frac{\partial y}{\partial x} = \rho(x, t) \ , \quad \frac{\partial y}{\partial t} = -(\rho u)(x, t) \ .$$

This system is consistent because $\rho_t = -(\rho u)_x$. Hence, it has a solution, $(y, t) = T(x, t)$, in the space of distributions. If $0 < \varepsilon \leq \rho(x, t) \leq M < \infty$ and $u \in L^\infty$ then T is a bi-Lipschitz homeomorphism from $\mathbb{R}\times\mathbb{R}^+$ onto itself.

The transformation proceeds via the change of variables formula for integrals. The Jacobian of the transformation is ρ. Consider one of the conservation laws (1.1), written as $D_t + F_x = 0$. The weak formulation of this equation, with initial conditions $D(x, 0) = D_0(x)$, is

$$\text{(2.1)} \qquad \iint_{t>0} \phi_t D + \phi_x F \, dxdt + \int_{t=0} \phi D_0 \, dx = 0 \ ,$$

for all C^1 test functions ϕ with compact support. We observe that for such weak solutions, (2.1) also holds for Lipschitz test functions. Indeed, if ϕ is Lipschitz with compact support, convolution with a standard smoothing kernel produces a C^∞ function with compact support, ϕ_ε , such that $\nabla\phi_\varepsilon \to \nabla\phi$ in L^1 as $\varepsilon \to 0$, so that

$$\iint_{t>0} \phi_{\varepsilon t} D + \phi_{\varepsilon x} F \, dxdt + \int_{t=0} \phi_\varepsilon D_0 \, dx$$
$$\to \iint_{t>0} \phi_t D + \phi_x F \, dxdt + \int_{t=0} \phi D_0 \, dx \ .$$

Change of variables transforms (2.1) to

$$0 = \iint_{t>0} \Big[(\phi_t \circ T^{-1})(D\circ T^{-1}) + (\phi_x \circ T^{-1})(F\circ T^{-1}) \Big]\cdot(\rho\circ T^{-1})^{-1} \, dydt$$
$$+ \int_{t=0} (\phi\circ T^{-1})(D_0\circ T^{-1})/(\rho_0\circ T^{-1}) \, dx$$
$$= \iint_{t>0} \Big[(\tilde\phi_t - (\widetilde{\rho u})\tilde\phi_y)\tilde D + (\tilde\phi_y \tilde\rho)\tilde F \Big]\cdot\tilde\rho^{-1} \, dydt + \int_{t=0} \tilde\phi \tilde D_0/\tilde\rho_0 \, dy$$

where $\tilde\phi = \phi\circ T^{-1}$, $\tilde D = D\circ T^{-1}$, etc. Since T is a bi-Lipschitz homeomorphism of $\mathbb{R}\times\mathbb{R}^+$ onto itself, the induced map $\phi \to \phi\circ T^{-1} = \tilde\phi$ is a bijection on the set of Lipschitz test functions on $\mathbb{R}\times\mathbb{R}^+$. Therefore we have that $(\tilde D/\tilde\rho)_t + (\tilde F - \tilde u \tilde D)_y = 0$ in the weak sense.

Thus (1.1) is transformed to:

$$
\begin{aligned}
1_t + 0_y &= 0 && \text{cons. of mass}\\
u_t + \tilde{p}(\tau, S)_y &= 0 && \text{cons. of momentum}\\
(u^2/2 + \tilde{e}(\tau, S))_t + (\tilde{p}u)_y &= 0 && \text{cons. of energy.}
\end{aligned}
$$

We seem to be missing (1.3a). This equation follows from the conservation of *volume*, $D = 1$, $F = 0$:

$$(1/\rho)_t - u_y = 0 \ .$$

One defines τ to be the specific volume $1/\rho$, and we have (1.2a).

The weak formulation of the increase of entropy, (1.1d), is

$$\iint_{t>0} \phi_t\rho S + \phi_x\rho u S \, dxdy + \int_{t=0} \phi\rho_0 S_0 \, dx \leq 0$$

for all non-negative C^1 test functions ϕ. Clearly this inequality is preserved in the transformation, and we obtain (1.2d). In addition, if (η, q) is any convex extension of (1.1), including $(-\rho S, -\rho uS)$, then any solution of (1.1) satisfies $\eta_t + q_x \leq 0$ if and only if the corresponding solution of (1.3) satisfies $(\eta/\rho)_t + (q-u\eta)_y \leq 0$. Furthermore, $\tilde{\eta}(\tau, u, e+u^2/2) = \tau\eta(1/\tau, u/\tau, (e+u^2/2)/\tau)$ is convex, as follows. Since η is convex, we know that the convex set $E = \{(z, U): z \geq \eta(U)\}$ is an intersection of half spaces. Consequently η is a supremum of affine functions:

$$\eta(\rho, \rho u, \rho e+\rho u^2/2) = \sup_{\alpha} (c_{0\alpha} + c_{1\alpha}\rho + c_{2\alpha}\rho u + c_{3\alpha}\rho(e+u^2/2)). \tag{2.2}$$

Then

$$
\begin{aligned}
\tilde{\eta}(1/\rho, u, e+u^2/2) &= \eta(\rho, \rho u, \rho e+\rho u^2/2)/\rho \\
&= \sup_{\alpha} (c_{0\alpha}/\rho + c_{1\alpha} + c_{2\alpha}u + c_{3\alpha}(e+u^2/2)),
\end{aligned} \tag{2.3}
$$

and thus is a convex function. Clearly this argument is reversible and generalizes for Theorem 2. Thus Lax's generalized entropy condition [7] holds for the transformed solution of (1.3) if and only if it holds for the corresponding solution of (1.1).

It is also clear from this argument, that in the theory of compensated compactness [3, 9], if the Young measure, associated with the weak convergence of a sequence of approximate solutions of (1.1), reduces to a point mass, so that one actually obtains pointwise convergence, then the corresponding Young measure for (1.3) also reduces to a point mass, *provided* that the corresponding sequence of approximate solutions converges in the same weak sense. Usually this means convergence in the weak-$*$ topology of L^∞. We warn the reader, however, that the L^∞ bounds that one usually needs to make this theory work, are difficult, if not impossible, to obtain for the equations of gas dynamics in Lagrangian coordinates.

In [6] it is shown that the convexity of $-S$ as a function of τ, u, and $E = e+u^2/2$, is equivalent to the convexity of E as a function of τ, u, and S. The above discussion is similar to a simple geometric explanation and proof of this fact, due to Andrew Majda; namely that a function is convex if and only if through each point on its graph there is a sub-tangent hyperplane such that the graph lies on one side of the hyperplane:

$$E-E_0 > c_1(\tau-\tau_0) + c_2(u-u_0) + c_3(S-S_0).$$

In this case $c_3 = (\partial E/\partial S)_{u,\tau} = T$ = temperature and is always positive. Thus

$$-(S-S_0) > (1/c_3)(c_1(\tau-\tau_0) + c_2(u-u_0) - (E-E_0)),$$

and we see that $-S$ is convex as a function of τ, u, and E. Essentially we are looking at the same graph, but from a different direction.

Remark. The level curves y = constant yield, rather easily, the "particle paths" of the solution.

3. IN CASE OF VACUUMS...

Liu and Smoller [8] have demonstrated solutions of the Eulerian isentropic gas dynamics equations, which contain vacuum states. In this section we show how our results extend to arbitrary bounded measurable solutions of (1.1), including those containing vacuum states, provided (1.4) holds, $p(0, S) = 0$, and $e(0, S)$ is finite. This condition is satisfied by ideal polytropic gases.

In this case the transformation T is still Lipschitz and, for fixed t, y is a monotone function of x. However, T is no longer one to one, and may in fact map sets of positive measure in (x, t) space, namely the vacuum regions, into sets of zero measure in (y, t) space. Hence we may no longer regard τ as a function; however it has a natural expression as a measure, namely $\tau = T_{\#}m_2$, where

$$T_{\#}m_2(E) = m_2(T^{-1}(E)).$$

The following example shows that in the presence of a vacuum, we must strengthen the definition of weak solution for (1.3b, c, d). Let $\tau_0 = 1 + \delta_0$, where δ_0 is the Dirac delta measure at $y = 0$, and let $u = 0$ and $S = 1$. Then p is a nonzero constant for $y \neq 0$, and is zero at $y = 0$. However since p is equal, a.e., to a constant, $p_y = 0$ in the sense of distributions. Hence we have described a *steady* solution to (1.3) or (1.12). This is clearly unphysical, since in Eulerian coordinates we have a vacuum of length one, and hence $p_x = (\text{const.})(\delta_1 - \delta_0)$.

One may view this example as showing that Lagrangian coordinates are unphysical when a vacuum is present. However, we now show in what way the two coordinate systems are equivalent. We motivate our new definition of weak solution by examining how Eulerian test functions pull back to Lagrangian coordinates. Given τ and u satisfying $\tau_t - u_y = 0$ in the usual weak sense, with τ a positive Radon measure, and $u \in L^\infty$, define $x(y, t)$ such that

$$\frac{\partial x}{\partial y} = \tau, \qquad \frac{\partial x}{\partial t} = u. \tag{3.1}$$

Since τ is a Radon measure, we have that $x \in BV_{loc}$, and for fixed t, x is a monotone increasing function of y. Thus Q, defined by $Q(y, t) = (x(y, t), t)$ has a unique monotone left inverse T, i.e., $T(Q(y, t)) = (y, t)$. If τ dominates m_2, in the sense that there is a positive constant K such that $\tau(E) \geq Km_2(E)$ for any subset E of $\mathbb{R}\times\mathbb{R}^+$, then T is Lipschitz continuous.

Let ϕ be an Eulerian test function. Then $\phi \circ Q$ is a Lagrangian test function which is discontinuous, but BV. By [10] we have $(\phi \circ Q)_t = (\phi_t \circ Q)\hat{} + (\phi_x \circ Q)\hat{}\, u$, and $(\phi \circ Q)_y = (\phi_x \circ Q)\hat{}\,\tau$, where $(\phi_t \circ Q)\hat{}$ and $(\phi_y \circ Q)\hat{}$ are defined at the regular points of Q. The regular points of Q are those points (y_o, t_o) at which the half-space approximate limits $l_{\pm a}x(y_o, t_o)$ exist for some $a \in \mathbb{R}^2$, such that

$$\lim_{r\to o+} m_2 \Big\{(y,t): \|(y-y_o,\ t-t_o)\| < r,\ |x(y,\ t) - l_{\pm a}x(y_o,\ t_o)| > \varepsilon,\ (y-y_o,\ t-t_o)\cdot(\pm a) > 0\Big\} \frac{1}{\pi r^2} = 0 .$$

Since x is locally bounded and of bounded variation, almost all (y, t), with respect to one-dimensional Hausdorff measure, are regular points of x, and of Q [10]. At such points, $(\nabla\phi \circ Q)\hat{}$ is defined by

$$(\nabla\phi \circ Q)\hat{}(y,\ t) = \int_0^1 \nabla\phi\big(l_{-a}Q(y,\ t)(1-s) + l_a Q(y,\ t)s\big)\, ds.$$

Since $(\phi \circ Q)_y = (\phi \circ Q)\hat{}\,\tau$, it is a measure which is absolutely continuous with respect to τ, while $(\phi \circ Q)_t$ is a function. This motivates the following definition.

Definition 2. We say that (τ, u, S) is a weak solution of (1.3), if τ is a Radon measure on $\mathbb{R}\times\mathbb{R}^+$, and u and S are bounded τ-measurable functions such that (1.3a) holds in the sense of distributions, and the weak formulation of (1.3b, c, d) holds with all test functions ϕ with compact support such that $\phi_y = f\tau$, and $\phi_t = g$, with $f, g \in L^\infty(\tau)$.

Using this definition, we can finish the proof of Theorem 1 for the case that vacuum states are present. The detail of this proof are in [11].

4. Other Coordinates

It should be clear that other transformations are possible. In particular, for (1.1), we could integrate the energy density $\rho e + \rho u^2$ instead of ρ to obtain a new coordinate, z. By Theorem 2, this yields an equivalent system of conservation laws, of which the non-trivial ones are the conservation of volume, mass, and momentum. One may also consider a transformation of both space and time. A simple sufficient condition, for the transformed system to be an autonomous system of conservation laws, is that the differential of the transformation T should be a function of the conserved unknown, U. If $T(x, t) = (x', t')$, and

$$DT = \left[\frac{\partial T}{\partial x},\ \frac{\partial T}{\partial t}\right] = \big[G(U),\ H(U)\big],$$

and $JT = \det(DT)$, then the transformed system is

$$\frac{\partial}{\partial t'}\left[\left(U\frac{\partial t'}{\partial t} + F(U)\frac{\partial t'}{\partial x}\right)\frac{1}{JT}\right] + \frac{\partial}{\partial x'}\left[\left(U\frac{\partial x'}{\partial t} + F(U)\frac{\partial x'}{\partial x}\right)\frac{1}{JT}\right] = 0.$$

Given a matrix [G(U), H(U)], a necessary condition for the existence of a transformation T such that DT = [G(U), H(U)] is that $G(U)_t = H(U)_x$ weakly. Thus [G(U), H(U)] must be of the form A•(U, -F(U)) + (b, c) , where A is a constant 2×n matrix and b and c are constant 2-vectors. Of course the transformation T must be one to one, Cauchy data must be prescribed on a space-like curve, and the forward time direction must be properly chosen to respect the entropy condition. Presumably these are the only transformations which produce an autonomous system of conservation laws which is equivalent, for weak solutions, to the original system.

The existence of many equivalent coordinate system suggests a manifold structure. The fact that affine functions of the conserved densities are transformed into other affine functions, and that convex functions are transformed into other convex functions (2.2, 2.3) suggests that this structure is intrinsically affine. The value of this insight, if any, remains to be determined.

References

1. Antman, S.S., Osborne, J.E., The Principle of Virtual Work and Integral Laws of Motion, Arch. Rat. Mech. Anal. 69, 1979, pp. 231-261.

2. Courant, R., and Friedrichs, K.O., Supersonic flow and shock waves (Pure Appl. Math., Vol 1). Interscience, New York, 1948.

3. Diperna, R.J., Convergence of the viscosity method for isentropic gas dynamics, Comm. Math. Phys., 91, (1983), 1-30.

4. Donoghue, W.F., Distributions and Fourier Transforms, Academic Press, New York and London, 1969.

5. Federer, H., Geometric measure Theory, Springer-Verlag, New York, 1969.

6. Friedrichs, K.O., and Lax, P.D., Systems of Conservation Equations with a Convex Extension, Proc. Nat. Acad. Sci. USA 68, 1686-1688, 1971.

7. Lax, P. D., Shock waves and entropy, in Contributions to nonlinear functional analysis (Zarantonello, E. A., ed.), Academic Press, 1971, 603-634.

8. Liu, T.P., and Smoller, J., The vacuum state in isentropic gas dynamics, Adv. Appl. Math., 1, 1980, 345-359.

9. Tartar, L., Compensated compactness and applications to partial differential equations. In: Research notes in mathematics, nonlinear analysis, and mechanics: Heriot-Watt Symposium, vol. 4, Knops, R. J. (ed.). New York, Pitman Press 1979.

10. Vol'pert, A.I., The spaces BV and quasilinear equations, Math. USSR-Sb., 2, 1967, pp. 225-267.

11. Wagner, D. H., Equivalence of the Euler and Lagrangian equations of gas dynamics for weak solutions, to appear in J. Differential Equations.

Two Existence Theorems For Systems of Conservation Laws with Dissipation

David Hoff*
Department of Mathematics
Indiana University
Bloomington, IN 47405 USA

§1. Introduction

In this paper we present two existence theorems for the Cauchy problem for systems of conservation laws with certain types of dissipation terms added. The first result applies to general systems in several space variables with the simplest (constant, diagonal) dissipation. The key hypothesis here is that the corresponding system of conservation laws should admit a so called entropy-entropy flux pair. This allows for certain energy-type estimates which serve to control local solutions pointwise. We then show by an explicit construction of the required entropy that our result applies to the standard systems modelling compressible fluid flow in several space variables. (These results represent joint work with Joel Smoller. Complete details may be found in [3] and [4].)

In section §3. we restrict attention to the Navier-Stokes equations for one-dimensional, isentropic flow. The main point of interest here is that the initial data is allowed to be discontinuous. We find that, owing to the parabolicity of the momentum equation, initial discontinuities in the velocity are smoothed out in positive time (and so are more or less irrelevant). However, initial discontinuities in the density persist for all time, and it is the analysis of the propogation of these discontinuities which is the most

*Research supported in part by the NSF under Grant No. MCS-8301141.

interesting aspect of this work. We show that jump discontinuities in the density and in the gradient of the velocity decay exponentially in time; this exponential decay then allows for time-independent, pointwise control of the nonlinear terms. Complete details of these results may be found in [2].

§2. Systems of parabolic conservation laws in several space variables

In this section we prove an existence theorem for systems of conservation laws with artificial viscosity terms added. Specifically, we consider the Cauchy problem for the system

$$\frac{\partial u}{\partial t} + \sum_{i=1}^{n} \frac{\partial}{\partial x_i} f_i(u) = D\Delta u \quad , \quad u = u(x,t),\ x \in \mathbb{R}^n \ , \ t > 0 \ , \tag{2.1}$$

with initial data

$$u(x,0) = u_0(x) \ . \tag{2.2}$$

Here $u = (u^1,\ldots,u^m)$, $f_i = (f_i^1,\ldots,f_i^m)$, and D is a constant, diagonalizable matrix with positive eigenvalues. The f_i are assumed to be defined and smooth in a neighborhood of a point $\bar{u}$, and the initial data will be assumed to be sufficiently close to $\bar{u}$ in $L^2 \cap L^\infty$. We shall show that our results apply to the equations of compressible fluid flow (conservation of mass, momentum, and energy) with artificial viscosity terms $D\Delta U$ added.

We first consider the case that D is a diagonal matrix: $D = \text{diag}(d_1,\ldots d_m) > 0$. Let K^j be the fundamental solution for the heat operator $\frac{\partial}{\partial t} - d_j\Delta$:

$$K^j(x,t) = (4\pi d_j t)^{-n/2} \exp(-|x|^2/4d_j t) .$$

The solution of (2.1)-(2.2) then formally satisfies the representation

$$(2.3) \qquad u^j(t) = K^j(t) * u_0^j - \sum_{i=1}^{n} \int_0^t K^j_{x_i}(t-s) * f_i^j(u(s)ds , \qquad j = 1,\dots,m ,$$

where $*$ denotes convolution in the spatial variable x, and the x-dependence has been surpressed. By iterating in (2.3) and using well-known properties of the heat kernel K^j, one can prove the following local existence result ([3], Lemma 2.2):

<u>Lemma 2.1</u>: Assume that $f = (f_1,\dots,f_n)$ is defined and smooth in a ball $\bar{B}_r(\bar{u})$ and that D is a constant, positive diagonal matrix. Then given $s < r$ there is a time $T > 0$ and a constant C_1 such that, if $u_0 - \bar{u}$ is in $L^2 \cap L^\infty$ with $\|u_0 - \bar{u}\|_\infty \le s < r$, then the Cauchy problem (2.1)-(2.2) has a solution defined on $\mathbb{R}^n \times [0,T]$. Moreover, u satisfies

$$\|u(\cdot,t) - \bar{u}\|_{L^\infty(\mathbb{R}^n)} \le r , \quad 0 \le t \le T$$

and

$$(2.4) \qquad \|u(\cdot,T) - \bar{u}\|_{H^p(\mathbb{R}^n)} \le C_1 \|u_0 - \bar{u}\|_{L^2(\mathbb{R}^n)} .$$

Here p is defined by

$$(2.5) \qquad p = \min\{m \in \mathbb{Z} : m > \frac{n}{2} + 1\} .$$

We can extend these local solutions to global solutions when the system (2.1) admits an entropy-entropy flux pair. These are defined as follows:

<u>Def</u>: The functions $\alpha:\bar{B}_r(\bar{u}) \to \mathbb{R}$ and $\beta = (\beta_1,\ldots,\beta_n):\bar{B}_r(\bar{u}) \to \mathbb{R}^n$ form an entropy-entropy flux pair for the system (2.1) if, for each $i=1,\ldots,n$ and u in $\bar{B}_r(\bar{u})$,

$$\nabla\alpha(u)^t f_i'(u) = \nabla\beta_i(u)^t \;, \quad i = 1,\ldots,n \;. \tag{2.6}$$

The entropy α will always be assumed to satisfy

$$\delta|u - \bar{u}|^2 \le \alpha(u) \le \delta^{-1}|u - \bar{u}|^2 \tag{2.7}$$

for some positive constant δ. Finally, α is said to be consistent with the diagonal matrix D if

$$w^t D\alpha''(u)w \ge 0 \tag{2.8}$$

holds for all u in $\bar{B}_r(\bar{u})$ and w in $\mathbb{R}^m$.

The existence of such a pair (α,β) enables us to derive the following a priori estimate for solutions of (2.1):

<u>Lemma 2.2</u>: Assume that there is an entropy-entropy flux pair (α,β) satisfying (2.6), (2.7) and (2.8), and that D is a constant, positive, diagonal matrix. Then there is a constnat C_2 depending only on the properties of α and f in $\bar{B}_r(\bar{u})$ such that, if u is any solution of (2.1) on $\mathbb{R}^n \times [0,t]$ with values in $\bar{B}_r(\bar{u})$, then

$$\|u(\cdot,t) - \bar{u}\|_{L^2(\mathbb{R}^n)} \le C_2\|u_0 - \bar{u}\|_{L^2(\mathbb{R}^n)} \;. \tag{2.9}$$

Sketch of proof: Without loss of generality, $\beta(\bar{u}) = 0$. We multiply (2.1) on the left by $\nabla\alpha(u)^t$ to obtain

$$\alpha(u)_t + \sum_{i=1}^{n} \beta_i(u)_{x_i} = \nabla\alpha(u)^t D\Delta u$$
$$= \sum_i (\nabla\alpha^t D u_{x_i})_{x_i} - \sum_i (D\alpha'' u_{x_i})^t u_{x_i} \ .$$

Integrating over $\mathbb{R}^n \times [0,t]$, we then have, using (2.8), that

$$\int_{\mathbb{R}^n} \alpha(x,\cdot)\Big|_0^t dx \le 0 \ .$$

The result then follows from the hypothesis (2.7).//

We can now state our global existence theorem for the case that the diffusion matrix D is diagonal. The proof will require the Sobolev inequality

$$(2.10) \qquad \|v\|_{L^\infty(\mathbb{R}^n)} \le C_3\|v\|_{H^{p-1}(\mathbb{R}^n)} \le C_3\|v\|_{H^p(\mathbb{R}^n)} \ .$$

(recall that p is defined in (2.5).)

<u>Lemma 2.3</u>: Assume that the hypotheses of lemmas 2.1 and 2.2 are in force, and let C_1, C_2, and C_3 be as defined in Lemma 2.1, Lemma 2.2, and (2.10) respectively. If $u_0 - \bar{u}$ is in $L^2 \cap L^\infty$ with $\|u_0 - \bar{u}\|_\infty \le s < r$ and $C_1C_2C_3\|u_0 - \bar{u}\|_{L^2} \le s$, then the Cauchy problem (2.1)-(2.2) has a unique global solution.

Sketch of proof: If a solution u has been constructed up to time $t \ge T$, then (2.10), (2.4), and (2.9) apply to show that

$$\|u(\cdot,t) - \bar{u}\|_{L^\infty} \le C_3\|u(\cdot,t)\ \bar{u}\|_{H^p}$$
$$\le C_1C_3\|u(\cdot,t - T) - \bar{u}\|_{L^2}$$
$$\le C_1C_2C_3\|u_0 - \bar{u}\|_{L^2} \le s \ .$$

Thus the solution u can be extended past time t .//

We can dispense with the requirement that D be a diagonal matrix by making a simple change of variable:

<u>Theorem 2.4</u>: Assume that f is smooth in $\bar{B}_r(\bar{u})$ and that the system (2.1) admits an entropy-entropy flux pair (α,β) satisfying (2.6) and (2.7). Let D be a diagonalizable matrix with positive eigenvalues, say

$$P^{-1}DP = \Lambda = \operatorname{diag}(d_1,\ldots,d_m) > 0$$

and assume that

$$\Lambda P^t \alpha''(u) P \geq 0 \quad \text{for} \quad u \in \bar{B}_r(\bar{u}) \ . \tag{2.10}$$

Then the Cauchy problem (2.1)-(2.2) has a unique global solution provided that

$$\|u_0 - \bar{u}\|_{L^\infty} < \frac{r}{\|P\| \ \|P^{-1}\|}$$

and that $\|u_0 - \bar{u}\|$ is sufficiently small.

Proof. Make the change of variables $v = P^{-1}u$, and apply Lemma 2.3 to the system satisfied by v .//

We now apply Theorem 2.4 to the equations of multidimensional compressible fluid flow. We consider the two systems

$$\begin{gathered} \rho_t + \operatorname{div}(\rho u) = 0 \\ (\rho u_i)_t + \operatorname{div}(\rho u_i u + P e_i) = 0 \ , \ i = 1,\ldots,n \\ (\rho S)_t + \operatorname{div}(\rho S u) = 0 \end{gathered} \tag{2.11}$$

and

$$\begin{aligned} &\rho_t + \operatorname{div}(\rho u) = 0 \\ (2.12) \qquad &(\rho u_i)_t + \operatorname{div}(\rho u_i u + Pe_i) = 0 \ , \ i = 1,\dots,n \\ &(\rho E)_t + \operatorname{div}(\rho E u + Pu) = 0 \end{aligned}$$

when appropriate (artificial) viscosity terms are added to the right hand-sides. These systems describe the flow of a compressible fluid in which ρ, $u = (u_1,\dots,u_n), P, S,$ and E are respectively the density, velocity, pressure, entropy, and energy. (e_i denotes the ith standard basis vector.) In (2.11) P is a smooth function of ρ and S, and is to satisfy

$$P_\rho > 0 \quad \text{for} \quad \rho > 0 \ .$$

In (2.12) $P = P(\rho,e)$, where $E = e + \frac{|u|^2}{2}$. System (2.12) can be derived formally from (2.11) and a fundamental thermodynamic relation involving $e, S,$ and ρ.

We first display an entropy-entropy flux pair for (2.11) when $n = 3$. The existence of the required entropy-entropy flux pair for (2.12) will then follow by a simple change of variable. The $n = 2$ case is similar.

Observe that all nonlinear functions appearing in (2.11) and (2.12) are defined and smooth in the region $\rho > 0$. We therefore fix a point $(\bar{\rho},\bar{u},\bar{S})$ with $\rho > 0$ and, without loss of generality, we take $\bar{u} = \bar{S} = 0$.

Define functions α and $\beta = (\beta_1, \beta_2, \beta_3)$ by

$$\alpha(\rho,u,S) = \frac{\rho|u|^2}{2} + \rho\int_{\bar{\rho}}^{\rho} \frac{P(\sigma,S) - \bar{P}}{\sigma^2}\, d\sigma + C\rho S^2$$

$$\beta_i(\rho,u,S) = \frac{\rho|u|^2u_i}{2} + \rho u_i\int_{\bar{\rho}}^{\rho} \frac{P(\sigma,S) - \bar{P}}{\sigma^2}\,d\sigma + u_i(P - \bar{P}) + C\rho u_i S^2$$

A somewhat tedious calculation ([3], sect. 3) shows that, when C is large, α and β satisfy (2.6) and (2.7), and that $\alpha'' > 0$. It follows that the hypotheses of Theorem 2.4 are satisfied when D is sufficiently close to a (positive) multiple of the identity matrix and the initial data is close to $(\bar{\rho},0,0)$ in $L^2 \cap L^\infty$.

A similar result for the energy formulation (2.12) of the compressible flow equations follows from the observation that (2.12) can be derived formally from (2.11) by the change of variables

$$(\rho,\rho u,\rho S) = h(\rho,\rho u,\rho E)$$

(see [1], pp. 15-16). It then follows that the functions $A = \alpha \circ h$ and $B = \beta \circ h$ from an entropy-entropy flux pair for (2.12) satisfying (2.6) and (2.7) in a neighborhood of $(\bar{\rho},0,0)$ in $(\rho,\rho u,\rho E)$ space, and that A'' is positive definite there. (See [2], Prop. 3.1). Thus A and B satisfy (2.6), (2.7), and (2.10) for system (2.12) when D is diagonalizable and close to a positive multiple of the identity.

We can now give a formal statement of the application of Theorem 2.4 to the compressible flow systems (2.11) and (2.12). In this statement we let the dependent variables be (ρ,v), where

$$v = \begin{bmatrix}\rho u\\ \rho S\end{bmatrix} \quad \text{and} \quad v = \begin{bmatrix}\rho u\\ \rho E\end{bmatrix}$$

in systems (2.11) and (2.12) respectively.

<u>Theorem 2.5</u>: Let $(\bar{\rho},\bar{v})$ be a given constant state in $\rho > 0$ and let d be a positive number. Then there is a number $r > 0$ such

that, if the initial data (ρ_0,v_0) satisfies $\|\rho_0,v_0) - (\bar{\rho},\bar{v})\|_{L^\infty} \le r$ and $\|(\rho_0,v_0) - (\bar{\rho},\bar{v})\|_{L^2} \le r$, and if $\|D - dI\| \le r$ then the systems (2.11) and (2.12), modified by the addition of terms $D\Delta u$ to the right-hand sides, have unique solutions defined in all of $t > 0$.

§3. An existence theorem for isentropic, compressible Navier-Stokes equations in one space dimension with nonsmooth initial data

In this section we consider the system of equations

$$\text{(3.1)} \qquad \begin{aligned} & v_t - u_x = 0 \\ & u_t + P(v)_x = (k(v)u_x)_x\,, \quad (v,u) = (v(x,t),u(x,t)), \\ & \qquad\qquad\qquad\qquad\qquad x \in \mathbb{R},\ t > 0\,, \end{aligned}$$

with initial data

$$\begin{aligned} v(x,0) &= v_0(x) \\ u(x,0) &= u_0(x)\,. \end{aligned}$$

(3.1) is the one-dimensional, isentropic version of the system (2.11) written in Lagrangean coordinates. (Thus the line x = constant is now a particle path.) v is the specific volume, which is the reciprocal of density, u is the velocity, and $P=P(v)$ is the pressure. For a polytropic gas, $P(v) = v^{-\gamma}$, $\gamma > 1$, and in the Navier-Stokes equations, $k(v) = \text{const.}/v$. We therefore assume that P and k satisfy

$$\text{(3.2)} \qquad P'(v) < 0 \quad \text{and} \quad k(v) > 0 \quad \text{for} \quad v > 0\,.$$

We are particularly interested in the Cauchy problem for (3.1) when v_0 and u_0 are not smooth functions. In particular, since the

Riemann problem for the corresponding conservation laws $(k = 0)$ is so well understood, it would be useful to solve (3.1)-(3.2) with Riemann initial data and compare the resulting solution with that of the conservation laws.

Existence results for (3.1)-(3.2) in the case of smooth initial data are obtained in [5] and [6], and for the case of nonsmooth data, in [7]. The latter result has some overlap with ours, but our approach is much more geometrical, and, we believe, gives an interesting insight into the character of the solution.

Theorem 3.1: Assume that P and k satisfy (3.2) and let $(\bar{v},\bar{u})$ be a constant state with $\bar{v} > 0$. If the initial data (v_0,u_0) satisfies $\operatorname{ess\,inf} v_0 > 0$, $u_0-\bar{u} \in L^2$, and $v_0-\bar{v} \in L^2 \cap BV$ with $\|u_0 - \bar{u}\|_{L^2}, \|v_0 - \bar{v}\|_{L^2}$, and $TV(v_0)$ all sufficiently small, then the Cauchy problem (3.1)-(3.2) has a unique weak solution defined in all of $t > 0$.

Sketch of proof: First, without loss of generality, we take $\bar{u} = 0$. Next, we assume for the time being that v_0 has a single discontinuity, located at $x = 0$. Specifically, we take

$$v_0 - \bar{v} \in H^1(\{x < 0\}) \cap H^1(\{x > 0\}) .$$

Now, owing to the parabolicity in the second equation in (3.1), we expect that u will become smoothed out in positive time. Indeed, a result of Hoff-Smoller ([4], Thm.4.2) not only comfirms this but also shows that the variable v *cannot* become smoothed out in $t > 0$. Let us suppose therefore that the discontinuity in v_0 at $x = 0$ propogates along a curve $x(t) = s$. The Rankine-Hugoniot condition applied to (3.1) then shows that

$$s[v] = [u]$$

and

$$s[u] = [P - ku_x] ,$$

where $[\cdot]$ denotes the jump across the discontinuity curve. However, $[u] = 0 \neq [v]$, so that $s = 0$. (This says that the discontinuity propogates along the particle path $x = 0$.) The other jump condition is then

$$[P] = [ku_x] . \tag{3.3}$$

As in §2., our strategy will be to extend local solutions to all of $t > 0$ by means of a priori energy-type estimates. However, rather than iterating in (3.1), (note that the second equation has discontinuous coefficients!) it is far simpler to build the heuristic jump condition (3.3) into a semidiscrete difference scheme. The latter consists of ode's which are automatically solvable locally in time. We then establish the a priori energy estimates at the semidiscrete level, extend the approximate solutions to all of $t > 0$, and then pass to the limit as the discretization tends to 0. We shall bypass these details in the present exposition and instead proceed in a heuristic manner at the continuous level. The complete, rigorous argument is presented in [2].

Now, the key role of the energy estimates is to insure that v remains bounded below, so that the nonlinear functions P and k are controlled. It will therefore be necessary to bound $v(\cdot,t) - \bar{v}$ in H^1. We first obtain an a priori L^2 estimate as follows. Define

$$\Psi(v) = \int_{\bar{v}}^{v} [P(\bar{v}) - P(s)]ds .$$

It then follows formally from (3.1) that

$$\left[\frac{u^2}{2} + \Psi(v)\right]_t + [(P(v) - P(\bar{v}))u]_x = (uku_x)_x - ku_x^{\,2} ,$$

so that

$$\text{(3.4)} \qquad \int\left[\frac{u^2}{2} + \Psi(v)\right]\Bigg|_0^T dx + \int_0^T \int ku_x^{\,2}dxdt = C_0 ,$$

where C_0 denotes a constant depending only on the initial data. Note that the jump condition (3.3) was used in the spatial integration by parts leading to (3.4).

Next, following Kanel [5], we define $K(v)$ by $K'(v) = k(v)$. It then follows from (3.1) that

$$\begin{aligned} K_{xt} &= (kv_t)_x = (ku_x)_x \\ &= u_t + P_x . \end{aligned}$$

We multiply this by K_x and formally take $\int_0^T\int dxdt \equiv \int_0^T\left[\int_{-\infty}^0 + \int_0^\infty\right]dxdt$ to obtain

$$\frac{1}{2}\int K(v)_x^2\Big|_0^T dx - \int_0^T \int P(v)_x K(v)_x dxdt = \int_0^T \int u_t K(v)_x dxdt .$$

(3.2) shows that $-\iint P_x K_x$ is positive, and

$$\begin{aligned} \iint u_t K_x &= \int uK_x\Big|_0^T dx - \iint uK_{xt} \\ &\le \int u(T)^2 dx + \frac{1}{4}\int k_x(T)^2 dx + C_0 \\ &\quad + \iint u_x K_t dxdt + \int_0^T u(0,t)[K(v(0,t))]_t dt . \end{aligned}$$

Substituting this into (3.4) and using the first energy estimate (3.4) and the fact that $u_x K_t = k u_x^2$, we then obtain

$$(3.5) \qquad \frac{1}{4}\int K(v(T))_x^2 dx \le C_0 + \int_0^T u(0,t)[K(v(0,t))]_t dt \ .$$

We need to bound the integral on the right independently of time. The crucial observation here is that, from the jump condition (3.3),

$$\begin{aligned} [K(v)]_t = [kv_t] &= [ku_x] \\ &= [P(v)] = \alpha(t)[K(v)] \end{aligned}$$

where

$$\alpha(t) = [P(v(0,t))]/[K(v(0,t))] = P'(\xi)/k(\eta) < 0$$

by (3.2). Thus

$$[K(v(0,t))]_t = [K(v(0,0))]\alpha(t)\exp[\int_0^t \alpha(s)ds]$$

decays exponentially in time. Using the first energy estimate (3.4) again, we can bound the integral on the right side of (3.5) by C_0 and conclude that

$$(3.6) \qquad \int K(v(x,T))_x^2 dx \le C_0 \ .$$

If bounds $\underline{v} \le v(x,t) \le \bar{v}$ are known, then (3.4) and (3.6) together given an a priori bound for $\|v(\cdot,t) - \bar{v}\|_{L^\infty}$ in terms of C_0 and properties of P and k in $[\underline{v},\bar{v}]$. By making C_0 sufficiently small, then, we can close the argument and show that v remains bounded away from 0 (as well as bounded above) for all time.

Of course, the argument given above can be made to accomodate a finite number of jumps, since discontinuities never interact in $t > 0$. Theorem 3.1 is then proved by approximating more general initial data by piecewise-H^1 data, applying the above argument with this approximate data, and then taking an appropriate limit.

There are two shortcomings in our result, Theorem 3.1. The first is that it requires the data to have the same limits at $x = +\infty$ as at $x = -\infty$. This precludes the treatment of Riemann initial data. The other deficiency is the requirement that the initial data be small. We believe that the positive viscosity in (3.1) should prevent the occurrence of vacuum states. Indeed, by a simple rearrangement of the estimates in [5], one can show, without making any smallness assumptions, that the Cauchy problem (3.1)-(3.2) is globally solvable, provided that $v_0 - \bar{v} \in H^1$ (i.e. that there are no initial discontinuities in v). It remains, therefore, to accommodate discontinuities in v in this observation.

REFERENCES

1. R. Courant and K.O. Friedrichs, Supersonic Flow and Shock Waves, Wiley-Interscience, New York, 1948.

2. David Hoff, Construction of solutions for compressible, isentropic Navier-Stokes equations in one space dimension with nonsmoothinitial data, to appear in Proc. Royal Soc. Edin.

3. David Hoff and Joel Smoller, Global existence for systems of parabolic conservation laws in several space variables, submitted to J. Diff. Eqns.

4. David Hoff and Joel Smoller, Solutions in the large for certain nonlinear parabolic systems, Ann. Inst. Henri Poincaré, 2, no. 3, (1985), pp. 213-235.

5. Ya. Kanel, On a model system of equations of one-dimensional gas motion, Diff. Eqns. 4, (1968), 374-380.

6. A Kazhikov and V. Shelukhin, Unique global solutions in time of initial boundary value problems for one dimensional equations of a viscous gas, PMMJ Appl. Math. Mech. 41, (1977), 273-283.

7. Jong Uhn Kim, Global existence of solutions of the equations of one dimensional thermoviscoelasticity with initial data in BV and L^1 ,Annali Scuola Normale Superiore Pisa X, no. 3 (1983), 357-427.

GLOBAL SOLUTIONS TO SOME FREE BOUNDARY PROBLEMS FOR QUASILINEAR HYPERBOLIC SYSTEMS AND APPLICATIONS

Li Ta-tsien (Li Da-qian)
Fudan University
Shanghai, China

INTRODUCTION

It is well known that for the Cauchy problem for first order quasilinear hyperbolic systems, in general classical solutions exist only locally in time and singularities of the solution may occur in a finite time, even if the initial data are sufficiently smooth and (even) small. However, there are certain examples of globally defined classical solutions, for example, solutions which represent the interaction of two rarefaction waves in gas dynamics. There fore, it is of great interest to determine the conditions which guarantee the global existence of classical solutions and the conditions which guarantee the development of singularities in a finite time.

Up to now, most of the studies concerning this problem are concentrated on the Cauchy problem for the reducible hyperbolic system

$$\frac{\partial r}{\partial t} + \lambda(r,s)\frac{\partial r}{\partial x} = 0,$$

$$\frac{\partial s}{\partial t} + \mu(r,s)\frac{\partial s}{\partial x} = 0.$$

For the boundary value problems, especially for the free boundary problems, however, there are only few results (cf.[1]-[23]).

In this paper I shall present some recent results obtained in collaboration with Zhao Yan-chun on global classical solutions to some free boundary problems for quasilinear hyperbolic systems and some applications in gas dynamics.

This paper is divided into two parts. In the first part, we consider a class of typical free boundary problems with characteristic boundary for system (1), and prove, under certain hypotheses of monotonicity, the existence or the non-existence of global classical solutions. This result can be used to consider a class of discontinuous initial value problems for the system of one-dimensional isentropic flow and obtain, in a class of piecewise continuous and piecewise smooth functions, the corresponding existence or non-existence theorem for global discontinuous solutions which contain only one shock and probably one centered wave. As simple examples, some interaction problems of a rarefaction wave with a shock will be discussed.

In the second part, we consider another class of typical free boundary problems for the same system (1) and prove, under certain hypotheses of dissipation, the global existence of classical solutions. Then two applications for the system of one-dimensional isentropic flow can be given. The first one is to get a unique global discontinuous solution containing only two shocks for a class of discontinuous initial value problems, and the second one is to prove the global existence of a discontinuous solution containing only one shock for a class of piston problems.

PART ONE

Suppose that on the domain under consideration, system (1) is strictly hyperbolic :

(2) $\lambda(r,s) < \mu(r,s)$

and genuinely nonlinear in the sense of P.D. Lax :

(3) $\dfrac{\partial\lambda}{\partial r}(r,s) > 0, \quad \dfrac{\partial\mu}{\partial s} > 0.$

On an angular domain

(4) $R = \{(t,x) \mid t \geq 0,\ x_1(t) \leq x \leq x_2(t)\},$

we consider the following typical free boundary problem with characteristic boundary :

$x = x_2(t)$ is a free boundary on which we prescribe the boundary conditions

(5) $r = g(t,x,s)$

and

(6) $x'_2(t) = G(t,x,s), \quad x_2(0) = 0.$

Here, (6) is an ordinary equation to be used to determine the position of the free boundary.

On the other hand, $x = x_1(t)$ is a backward characteristic curve passing through the origin, on which we prescribe the boundary condition

(7) $s = s_0(t)$.

Moreover, on $x = x_1(t)$ we have

(8) $r = r_0 = g(0,0,S_0)$, with $S_0 = s_0(0)$

and

(9) $x'_1(t) = \lambda(r_0, s_0(t)), x_1(0) = 0$.

We give the following hypotheses :

(H1). On the domain under consideration, λ, μ, s_0, g and $G \in C^1$, $x_1(t) \in C^2$.

(H2). On $x = x_2(t)$, we have

(10) $\lambda(r,s) < G(t,x,s) < \mu(r,s)$

and

(11) $G(t,x,s) - \lambda(r,s) \geq a(T_0,A,B) > 0, \; \forall \; 0 \leq t \leq T_0, \; r \leq A, \; s \leq B$,

where a (T_0,A,B) is a constant depending only on T_0, A and B.

(H3). We have

(12) $s'_0(t) \leq 0, \forall \; t \geq 0$

and on $x = x_2(t)$:

(13) $\frac{\partial g}{\partial t}(t,x,s) + \frac{\partial g}{\partial x}(t,x,s)G(t,x,s) \geq 0$,

and on $x = x_2(t)$:

(14) $\frac{\partial g}{\partial s}(t,x,s) \leq 0$.

(H4). On $x = x_2(t)$ we have

(15) $|G(t,x,s)| \leq a_1(T_0,B) + a_2(T_0,B) \; |x| \; \forall \; 0 \leq t \leq T_0, |s| \leq I$

where $a_1(T_0,B)$ and $a_2(T_0,B)$ are constants depending only on T_0 and B.

Theorem 1 : Under hypotheses (H1) - (H4), the preceding free boundary problem admits on the angular domain R a unique global classical solution : $(r(t,x),s(t,x)) \in C^1(R)$ and $x_2(t) \in C^2, \forall \; t \geq 0$. Moreover, we have

(16) $\frac{\partial r}{\partial x}(t,x) \geq 0, \; \frac{\partial s}{\partial x}(t,x) \geq 0 \; , \forall \; (t,x) \in R$.

□

The idea of the proof is as follows. According to the corresponding local existence theorem (cf. [24] -[25]), this problem always admits a local classical solution on an angular domain

(17) $R(\delta_0) = \{(t,x) | 0 \leq t \leq \delta_0, x_1(t) \leq x \leq x_2(t)\}$.

where $\delta_0>0$ is suitably small, and then, in order to get a global classical solution, it is sufficient to establish the following uniform a priori estimates :

For any given $T_0>0$, if on a domain

(18) $R(T) = \{(t,x) | 0 \leq t \leq T, x_1(t) \leq x \leq x_2(t)\}$

with $0 < T \leq T_0$, this free boundary problem possesses a classical solution : $(r(t,x)s(t,x)) \in C^1$ and $x_2(t) \in C^2$, then the C^1 norm of $(r(t,x), s(tx,))$ has an upper bound depending only on T_0.

Remark 1 : If

(19) $s_0(t) \equiv S_0$,

then the solution is a backward rarefaction wave : $s(t,x) \equiv S_0$. In this case, $x = x_1(t)$ must be a straight line and Theorem 1 is still valid if we omit hypotheses (11) and (14) in (H2) and in (H3) respectively.

Remark 2 : If the free boundary problem under consideration admits a global bounded classical solution on the angular domain R, then the value of r on the free boundary

(20 $r_0(t) = r(t,x_2(t))$

must be an non-decreasing function.

As an application, we consider the system of one-dimensional isentropic flow in Lagrangian representation. Taking the Riemann invariants as unknown functions, the system is actually of the preceding form (1) with

(21) $- \lambda(r,s) = \mu(r,s) = \sqrt{-p'(\tau(s-r))} = a(s-r)^{\frac{\gamma+1}{\gamma-1}}$

(a > 0 constant)

for polytropic gases, where τ is the specific volume, p is the pressure :

(22) $p(\tau) = A\tau^{-\gamma}$(A >0 constant)

and $\gamma>1$ is the adiabatic exponent.

It is easy to see that the system is strictly hyperbolic and genuinely nonlinear in the sense of P.D.Lax, if and only if $s - r > 0$, i.e. there never exists the vacuum state.

Consider the following discontinuous initial value problem with the piecewise continuous and piecewise smooth initial data :

$$(23)\quad t = 0 : r = \begin{cases} r_0^-(x), \\ r_0^+(x), \end{cases} \qquad s = \begin{cases} s_-, x \leq 0, \\ s_+, x \geq 0, \end{cases}$$

where $s_\pm$ are constants. Setting

$$(24)\quad r_\pm = r_0^\pm(0),$$

suppose that the similarity solution of the corresponding Riemann problem with the piecewise constant initial data :

$$(25)\quad t = 0 : r = \begin{cases} r_-, \\ r_+, \end{cases} \qquad s = \begin{cases} s_-, & x \leq 0, \\ s_+, & x \geq 0, \end{cases}$$

is composed of constant states with a backward centered rarefaction wave and a forward typical shock, using Theorem 1 and Remark 1 we can prove the

Theorem 2 : Suppose that $r_0^\pm(x)$ are bounded, C^1 functions on $x \geq 0$ and on $x \leq 0$ respectively, and

$$(26)\quad r_0^{+'}(x) \geq 0, \forall\ x \geq 0, \quad r_0^{-'}(x) \geq 0, \forall\ x \leq 0.$$

Suppose furthermore that there is no vacuum state at the initial time i.e.

$$(27)\quad s_+ - r_0^+(x) > 0, \forall\ x \geq 0, \quad s_- - r_0^-(x) > 0, \forall x \leq 0.$$

Then the discontinuous initial value problem (1), (21), (23) admits a unique global discontinuous solution in a class of piecewise continuous and piecewise smooth functions. This solution contains only a backward centered wave with the origin as its center and a forward shock $x = x_2(t)$ passing through the origin. Moreover, on both the right side and the left side of the shock the solutions are all the backward rarefaction waves $s \equiv s_+$ and $s \equiv s_-$ respectively, and there never exists any vacuum state on $t \geq 0$. □

The following well-known result comes directly from Theorem 2.

Corollary 1 : The problem of interaction of a forward typical shock with a backward rarefaction wave $s \equiv s_+$ admits a unique global discontinuous solution containing only one forward shock $x = x_2(t)$ on $t \geq 0$ in a class of piecewise continuous and piecewise smooth functions. Moreover, after the collision, the original backward rarefaction wave $s \equiv s_+$ becomes another backward rarefaction wave $s \equiv s_-$. □

Now we consider the general case with the initial data :

$$(28)\quad t = 0 : r = \begin{cases} r_0^-(x), \\ r_0^+(x), \end{cases} \qquad s = \begin{cases} s_0^-(x), & x \leq 0 \\ s_0^+(x), & x \geq 0 \end{cases}$$

for the system of one dimensional isentropic flow.

Setting

(29) $r_\pm = r_0^\pm(0)$, $s_\pm = s_0^\pm(0)$,

we still suppose that the corresponding Riemann problem has a unique similarity solution composed of a backward centered rarefaction wave and a forward typical shock

(30) $$(r,s) = \begin{cases} (r_+,s_+), & x \geq Vt, \\ (r_0,s_-), & x \leq Vt, \end{cases}$$

where

(31) $$r_0 > r_-,$$

V is the propagation speed of shock, and (r_+,s_+), (r_0, s_-) satisfy the corresponding Rankine-Hugoniot condition :

(32) $$r_0 = g(r_+,s_+,s_-)$$

with

(33) $$\frac{\partial g}{\partial r_+} > 0, \quad \frac{\partial g}{\partial s_+} < 0, \quad 0 < \frac{\partial g}{\partial s_-} < 1.$$

Using Remark 2, we can prove the following

Theorem 3 : If

$$\frac{\partial g}{\partial r_+}(r_+,s_+,s_-)(V-\lambda(r_+,s_+))r_0^{+'}(0) +$$

(34) $$\frac{\partial g}{\partial s_+}(r_+,s_+,s_-)(V-\mu(r_+,s_+))s_0^{+'}(0)+$$

$$\frac{\partial g}{\partial s_-}(r_+,s_+,s_-)(V-\mu(r_0,s_-))s_0^{-'}(0) < 0$$

then, in a class of piecewise continuous and piecewise smooth functions, the discontinuous initial value problem (1), (21), (28) never admits a global discontinuous solution containing only one forward shock and a backward centered wave on $t \geq 0$. □

By means of Theorem 3, we can easily obtain the following

Corollary 2 : The interaction problem either of a forward rarefaction wave catching up with a forward typical shock or a forward shock catching up with a forward rarefaction wave never exists a global discontinuous solution containing only one forward shock on $t \geq 0$. □

In contrary, however, we can prove the following

Theorem 4 : For the interaction problem either of a forward compression wave catching up with a forward typical shock or of a forward typical shock catching up with a forward compression wave, if the "strength" of the compression wave, i.e.

$a \max_{0\leq x\leq a} \left| s_0^{+\prime}(x) \right|$ or $a \max_{-a\leq x\leq 0} \left| s_0^{-\prime}(x) \right|$ is suitably small, where a is the initial width of the compression wave, then on $t \geq 0$ there exists a unique global discontinuous solution containing only one forward shock in a class of piecewise continuous and piecewise smooth functions. Moreover, after the interaction, the forward compression wave becomes a backward rarefaction wave. □

PART TWO

In this part we consider the same hyperbolic system (1), but only suppose that the system is strictly hyperbolic.

On an angular domain

(35) $R = \{(t,x) \mid t \geq 0, x_1(t) \leq x \leq x_2(t)\}$,

where $x = x_1(t)$ and $x = x_2(t)$ are all free boundaries, we consider the following typical free boundary problem :

(36) on $x = x_1(t)$: $s = f(a(t,x),r)$,

(37) $\dfrac{dx_1(t)}{dt} = F(\alpha(t,x),r,s), x_1(0) = 0,$

(38) on $x = x_2(t)$: $r = g(b(t,x),s)$,

(39) $\dfrac{dx_2(t)}{dt} = G(\beta(t,x),r,s),\ x_2(0) = 0.$

We give the following hypotheses :

(H1). On the domain under consideration, all given functions $\lambda,\mu,a,b,\alpha,\beta,f,g,F$ and $G \in C^1$.

(H2). There exists a state (r_0,s_0) such that

(40) $s_0 = f(a(0,0)r_0),\ r_0 = g(b(0,0),s_0).$

(H3). There is no characteristic passing through the origin and entering into the angular domain R, i.e.

(41) $\lambda(r_0,s_0) < F(\alpha(0,0),r_0,s_0) < G(\beta(0,0),r_0,s_0)<\mu(r_0,s_0)$.

(H4). The so-called minimal characterizing number of this problem (see [25]) is less than 1, namely

(42) $$\left|\frac{\partial f}{\partial r}(a(0,0)r_0)\ \frac{\partial g}{\partial s}(b(0,0),s_0)\right| < 1.$$

Then we have the following

<u>Theorem 5</u> : Under hypotheses (H1) - (H4), if it holds that

(43) on $x = x_1(t)$: $|a(t,x) - a(0,0)|$, $|\alpha(t,x) - \alpha(0,0)| \leq \varepsilon, \forall t \geq 0$

(44) $$\left|\frac{\partial a}{\partial t}(t,x)\right| , \left|\frac{\partial a}{\partial x}(t,x)\right| \leq \frac{\eta}{t} , \forall\ t > 0,$$

(45) on $x = x_2(t)$: $|b(t,x)-b(0,0)|, |\beta(t,x)-\beta(0,0)| \leq \varepsilon, \forall\ t \geq 0$

(46) $$\left|\frac{\partial b}{\partial t}(t,x)\right|, \left|\frac{\partial b}{\partial x}(t,x)\right| \leq \frac{\eta}{t} , \forall\ t > 0,$$

where $\varepsilon > 0$ and $\eta>0$ are suitably small, then this typical free boundary problem admits on the angular domain R a unique global classical solution : $(r(t,x),s(t,x)) \in C^1$ and $x_1(t)$, $x_2(t) \in C^2$ with $(r(0,0),s(0,0)) = (r_0,s_0)$. Moreover, we have

(47) $|r(t,x) - r_0|, |s(t,x) - s_0| \leq K_0\ \varepsilon, \forall\ (t,x) \in R,$

(48) $$\left|\frac{\partial r}{\partial x}(t,x)\right|, \left|\frac{\partial r}{\partial t}(t,x)\right|, \left|\frac{\partial s}{\partial x}(t,x)\right|, \left|\frac{\partial s}{\partial t}(t,x)\right| \leq \frac{K_1\eta}{1+t} , \forall (t,x) \in R$$

and

(49) $|x_1'(t) - x_1'(0)|$, $|x_2'(t) - x_2'(0)| \leq K_2\ \varepsilon,\ \forall\ t \geq 0,$

where

(50) $$\begin{cases} x_1'(0) = F(\alpha(0,0),r_0,s_0), \\ x_2'(0) = G(\beta(0,0),r_0,s_0), \end{cases}$$

and K_0, K_1, K_2 are positive constants.

□

<u>Remark 3</u> : If we replace the boundary conditions by the following simpler ones :

(51) on $x = x_1(t)$: $s = f(a(0,0),r)$,

(52) $$\frac{dx_1(t)}{dt} = F(\alpha(0,0),r,s), x_1(0)= 0$$

(53) on $x = x_2(t)$: $r = g(b(0,0),s)$,

$$(54)\quad \frac{dx_2(t)}{dt} = G(\beta(0,0),r,s), x_2(0) = 0,$$

the corresponding typical free boundary problem has obviously a trivial global solution in which $(r,s) \equiv (r_0,s_0)$ is a constant state and the free boundaries are all straight lines :

$$(55)\quad \begin{cases} x_1(t) = F(\alpha(0,0),r_0,s_0)t, \\ x_2(t) = G(\beta(0,0),r_0,s_0)t. \end{cases}$$

the typical free boundary problem considered in Theorem 5 can be regarded as a perturbation of this simpler problem and Theorem 5 shows that this perturbated problem still possesses a global classical solution which can be also regarded as a perturbation of the trivial one.

The idea of the proof is as follows : According to the local existence theorem (cf. [25]-[26]), this typical free boundary problem always admits a local classical solution on an angular domain

$$(56)\quad R(\delta_0) = \{(t,x) \mid 0 \leq t \leq \delta_0, x_1(t) \leq x \leq x_2(t)\}.$$

Moreover, we can show that, under the hypotheses of Theorem 5, on any domain where there exists a classical solution, the uniform estimates (47) (48) hold with some positive constants K_0 and K_1, provided that $\varepsilon > 0$ and $\eta > 0$ are suitably small.

Remark 4 : If we suppose furthermore in Theorem 5 that

$$(57)\quad \text{on } x = x_1(t) : \left|\frac{\partial\alpha}{\partial t}(t,x)\right|, \left|\frac{\partial\alpha}{\partial x}(t,x)\right| \leq \frac{\eta}{t}, \forall\, t > 0,$$

$$(58)\quad \text{on } x = x_2(t) : \left|\frac{\partial\beta}{\partial t}(t,x)\right|, \left|\frac{\partial\beta}{\partial x}(t,x)\right| \leq \frac{\eta}{t}, \forall\, t > 0,$$

then we have

$$(59)\quad |x_1''(t)|, |x_2''(t)| \leq \frac{K_3\eta}{1+t}, \forall\, t > 0,$$

where K_3 is a positive constant.

Remark 5 : Set

$$(60)\quad \alpha_0 = \left|\frac{\partial f}{\partial r}(a(0,0)r_0)\,\frac{\partial g}{\partial s}(b(0,0),s_0)\right| < 1,$$

$$(61)\quad \beta_0 = \frac{\eta(r_0,s_0) - G(\beta(0,0),r_0,s_0)}{\eta(r_0,s_0)-F(\alpha(0,0),r_0,s_0)} \cdot \frac{F(\alpha(0,0),r_0,s_0)-\lambda(r_0;s_0)}{G(\beta(0,0),r_0,s_0)-\lambda(r_0,s_0)} < 1$$

and

(62) $c_0 = \dfrac{\ln\alpha_0}{\ln\beta_0} > 0.$

For any given c with $0 < c < c_0$, if in Theorem 5 we suppose furthermore that

(63) on $x = x_1(t)$: $|a(t,x) - a(0,0)|$, $|\alpha(t,x) - \alpha(0,0)| \le \dfrac{\varepsilon}{t^c}$, $\forall\ t > 0$,

(64) on $x = x_2(t)$: $|b(t,x) - b(0,0)|$, $|\beta(t,x) - \beta(0,0)| \le \dfrac{\varepsilon}{t^c}$, $\forall\ t > 0$,

then the global classical solution itself satisfies the following decay estimates :

(65) $|r(t,x) - r_0|$, $|s(t,x) - s_0| \le \dfrac{K_4\varepsilon}{(1+t)^c}$,$\forall\ (t,x) \in R$

and

(66) $|x_1'(t) - x_1'(0)|$, $|x_2'(t) - x_2'(0)| \le \dfrac{K_4\varepsilon}{(1+t)^c}$,$\forall\ t \ge 0$,

where K_4 is a positive constant.

Now we consider a class of discontinuous initial value problems for the system of one-dimensional flow in Lagrangian representation with the piecewise continuous and piecewise smooth initial data (28). Suppose that the corresponding Riemann problem admits a similarity solution composed of constant states (r_-,s_-), (r_0,s_0) and (r_+,s_+) with a forward typical shock

(67) $(r,s) = \begin{cases} (r_0,s_0), & x \le Vt, \\ (r_+,s_+), & x \ge Vt \end{cases}$

and a backward typical shock

(68) $(r,s) = \begin{cases} (r_-,s_-), & x \le Ut, \\ (r_0,s_0), & x \ge Ut; \end{cases}$

where U and V denote the propagation speeds of these typical shocks.

We can use Theorem 5 to prove the following result of perturbation.

<u>Theorem 6</u> : Suppose that $r_0^\pm(x)$, $s_0^\pm(x) \in C^1$ and there is no vacuum state at the initial time :

(69) $s_0^-(x) - r_0^-(x) > 0$,$\forall\ x \le 0$, $s_0^+(x) - r_0^+(x) > 0$, $\forall\ x \ge 0$.

If it holds that

$$(70)\quad \begin{aligned} &|r_0^-(x) - r_-| \ , |s_0^-(x) - s_-| \le \varepsilon, \forall\ x \le 0, \\ &|r_0^+(x) - r_+| \ , |s_0^+(x) - s_+| \le \varepsilon, \forall\ x \ge 0 \end{aligned}$$

and

$$(71)\quad \begin{aligned} &0 \le r_0^{-'}(x) \ , \ s_0^{-'}(x) = \frac{\eta}{x} \ , \quad \forall\ x < 0, \\ &0 \le r_0^{+'}(x) \ , \ s_0^{+'}(x) = \frac{\eta}{x} \ , \quad \forall\ x > 0, \end{aligned}$$

where $\varepsilon > 0$ and $\eta > 0$ are suitably small, then this discontinuous initial value problem admits a unique global discontinuous solution $(r(t,x),s(t,x))$ on $t \ge 0$ in a class of piecewise continuous and piecewise smooth functions. This solution contains only a backward shock $x=x_1(t)$ $(x_1(0) = 0)$ and a forward shock $x = x_2(t)$ $(x_2(0) = 0)$ and there never exists vacuum states on $t \ge 0$, i.e. we always have

$$(72)\quad s(t,x) - r(t,x) > 0 \text{ and } t \ge 0.$$

Moreover, we have that

$$(73)\quad |r(t,x) - r_-| \ , \ |s(t,x) - s_-| \le \varepsilon$$

on the left side of $x = x_1(t)$,

$$(74)\quad |r(t,x) - r_+| \ , \ |s(t,x) - s_+| \le \varepsilon$$

on the right side of $x = x_2(t)$, and on the angular domain between two shocks :

$$(75)\quad R = \{(t,x) \mid t \ge 0, \ x_1(t) \le x \le x_2(t) \}$$

$$(76)\quad |r(t,x) - r_0| \ , \ |s(t,x) - s_0| \le K_0\varepsilon,$$

$$(77)\quad \left|\frac{\partial r}{\partial x}(t,x)\right| , \left|\frac{\partial r}{\partial t}(t,x)\right| , \left|\frac{\partial s}{\partial x}(t,x)\right| , \left|\frac{\partial s}{\partial t}(t,x)\right| \le \frac{K_1\eta}{1+t}$$

and

$$(78)\quad |x_1'(t) - U|, |x_2'(t) - V| \le K_2\varepsilon \ , \forall\ t \ge 0,$$

$$(79)\quad |x_1''(t)|, |x_2''(t)| \le \frac{K_3\eta}{1+t} \ , \forall\ t \ge 0,$$

where K_0, K_1, K_2 and K_3 are positive constants. □

<u>Remark 6</u> : If we suppose furthermore that

$$(80)\quad \begin{aligned} &|r_0^-(x) - r_-| \ , |s_0^-(x) - s_-| \le \frac{\varepsilon}{x^c} \ , \ \forall\ x < 0, \\ &r_0^+(x) - r_+ \ , \ s_0^+(x) - s_+ \le \frac{\varepsilon}{x^c} \ , \ \forall\ x > 0, \end{aligned}$$

where c is defined as before, then on the angular domain R, the solution itself satisfies the following decay estimates :

(81) $\left|r(t,x) - r_0\right|$, $\left|s(t,x) - s_0\right| \leq \frac{K_4\varepsilon}{(1+t)^c}$, $\forall$ $(t,x) \in R$

and

(82) $\left|x_1'(t) - U\right|$, $\left|x_2'(t) - V\right| \leq \frac{K_4\varepsilon}{(1+t)^c}$, $\forall$ $t \geq 0$.

Now we use Theorem 5 to discuss a class of piston problems. Suppose that a piston originally located at the origin at $t = 0$ moves with the speed $u = \varphi(t)$ in a tube, we want to determine the state of the gas as the right side of this piston. In Lagrangian representation this piston problem asks us to solve a mixed initial-boundary value problem for the system of isentropic flow mentioned above with the following conditions :

(83) $t = 0 : r = r_0^+(x)$, $s = s_0^+(x)$ $(x \geq 0)$,

(84) $x = 0 : s = -r + \varphi(t)$ $(t \geq 0)$.

Suppose that

(85) $\varphi(0) > r_+ + s_+ \overset{\Delta}{=} u_0$,

where $r_+ = r_0^+(0)$, $s_+ = s_0^+(0)$ and u_0 denotes the velocity of the gas at the origin at $t = 0$.

It is well known that if $\varphi\ (t) \equiv \varphi(0)$, $r_0^+(x) \equiv r_+$ and $s_0^+(x) \equiv s_+$, then this piston problem has a trivial global solution :

(86) $(r,s) = \begin{cases} (r_0,s_0), & 0 \leq x \leq Vt, \\ (r_+,s_+), & x \geq Vt, \end{cases}$

which is a tipical forward shock, where V is the propagation speed of the shock and $r_0 + s_0 = \varphi(0)$.

As a perturbation result, we have

<u>Theorem 7</u> :Suppose that $r_0^+(x)$, $s_0^+(x)$ and $\varphi(t) \in C^1$, and there is no vacuum state at the initial time :

(87) $s_0^+(x) - r_0^+(x) > 0$,$\forall$ $x \geq 0$.

If it holds that

(88) $\left|r_0^+(x) - r_+\right|$, $\left|s_0^+(x) - s_+\right| \leq \varepsilon$, $\forall$ $x \geq 0$,

(89) $\left|\varphi(t) - \varphi(0)\right| \leq \varepsilon$, $\forall$ $t \geq 0$,

(90) $0 \leq r_0^{+'}(x)$, $s_0^{+'}(x) \leq \frac{\eta}{x}$, $\forall$ $x > 0$,

(91) $|\varphi'(t)| \leq \frac{\eta}{t}, \forall\, t > 0,$

where $\varepsilon > 0$ and $\eta > 0$ are suitably small. Then this piston admits, on the whole domain $\{(t,x) \mid t \geq 0,\ x \geq 0\}$, a unique global discontinuous solution $(r(t,x),s(t,x))$ in a class of piecewise continuous and piecewise smooth functions. This solution contains only one forward shock $x = x_2(t)$ and satisfies the following estimates :

(92) $|r(t,x) - r_+|,\ |s(t,x) - s_+| \leq \varepsilon$ on $R_+ = \{(t,x) \mid t \geq 0;\ x \geq x_2(t)\}$.

(93) $|r(t,x) - r_0|,\ |s(t,x) - s_0| \leq K_0\varepsilon$

and

(94) $\left|\frac{\partial r}{\partial x}(t,x)\right|,\ \left|\frac{\partial r}{\partial t}(t,x)\right|,\ \left|\frac{\partial s}{\partial x}(t,x)\right|,\ \left|\frac{\partial s}{\partial t}(t,x)\right| \leq \frac{K_1\eta}{1+t}$

on $R = \{(t,x) \mid t \geq 0,\ 0 \leq x \leq x_2(t)\}$, and

(95) $|x_2'(t) - V| \leq K_2\varepsilon, \forall\, t \geq 0,$

(96) $|x_2''(t)| \leq \frac{K_3\eta}{1+t}, \forall\, t \geq 0,$

where K_0, K_1, K_2 and K_3 are positive constants. Moreover, there never exists vacuum states on $t \geq 0$. □

<u>Remark 7</u> : If we suppose furthermore that

(97) $|r_0^+(x) - r_+|,\ |s_0^+(x) - s_+| \leq \frac{\varepsilon}{x^c}, \forall\, x > 0,$

(98) $|\varphi(t) - \varphi(0)| \leq \frac{\varepsilon}{t^c}, \forall\, t > 0,$

where c is still defined as before, then we can obtain the following decay estimates for the solution itself :

(99) $|r(t,x) - r_0|,\ |s(t,x) - s_0| \leq \frac{K_4\varepsilon}{(1+t)^c}, \forall\, (t,x) \in R$

and

(100) $|x_2'(t) - V| \leq \frac{K_4\varepsilon}{(1+t)^c}, \forall\, t \geq 0.$

REFERENCES

1. Gu Chao-hao, Collections of Mathematical Papers of Fudan Univ., (1960), 36-39.

2. Lin Long-wai, Journal of Jilin Univ., 4(1963), 83-96.

3. P.D. Lax, J. Math. Phys., 5(1964), 611-613.

4. J.B.Keller and L.Ting, Comm. Pure Appl. Math., 19(1966), 371-420.

5. J.L.Johnson, Bull. Amer. Math. Soc;, 73(1967), 639-641.

6. M.Yamaguti and T.Nishida, Funkcial Ekvac., 11(1968), 51-57.

7. A.Jeffrey and A.Donato, Wave motion, 1(1979), 177-185.

8. B.L. Rozdestvenskii and N.N.Yanenko, Systems of Quasilinear Hyperboli Equations, Izd. Nauka, Moskva, 1968.

9. Li Ta-tsien and Shi Jia-hong, Chin. Ann. of Math., 5B (1984), 241-248

10. Li Ta-tsien, Proceedings of the Royal Society of Edinburgh, 87A (1981 255-261.

11. Li Ta-tsien and Yu Wen-ci, J. Math. Pures et Appl., 61(1982), 401-409

12. A.Majda and S.Klainerman, Comm. Pure Appl. Math., 33(1980), 241-263.

13. Li Ta-tsien and Shi Jia-hong, Proceedings of the Royal Society of Edinburgh, 94 A(1983), 137-147.

14. Li Ta-tsien and Qin Tie-hu, Acta Math. Sinica, 28(1985), 606-613.

15. Li Ta-tsien and Li Cai-zhong, Acta Math. Scientia, 4(1984), 497-507.

16. Yu Jin-guo and Zhao Yan-chun, Chin. Ann. of Math., 6A (1985), 595-609

17. F. John, Comm. Pure Appl. Math., 27 (1974), 377-405.

18. T. Nishida, Publications Mathématiques d'Orsay, 1978.

19. Hsiao Ling and Li Ta-tsien, Chin. Ann. of Math., 4B (1983), 109-115.

20. Li Ta-tsien and Qin Tie-hu, Chin. Ann. of Math., 6B (1985), 190-115.

21. J.M. Greenberg and Li Ta-tsien, J. Diff. Equa., 52(1984), 66-75

22. Qin Tie-hu, Chin. Ann. of Math., 6B (1985)

23. Lin Long-wai, Journal of Huachaw Univ., 2(1984), 1-4.

24. Gu Chao-hao, Li Ta-tsien and Ho Zon-y, Acta Math. Sinica, 4(1961), 324-327.

25. Li Ta-tsien and Yu Wen-ci, Boundary Value Problems for Quasilinear

hyperbolic Systems, Duke University Mathematics Series V, 1985.

26. Gu Ghao-hao, Li Ta-tsien and Ho Zon-ỳ, Acta Math. Sinica, 2(1962), 132-143

UN THEOREME D'EXISTENCE GLOBALE EN
ELASTICITE NON LINEAIRE MONO-DIMENSIONNELLE

Michel Rascle
Analyse Numérique, Université de St-Etienne,
42023 Saint-Etienne Cédex 2, FRANCE

A partir de Septembre 1986 : Mathématiques, Université de Nice, Parc Valrose, 06034 Nice Cédex, France

Résumé : On montre comment passer à la limite faible sur le produit des invariants de Riemann d'un système strictement hyperbolique (non linéaire) de deux lois de conservation. On explique ensuite comment ce résultat permet de démontrer aisément la convergence d'une suite de solutions approchées du système de l'élasticité non linéaire. Ce résultat s'applique aussi au système de la dynamique des gaz isentropiques, en coordonnées eulériennes, s'il n'y a pas cavitation.

Abstract : We show how to pass to the weak limit on the product of the Riemann invariants of a 2 × 2 strictly hyperbolic (non linear) system of conservation laws.Then we show how this result enables us to prove easily the convergence of a sequence of approximate solutions to the nonlinear elasticity system. This result also applies to the isentropic gas dynamics system, in eulerian coordinates, if there is no cavitation.

1- Introduction :

On considère le problème de Cauchy pour un système 2×2 strictement hyperbolique :

$$\begin{cases} u_t + (f(u))_x = 0, \quad u : Q = \mathbb{R} \times \mathbb{R}_+ \to \mathbb{R}^2, \quad f : \mathbb{R}^2 \to \mathbb{R}^2 & (1.1) \\ u(x,0) = u_0(x) & (1.2) \end{cases}$$

qu'on approche (par exemple) par "viscosité" :

$$\begin{cases} u_t^\varepsilon + (f(u^\varepsilon))_x = \varepsilon\,(Du_x^\varepsilon)_x, \ D \text{ (semi)-définie positive} & (1.3) \\ u^\varepsilon\,(x,0) = u_0(x)\ , \qquad \varepsilon \downarrow 0 & (1.2) \end{cases}$$

On note $\lambda_1(u) < \lambda_2(u)$ les valeurs propres de la matrice jacobienne $f'(u)$, $w(u)$, $z(u)$ les invariants de Riemann, et $(\varphi(u), \Psi(u))$ un couple entropie-flux (e-f) : pour des solutions régulières, on a :

$$\begin{cases} \partial_t w(u) + \lambda_2(w(u),z(u))\, \partial_x w(u) = 0 \\ \partial_t z(u) + \lambda_1(w(u), z(u))\, \partial_x z(u) = 0 \end{cases} \tag{1.4}$$

et

$$\partial_t \varphi(u) + \partial_x \Psi(u) = 0 \tag{1.5}$$

(On suppose que l'application $u \mapsto (w,z)$ est bijective).

On suppose dans toute la suite avoir une estimation L^∞ indépendante de ε sur la suite (u^ε) :

$$\|u^\varepsilon\|_{L^\infty(Q)} \leq c \tag{1.6}$$

(en général, cette estimation est très difficile à obtenir).

On suppose aussi qu'il existe une entropie Φ uniformément convexe (l'énergie) ce qui entraîne :

$$\|\sqrt{\varepsilon}\, u_x^\varepsilon\|_{L^2} \leq c$$

. Alors, en utilisant le Lemme de Murat [3], on obtient aisément :

$$\begin{cases} \partial_t w^\varepsilon + \lambda_2\, (w^\varepsilon,z^\varepsilon)\, \partial_x w^\varepsilon \in K \\ \partial_t z^\varepsilon + \lambda_1\, (w^\varepsilon,z^\varepsilon)\, \partial_x z^\varepsilon \in K' \end{cases} \tag{1.7}$$

$$\partial_t \varphi^\varepsilon + \partial_x \Psi^\varepsilon \in K'' \tag{1.8}$$

où K, K', K'' sont fortement compacts dans l'espace de Sobolev $H^{-1}_{loc}(Q)$, avec $w^\varepsilon = w(u^\varepsilon)$, $z^\varepsilon = z(u^\varepsilon)$ etc...

Alors, si $(\hat{\varphi}, \hat{\psi})$ est un autre couple e - f, la compacité par compensation entraîne

$$\varphi^\varepsilon\, \hat{\Psi}^\varepsilon - \Psi^\varepsilon\, \hat{\varphi}^\varepsilon \rightharpoonup \varphi^*\, \hat{\Psi}^* - \Psi^*\, \hat{\varphi}^* \tag{1.9}$$

(où $\rightharpoonup$ désigne la convergence dans L^∞ faible -étoile)

Soit alors $\{\nu_{x,t}\}_{(x,t)\in Q}$ la famille de mesures d'Young associée à la suite $(w^\varepsilon, z^\varepsilon)$ (voir e.g. [10]) : pour toute fonction continue $F : \mathbb{R}^2 \to \mathbb{R}$, une suite extraite

$F(u^\varepsilon) \rightharpoonup F^*$, avec

$$F^*(x,t) = \langle \nu_{x,t}, F(.) \rangle \qquad (1.10)$$

On rappelle que la suite (u^ε) converge <u>fortement</u> dans $L^p_{loc}(Q)$, $\forall\, p \in [1,+\infty]$ si et seulement si

$$\nu_{x,t} = \delta_{u(x,t)} \text{ pour presque tout } (x,t) \in Q \qquad (1.11)$$

(i.e. $\nu_{x,t}$ est une <u>masse de Dirac</u>).

Alors on peut réécrire (1.9) sous la forme :

$$\begin{cases} \text{pour tous couples e-f } (\varphi, \Psi) \text{ et } (\hat{\varphi}, \hat{\Psi}), \text{ et p.p. dans } Q \\ \langle \nu_{x,t}, \varphi\hat{\Psi} - \Psi\hat{\varphi} \rangle = \langle \nu_{x,t}, \varphi \rangle . \langle \nu_{x,t}, \hat{\Psi} \rangle - \langle \nu_{x,t}, \Psi \rangle . \langle \nu_{x,t}, \hat{\varphi} \rangle \end{cases} \qquad (1.12)$$

Cette équation, introduite (et résolue dans le cas scalaire) par L. Tartar [10] <u>permet d'identifier la mesure $\nu_{x,t}$</u> ($\forall (x,t)$fixé) en utilisant <u>suffisamment</u> de couples e-f : R.J. Di Perna [1] a utilisé les entropies de Lax (où $k \to \pm \infty$)

$$\varphi_k = e^{kw}\,(V_0 + k^{-1}V_1 + k^{-2}A_k) \;;\; \Psi_k = e^{kw}(\lambda_2 V_0 + k^{-1}H_1 + k^{-2}B_k) \qquad (1.12)$$

(et les couples analogues associés à l'autre invariant de Riemann z). Les termes dominants V_0 (resp. $\hat{V}_0$) vérifient :

$$(1.13) \qquad (\lambda_2 - \lambda_1)\,\frac{\partial V_0}{\partial z} + V_0\,\frac{\partial \lambda_2}{\partial z} = 0 \;\; (\text{resp.} (\lambda_1 - \lambda_2)\,\frac{\partial \hat{V}_0}{\partial w} + \hat{V}_0\,\frac{\partial \lambda_1}{\partial w} = 0), V_0, \hat{V}_0 > 0$$

Avec ces couples e-f, R.J. Di Perna a montré que, si le système est <u>vraiment non linéaire</u> (V.N.L.), alors (1.12) entraîne que $\nu_{x,t}$ est une masse de Dirac : si le système est V.N.L., <u>aucune oscillation ne peut apparaître</u> ou <u>se propager</u> dans une solution (entropique) limite de solutions "visqueuses" : même s'il y avait des oscillations dans la condition initiale, elles sont "tuées" par la condition d'entropie.

Au contraire, il est bien connu que, dans le cas linéairement dégénéré (L.D.), il n'y a pas de condition d'entropie : on peut construire une suite de solutions oscillantes (périodiques) ne prenant que 2 valeurs a et b reliées par une discontinuité de contact. Une telle suite de solutions ne converge pas fortement : la mesure $\nu_{x,t}$ n'est pas une masse de Dirac. Dans ce cas, on soupçonne fortement qu'aucune oscillation ne peut apparaître dans la solution, mais ce n'est pas (encore) démontré. Par contre, cf D. Serre [9], des oscillations dans la condition initiale peuvent se propager.

Le résultat général présenté ici s'applique (pour l'instant) au système "de l'élasticité" :

$$\begin{cases} u_t - (\sigma(v))_x = 0, & \sigma'(v) > 0 \\ v_t - u_x = 0, & u : \text{vitesse}, \ \sigma(v) : \text{contrainte} \end{cases} \tag{1.14}$$

qu'on perturbe (par exemple) par visco-élasticité

$$\begin{cases} u_t^\varepsilon - (\sigma(v^\varepsilon))_x = \varepsilon u_{xx}^\varepsilon & , \ \varepsilon \downarrow 0 \\ v_t^\varepsilon - u_x^\varepsilon = 0 \end{cases} \tag{1.15}$$

ainsi qu'au système de la dynamique des gaz isentropiques :

$$\begin{cases} \rho_t + (\rho u)_x = 0 \ , & p'(\rho) > 0 \\ (\rho u)_t + (\rho u^2 + p(\rho))_x = 0, & \rho : \text{densité}, \ p : \text{pression}, \ u : \text{vitesse} \end{cases} \tag{1.16}$$

en supposant dans ce dernier cas qu'il n'y a pas de cavitation, i.e. que le système reste strictement hyperbolique ; pour les difficultés supplémentaires liées à la cavitation, voir R.J. Di Perna [2].

Pour l'un ou l'autre de ces deux systèmes, on étend le résultat de Di Perna à tous les cas intermédiaires entre le cas V.N.L. et le cas L.D., par exemple :

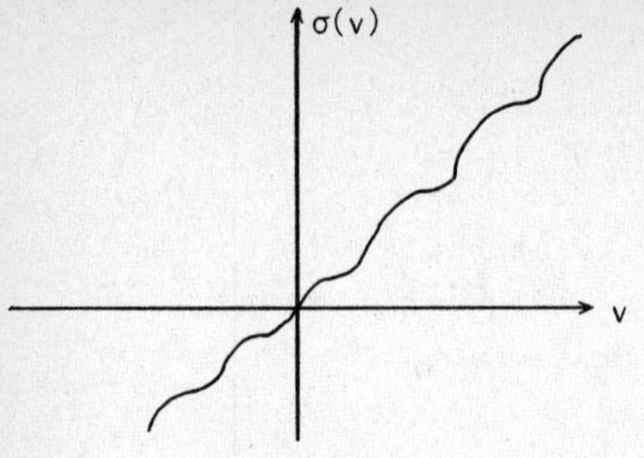

Figure 1

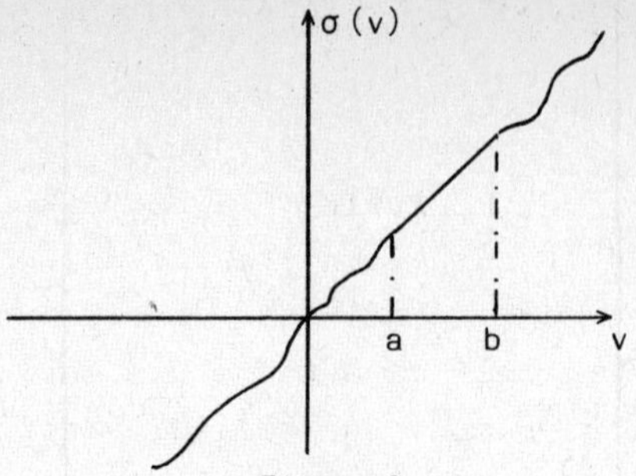

Figure 2

Par exemple, pour le système de l'élasticité, dans le cas de la figure 1, il n'y a pas d'oscillation dans la solution (pour tout $t > 0$). Dans le cas de la figure 2, il y a éventuellement des oscillations pour $v \in [a,b]$, mais, même dans ce cas, une sous-suite extraite de la suite $(u^{\varepsilon}, v^{\varepsilon})$ converge faiblement vers une solution faible du problème de Cauchy.

Pour démontrer ceci, le point crucial est de passer à la limite faible sur le produit des invariants de Riemann w^{ε}, z^{ε}, ce qui, au vu de (1.7) est un résultat de compacité par compensation à coefficients variables.

Dans le paragraphe 2, on énonce ce résultat, avec quelques commenraires, et une brève indication sur la démonstration, et dans le paragraphe 3 on l'applique au système de l'élasticité.

La conclusion constitue le paragraphe 4. Pour les détails de la démonstration du Théorème 1, on renvoie à M. Rascle [6], et à un travail à paraître.

2- Le Théorème général :

Dans M. Rascle [4], en 1983, on émettait la Conjecture suivante, pour le système de l'élasticité (1.15) :

$$\text{(C)} \qquad w^{\varepsilon} \,.\, z^{\varepsilon} \rightharpoonup w^{*} \,.\, z^{*}$$

Vers la même époque, D. Serre et L. Tartar donnaient un contre-exemple montrant que (C) est en général fausse pour un système hyperbolique 2×2 quelconque. Enfin, fin 1983, D. Serre émettait la Conjecture plus générale suivante (voir [8]) :

$$(C')\begin{cases} \forall\ f,g \text{ continues} \\ (\underset{\varepsilon\to 0}{\text{w-lim}}\ p^\varepsilon f(w^\varepsilon)g(z^\varepsilon)).(\underset{\varepsilon\to 0}{\text{w-lim}}\ p^\varepsilon)=(\underset{\varepsilon\to 0}{\text{w-lim}}\ p^\varepsilon f(w^\varepsilon)).(\underset{\varepsilon\to 0}{\text{w-lim}}\ p^\varepsilon g(z^\varepsilon)) \end{cases}$$

où le poids $p^\varepsilon = p(w^\varepsilon,z^\varepsilon)$ vérifie

$$p = \exp(-H)\ ;\ \frac{\partial^2 H}{\partial w \partial z} = (\lambda_2-\lambda_1)^{-2}\ \frac{\partial(\lambda_1,\lambda_2)}{\partial(w,z)} \tag{2.1}$$

où $\frac{\partial(\lambda_1,\lambda_2)}{\partial(w,z)}$ désigne le jacobien de l'application : $(w,z) \to (\lambda_1,\lambda_2)$.

Par définition des mesures d'Young, on peut réécrire (C') sous la forme

$$\nu = \nu_{x,t} = p^{-1}\ \mu_1 \otimes \mu_2 = \exp\ (H)\mu_1 \otimes \mu_2 \tag{C'}$$

où μ_1 (resp. μ_2) opère sur les fonctions de w(resp. z).

Remarquons d'abord que pour le système de l'élasticité (1.14), $\lambda_1 = -\lambda_2$, donc $\partial(\lambda_1,\lambda_2)/\partial(w,z) = 0$, donc on peut choisir $p = \exp\ (H) \equiv 1$: la Conjecture (C') contient le cas particulier de la Conjecture (C).

Théorème 1 :

On suppose que le système (1.1) est strictement hyperbolique, admet une entropie uniformément convexe Φ, et que (1.6) est vérifiée. Alors la Conjecture générale (C') est vraie : en d'autres termes, (1.7) entraine (C').

Remarque 1 :

Le poids p(w,z) dépend de $\partial(\lambda_1,\lambda_2)/\partial(w,z)$, donc du couplage du système et en particulier de la manière dont les courbes caractéristiques sont déformées dans l'interaction de deux ondes. Par exemple , dans le cas du système étudié par Serre et Tartar, les solutions régulières vérifient

$$w_t + zw_x = 0\ ;\ z_t + wz_x = 0 \tag{2.2}$$

i.e. $\lambda_2 \equiv z$ et $\lambda_1 \equiv w$ sont linéairement dégénérées : on peut donc construire une suite de solutions oscillantes $(w^\varepsilon, z^\varepsilon)(x,t) \equiv (w,z)(\frac{x}{\varepsilon}, \frac{t}{\varepsilon})$ où (w,z) est une solution périodique

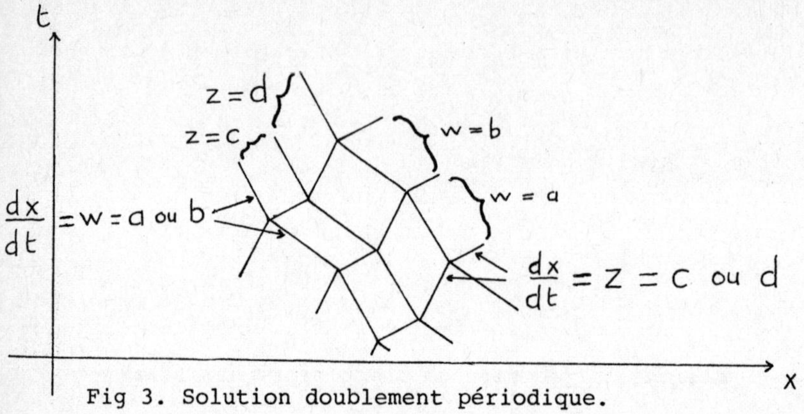

Fig 3. Solution doublement périodique.

En calculant les limites faibles de $w^\varepsilon, z^\varepsilon$, et donc leurs valeurs moyennes sur une période, on trouve aisément que dans ce cas, $p(w,z) = (\lambda_2 - \lambda_1)^{-1} = (z-w)^{-1}$ (on suppose $z > w$).

<u>Indications sur la démonstration</u> :

a) On construit d'abord les couples e-f suivants

$$\begin{cases} \varphi_k = e^{ik(w-w^*)^2}(V_0 + k^{-1}R_k) \ ; \ \Psi_k = e^{ik(w-w^*)^2}(\lambda_2 V_0 + k^{-1}S_k) \\ \hat{\varphi}_l = e^{il(z-z^*)^2}(\hat{V}_0 + l^{-1}\hat{R}_l) \ ; \ \hat{\Psi}_l = e^{il(z-z^*)^2}(\lambda_1 \hat{V}_0 + l^{-1}\hat{S}_l) \end{cases} \tag{2.3}$$

Grâce au théorème de la phase stationnaire,

$$\forall f, \forall w, \ f(w) = \lim_{k\to\infty} k^{\frac{1}{2}} \int e^{ik(w-w^*)^2} f(w^*)dw^*$$

ces couples se concentrent sur les droites $\{w=w^*\}$ ou $\{z=z^*\}$ quand k ou $l \to \pm\infty$, et les termes correcteurs $k^{-1}R_k, l^{-1}\hat{S}_l \ldots$ sont majorés par $O(k^{-\frac{1}{2}})$, ou $O(l^{-\frac{1}{2}})$, uniformément sur tout rectangle borné dans le plan (w,z). Les fonctions V_0, $\hat{V}_0$ sont encore solutions de (1.13).

b) On vérifie ensuite que, à un produit tensoriel près

$$p(w,z) = (\lambda_2 - \lambda_1)\, V_0\, \hat{V}_0 = \lambda_2\, V_0 . \hat{V}_0 - V_0 \, . \, \lambda_1\, \hat{V}_0 \tag{2.4}$$

(le second membre est relié à $\Psi_k\, \hat{\varphi}_l - \varphi_k\, \hat{\Psi}_l$, ce qui permet d'utiliser la compacité par compensation).

c) La difficulté est ensuite d'appliquer la méthode de la phase stationnaire en intégrant par rapport à la mesure d'Young, qui peut être singulière : dans ce cas, cette méthode ne s'applique pas. On tourne la difficulté en intégrant une fois de plus, par rapport au point w* (ou z*) où la phase est stationnaire. On applique alors le Théorème de Fubini pour faire porter la méthode de la phase stationnaire sur cette nouvelle intégration. On obtient ainsi, en appliquant (1.12) aux couples (φ_k, Ψ_k) et $(\hat{\varphi}_1, \hat{\Psi}_1)$:

$$\begin{cases} \forall\, f,g, < \nu, (\lambda_2-\lambda_1)\, V_0 \hat{V}_0\, f(w)\, g(z) > \; = \; <\nu, p.f(w)g(z)> : \\ \qquad = <\nu,\; \lambda_2 V_0 f(w)>.<\nu, \hat{V}_0 g(z)> - <\nu, V_0 f(w)>.<\nu, \lambda_1 \hat{V}_0 g(z)> \end{cases} \tag{2.5}$$

ce qui exprime que $\nu.p$ est une <u>différence</u> de deux produits tensoriels.

En appliquant encore la compacité par compensation, i.e. (1.12), avec deux fonctions $f_1(w)$ et $f_2(w)$, on obtient :

$$0 = <\nu,\; \lambda_2 V_0 f_1(w)\; >.<\nu,\; V_0 f_2(w)> - <\nu, V_0 f_1(w)>.<\nu, \lambda_2 V_0 f_2(w)>$$

et une relation analogue pour l'autre famille, ce qui montre que les deux produits tensoriels impliqués dans (2.5) sont proportionnels. Donc $\nu.p$ se réduit à un seul produit tensoriel, ce qui achève la démonstration.

3- Application au système de l'élasticité

<u>Théorème 2</u> :

<u>On suppose que la suite $(u^\varepsilon, v^\varepsilon)$ de solutions du problème de Cauchy pour le système (1.15) vérifie</u> : $\|(u^\varepsilon, v^\varepsilon)\|_{L^\infty(Q)} \leq C$. <u>Alors, une sous-suite $(u^\varepsilon, v^\varepsilon)$ converge dans L^∞ faible-étoile vers une solution faible (u,v) du problème de Cauchy. De plus, si σ n'est affine sur aucun intervalle non trivial, $(u^\varepsilon, v^\varepsilon) \to (u,v)$ fortement dans</u> $L^p_{loc}(Q)$, $\forall p < +\infty$ et (u,v) satisfait la condition d'entropie de Lax : <u>pour toute entropie convexe</u> φ, $\partial_t\, \varphi(u) + \partial_x\, \varphi(u) \leq 0$ (3.1)

<u>Idée de la démonstration</u> :

Pour plus de détails (avec des hypothèses un peu plus faibles), voir M. Rascle [4] [5]. On montre aisément que $\|\sqrt{\varepsilon}\, u^\varepsilon_x\|_{L^2(Q)} + \|\sqrt{\varepsilon}\, v^\varepsilon_x\|_{L^2(Q)} \leq c$ (3.2)

On peut donc appliquer la Conjecture (C'), qui se réduit ici à (C) :

(3.3) $$w^\varepsilon . z^\varepsilon = (u^\varepsilon - g(v^\varepsilon)).(u^\varepsilon + g(v^\varepsilon)) \rightharpoonup w^*.z^* = (u-g^*)(u+g^*) = u^2-(g^*)^2$$

(on a posé : $g(v) = \int_0^v (\sigma'(s))^{\frac{1}{2}} \, ds$)

Par ailleurs, la compacité par compensation appliquée aux deux équations de (1.15) s'écrit, avec des notations évidentes

$$(u^\varepsilon)^2 - v^\varepsilon \, \sigma(v^\varepsilon) \rightharpoonup u^2 - v \, . \, \sigma^* \qquad (3.4)$$

Soustrayons (3.3) de (3.4), il vient :

$$(g(v^\varepsilon))^2 - v^\varepsilon \sigma(v^\varepsilon) \rightharpoonup (g^*)^2 - v \, \sigma^* \qquad (3.5)$$

Posons maintenant

$$\chi^\varepsilon \underset{\text{déf}}{=} (g(v^\varepsilon) - g(v))^2 - (v^\varepsilon - v)\,(\sigma(v^\varepsilon) - \sigma(v))$$

Développons χ^ε et passons à la limite : il vient, compte tenu de (3.5)

(3.6) $$\chi^\varepsilon \rightharpoonup (g^*)^2 - 2g^* g(v) + (g(v))^2 - v\sigma^* + v\sigma(v) - v\sigma(v) + v\sigma^* = (g^* - g(v))^2 \geq 0$$

D'autre part, grâce à l'inégalité de Cauchy-Schwarz :

$$(g(v^\varepsilon) - g(v))^2 = \left(\int_v^{v^\varepsilon} 1. \, (\sigma'(s))^{\frac{1}{2}} ds\right) \leq (v^\varepsilon - v)(\sigma(v^\varepsilon) - \sigma(v)) \qquad (3.7)$$

χ^ε est une suite de fonctions négatives ou nulles.

Donc $\chi^\varepsilon \to g^* - g(v) = 0$ dans $L^1_{loc}(Q)$, et p.p. dans Q (on a au besoin extrait une sous-suite).

De plus, $\chi^\varepsilon = 0$ dans deux cas seulement : ou $v^\varepsilon = v$ (ce qui correspond à la convergence forte de (v^ε) vers v quand $\varepsilon \to 0$), ou $\sigma'(.)$ est constante sur l'intervalle $[v, v^\varepsilon]$, ce qui correspond au cas <u>linéairement dégénéré</u> : le cas <u>d'égalité dans l'inégalité de Cauchy-Schwarz</u> (3.7) correspond à la possibilité <u>d'oscillations</u> dans un système linéairement dégénéré. Cette situation ne peut se produire que dans le cas de la figure 2. Au contraire, dans le cas de la figure 1, il y a convergence forte.

Remarque 2 :

La même méthode s'applique au cas du système de la dynamique des gaz isentropiques (1.16), en supposant qu'il n'y a pas cavitation, voir M. Rascle, D. Serre [7].

Dans ce cas, le poids p de la Conjecture (C') (à ne pas confondre avec la pression!) est en fait la densité ρ. On construit encore une suite de fonctions $\chi^\varepsilon \leq 0$ qui converge faiblement vers une limite ≥ 0. Cette fois, l'inégalité de Cauchy-Schwarz est la suivante

$$\left(\int_{\rho^*}^{\rho^\varepsilon} s^{-1} \left(p'(s)\right)^{\frac{1}{2}} ds\right) \leq \frac{\rho^\varepsilon - \rho^*}{\rho^\varepsilon \rho^*} \cdot \left(p(\rho^\varepsilon) - p(\rho^*)\right) \tag{3.8}$$

(le premier membre intervient dans l'expression des invariants de Riemann). Ici encore, le cas d'égalité dans (3.8) correspond au cas linéairement dégénéré (irréaliste physiquement) où $p(\rho) = A - \frac{B}{\rho}$.

4- Conclusion :

En conclusion, le Théorème 1, qui ne suppose <u>aucune hypothèse</u> sur le caractère <u>V.N.L. ou L.D.</u> du système, permet d'étudier dans tous les cas (où le système est strictement hyperbolique) la structure de produit tensoriel de la famille de mesures d'Young associée à une suite de solutions approchées de (1.1).

Dans le cas particulier du système de l'élasticité (1.14) ou du système de la dynamique des gaz (1.16), ce résultat permet de démontrer aisément la convergence d'une suite de solutions approchées (méthode de viscosité, certains schémas numériques...) vers une solution faible "entropique" du système. Cette solution est alors définie globalement en temps.

Bibliographie :

[1] R.J. Di Perna : Convergence of Approximate Solutions to conservation laws, Arch. Rat. Mech. Anal., 82 (1983), 27-70.

[2] R.J. Di Perna : Convergence of the viscosity method for isentropic gas dynamics, Comm. Math. Phys. 91 (1983), 1-30.

[3] F. Murat : L'injection du cône positif de H^{-1} dans $W^{-1,q}$ est compacte pour tout $q < 2$, preprint.

[4] M. Rascle : Perturbations par viscosité de certains systèmes hyperboliques non linéaires, Thèse, Université Lyon 1, mai 1983.

[5] M. Rascle : On the convergence of the viscosity method for the system of nonlinear (1-D) elasticity, Workshop on oscillations and compensated compactness, Minneapolis (1985), Lectures in Applied Math. vol. 23.

[6] M. Rascle : Un résultat de compacité par compensation à coefficients variables. Application à l'élasticité non linéaire, C.R. Acad. Sc. Paris, t. 302, série I, n° 8 (1986), 311-314.

[7] M. Rascle, D. Serre : Compacité par compensation et systèmes hyperboliques de lois de conservation, C.R. Acad. Sc., Paris, t. 299, série I (1984), 673-676.

[8] D. Serre : Compacité par compensation et systèmes hyperboliques, C.R. Acad. Sc., Paris, t. 299, série I (1984), 555-558.

[9] D. Serre : cf ce volume

[10] L. Tartar : Compensated compactness and applications to P.D.E., in Research Notes in Math., Nonlinear Analysis and mechanics, Heriot-Watt Symposium, 4 (1979), R.J. Knops editor, Pitman.

A NONHOMOGENEOUS SYSTEM OF EQUATIONS OF NONISENTROPIC GAS DYNAMICS

WANG JINGHUA
The Institute of Systems Science
Academia Sinica,
China

1. **Introduction** : In this paper we consider the nonhomogeneous system of equations of nonisentropic gas dynamics in the Lagragian coordinates

$$
\begin{aligned}
&u_t + p_x = U(u,v) \\
&v_t - u_x = V(u,v) \\
(1.1)\qquad &(e+\frac{u^2}{2})_t + (pu)_x = E(u,v,s) \\
&p = \frac{a^2}{v}, \quad e = a^2 (\log p + \frac{S}{R})
\end{aligned}
$$

where $u, v > 0$, $p > 0$, e and s are velocity, specific volume, pressure, internal energy and entropy respectively and $a > 0$, $R > 0$ are constants. Without loss of generality we assume that $a = 1$. We are interested in the initial value problem for (1.1) with the initial value

$$(1.2)\qquad (u,v,s)(x,o) = (u_o,v_o,s_o)(x)$$

Here $u_o(x)$, $v_o(x)$, $s_o(x)$ are bounded and have bounded total variation and $0 < \delta \leq v_o(x)$. Without loss of generality we may assume that

$$(u_o,v_o)(x) = \begin{cases} (u_o, v_o)(-L) & x \leq -L \\ (u_o,v_o)(L) & x \geq L \end{cases}$$

where L is a constant.

System (1.1) is hyperbolic provided $u > 0$, the eigenvalues and Riemann invariants are given by the following

$$(1.3)\qquad \lambda = \frac{1}{v} \qquad \nu = 0, \qquad \mu = \frac{1}{v}$$

$$(1.4)\qquad w=u-q, \qquad z = u+q$$

where

(1.5) $q = \log p = -\log v$

The mapping from (u,q,s) to (w,z,s) and (u,v,s,), v > 0 to (w,z,s) are one to one and onto, we may use (u,q,s), (u,v,s) or (w,z,s) as basic unknown variables according to our convenience. Sometimes we let P_i denotes (u_i, q_i, s_i), (u_i, v_i, s_i) or (w_i, z_i, s_i).

If (u,v,s) is a smooth solution of (1.1), the system (1.1) can be rewritten as follows.

$$(1.6)\quad \begin{cases} w_t + \lambda w_x = W(w,z) \\ z_t + \mu z_x = Z(w,z) \\ s_t = S(w,z,s) \end{cases}$$

Here

$$(1.7)\quad \begin{cases} W(w,z) = U(u,v) + \dfrac{V(u,v)}{v} \\ Z\ (w,z) = U(u,v) - \dfrac{V(u,v)}{v} \\ S(w,z,s) = R(E(u,v,s) + v\ V(u,v) - u\ U(u,v)) \end{cases}$$

In this papers we assume that for the nonhomogeneous terms in the right side of system (1.1) the following Lipschitz condition holds

$$|W\ (w_1,z_1) - W\ (w_2,z_2)| \le K\ (|w_1 - w_2| + |z_1 - z_2|)$$
$$|Z\ (w_1,z_1) - Z\ (w_2,z_2)| \le K(|w_1 - w_2| + |z_1 - z_2|)$$
$$|S\ (w_1,z_1,s_1) - S(w_2,z_2,s_2)| \le K(|w_1 - w_2| + |z_1 - z_2| + |s_1 - s_2|)$$

The homogeneous system corresponding to system (1.1) is

$$(1\text{-}1)'\quad \begin{cases} u_t + p_x = 0 \\ v_t - u_x = 0 \\ (e + \frac{u^2}{2})_x + (pu)_x = 0 \\ p = \frac{1}{v},\ e = \log p + \frac{s}{R} \end{cases}$$

The existence of the global solution of the initial value problem (1.1)' and (1.2) was proved by Nishida [7].

The Riemann problem (1.1)' with the initial value

$$(1.9)\qquad (u,v,s)\,(x,0) = \begin{cases} (u_-,v_-,s_-), & x < 0 \\ (u_+,v_+,s_+), & x > 0 \end{cases}$$

, where $u_{\mp}$, $v_{\mp}$, > 0 $s_{\mp}$ are constants, has a piecewise continuous solution in the whole $t > 0$ half plane. The solution of the Riemann problem consits of at most four constant states (u_-,v_-,s_-), (u,v,s_1), (u,v,s_2) and (u_+,v_+,s_+) separated by 1-wave (1-shock wave or 1-rarefaction wave), 0-wave (contact discontinuity) and 2-wave (2-shock wave or 2-rarefaction wave) respectively.

For the shock curves, rarefaction wave curves and contact discontinuity curves the following fact are well known [7] :

The 1-shock curve starting from the state $P_- = (u_-,q_-,s_-)$ is

$$S_1(P_-)=\{P=(u,q,s),u-u_-=-2\mathrm{sh}\frac{q-q_-}{2},s-s_-=R(-(q-q_-)+\mathrm{sh}(q-q_-)),q>q_-\}=$$

$$(1.10)\qquad =\{P=(w,z,s),z_--z=f(w_--w),s-s_-=g(w_--w),w<w_-\},$$

The 1-rarefaction wave curves starting from the state $P_-=(u_-,q_-,s_-)$ is

$$R_1(P_-)=\{P=(u,q,s):u-u_-=-(q-q_-),s=s_-\ q<q_-\}=\{P=(w,z,s),z=z_-,$$

$$(1.11)\qquad s=s_-,w>w_-\}$$

The 2-shock curve starting from the state $P_+=(u_+,q_+,s_+)$ is

$$S_2(P_+)=\{P=(u,q,s),u-u_+=2\mathrm{sh}\frac{q-q_+}{2}\ ,s-s_+=R(-(q-q_+)+\mathrm{sh}(q-q_+)),q>q_+\}$$

$$(1.12)\qquad =\{P=(w,z,s),\ w-w_+=f(z-z_+),s-s_+=g(z-z_+);z>z_+\}$$

The 2-rarefaction wave curve starting from the state $P_+=(u_+,q_+,s_+)$ is

$$R_2(P_+)=\{P=(u,q,s),u-u_+=q-q_+,s=s_+,q<q_+\}=\{P=(w,z,s),w=w_+,s=s_+,$$

$$(1.13)\qquad z<z_+\}$$

The 0-wave (contact discontinuity) curve starting from the state $P_1=(u_1,q_1,s_1)$ is

$$(1.14)\qquad C(P_1) = \{P=(u,q,s),\ u = u_1,\ q = q_1\ s \gtrless s_1\}$$

Here the function f and g have the following properties :

$$(1.15)\qquad f(0) = g(0) = f'(0) = g'(0)=f''(0) = g''(0) = 0$$

$$(1.16)\qquad 0<f(y),\ 0<g(y),0<g'(y),0<f''(y),0<g''(y),\text{for } y > 0$$

$$0 < f'(y) < 1 \qquad \text{for } y > 0$$

(1.17) $$\lim_{y \to +\infty} f'(y) = 1$$

(1.18) $$0 \leq f'''(0),\ 0 < g'''(0)$$

The letters α, β, δ, ξ, η denote the 1-shock wave, 2-shock wave, contact discontinuity, 1-rarefaction wave and 2-rarefaction wave. The strength of the above elementary waves are defined in the two following ways :

The first way is that the strenght of shock wave and rarefaction wave is measured by the absolute value of the jump of the quantity q across the corresponding wave as follows

(1.19)
$$\begin{aligned} |\alpha| &= q-q_0 && \text{for } P \in S_1(P_0) \\ |\beta| &= q-q_0 && \text{for } P \in S_2(P_0) \\ |\xi| &= q_0-q && \text{for } P \in R_1(P_0) \\ |\eta| &= q_0-q && \text{for } P \in R_2(P_0) \end{aligned}$$

The increase of the entropy across a contact discontinuity, a 1-shock wave and a 2-shock wave is denoted by

(1.20)
$$\begin{aligned} |\delta| &= |s-s_0| && \text{for } P \in C(P_0) \\ \delta_\alpha &= s - s_0 && \text{for } P \in S_1(P_0) \\ \delta_\beta &= s - s_0 && \text{for } P \in S_2(P_0) \end{aligned}$$

The second one is that the strength of shock wave and rarefaction wave is measured by the absolute value of the jump of the quantity of the Riemann invariant w (resp. z) cross 1-wave (resp. 2-wave) as follows

(1.21)
$$\begin{aligned} \|\alpha\| &= w_0-w, && \text{for } P \in S_1(P_0) \\ \|\beta\| &= z - z_0, && \text{for } P \in S_2(P_0) \\ \|\xi\| &= w - w_0, && \text{for } P \in R_1(P_0) \\ \|\eta\| &= z_0-z, && \text{for } P \in R_2(P_0) \end{aligned}$$

Nonhomogeneous quasilinear hyperbolic systems of equations are reduced from conservation laws with physical or geometrical effects. This field has had atracted the attention of some authors such as Liu [3], [4], Ying and Wang [9], [10], Wang [8] , Luskin and Temple [5] Dafermos and Hsiao [1] and many others.

2. <u>The difference scheme</u> We use a difference scheme to construct the approximate solution $(u^l, v^l, s^l)(x,t)$ of the initial value problem

(1.1), (1.2). The scheme is a modification of Glimm's scheme [2] for dealing with the nonhomogeneous terms in (1.1). It has been used in [8], [9] and [10].

We define the time mesh length h and the space mesh length l such that the Courant-Friedrich-Lewy condition $\frac{l}{h} = \text{const.} \geq (v^l(x,t))^{-1}$ holds and choose a sequence $\{\alpha_i, i \geq 0\}$ equidistributed in (0,1). Assuming that the approximate solution (u^l,v^l,s^l) (x,t) has been defined for $0 \leq t < nh$. We define $(\hat{u}^l, \hat{v}^l, \hat{s}^l)$ in $Y_{m,n} = \{(x,t);(m-1)l \leq x < (m+1)l, nh \leq t < (n+1)h\}$, m+n even as the restriction on $Y_{m,n}$ of the solution of the Riemann problem (1.1)' with the initial value

$$(\hat{u}^l,\hat{v}^l,\hat{s}^l)(x,nh) = \begin{cases} (u^l,v^l,s^l)((m-2)l+2\alpha_n l, nh-0), & x > ml \\ (u^l,v^l,s^l)(ml + 2\alpha_n l, nh - 0), & x > ml. \end{cases}$$

Then by (1.4) and (1.5) we define

$$(\hat{w}^l,\hat{z}^l)(x,t) = (\hat{u}^l + \log \hat{v}^l, \hat{u}^l - \log \hat{v}^l)(x,t), \quad nh \leq t < (n+1)h$$

Finally we construct the approximate solution $(w^l,z^l,s^l)(x,t)$ for $nh \leq t < (n+1)$ h as follows

$$(w^l,z^l,s^l)(x,t)=(\hat{w}^l,\hat{z}^l,\hat{s}^l)(x,t)+(t-nh)(\bar{W}(\hat{w}^l,\hat{z}^l),S(\hat{w}^l,\hat{z}^l,\hat{s}^l))(x,t),$$

$$\text{for } nh \leq t \leq (n+1)\, h \tag{2.1}$$

By (1.4) and (1.5) again, we have

$$(u^l,v^l)(x,t) = \frac{w^l + z^l}{2}, \exp \frac{w^l - z^l}{2}))(x,t), \text{ for } nh \leq t < (n+1)h \tag{2.2}$$

To start out, at n = 0, we define (u^l,v^l,s^l) $(x,0 - 0) = (u_0, v_0,s_0)$ (x).

3. Estimations for approximate solution

In this section we derive certain estimates on the approximate solution $(u^l, v^l, s^l)(x,t)$. First we show that for any given $T > 0$, the total variation of $u^l(x,t)$ and $v^l(x,t)$ is bounded and $v^l(x,t)$ is bounded from a positive constant below.

Using the terminology introduced in section 1, for the four constant states (u_-, q_-, s_-), (u, q, s_1), (u, q, s_2) and (u_+, q_+, s_+) in the solution of the Riemann problem (1.1)', (1.9) we have $(u, q, s_1) \in S_1(u_-, q_-, s_-)$ or $R_1(u_-, q_-, s_-)$, $(u, q, s_2) \in C\,(u, q, s_1)$ and $(u, q, s_2) \in S_2(u_+, q_+, s_+)$ or $R_2(u_+, q_+, s_+)$. Then we define

$$
(3.1) \quad \begin{cases}
Q\,(P_-, P_+) = \max\,(0, w_- - w) + \max\,(0, z - z_+) \\[2ex]
Q_1(P_-, P_+) = |q_- - q| + |q - q_+| \\[2ex]
Q_2(P_-, P_+) = |S_1 - S_2| - (S_1 - S_-) - (S_2 - S_-) \\[2ex]
Q_3\,(P_-, P_+) = |S_1 - S_2| + (S_1 - S_-) + (S_2 - S_-)
\end{cases}
$$

i.e. $Q\,(P_-, P_+)$ is the sum of the strength, defined in (1.21) of shock waves in Riemann problem (1.1)', (1.9). $Q_1\,(P_-, P_+)$ is the sum of the strength, defined in (1.19), of 1-wave and 2-wave in the Riemann problem (1.1)', (1.9). Similar to Nishida [7] we have

$$
(3.2) \qquad Q(P_-, P_+) \le Q\,(P_-, P_m) + Q(P_m, P_+)
$$

where $P_m = (u_m, q_m, s_m)$ is a constant state.

From the definition of Q in (3.1) and the fact that the shock curve and rarefaction wave curve in terms of (u,q,s) or (w,z,s,) are independent of their initials respectively, we may use $Q\,((w_- - w_+), (z_- - z_+))$ to denote $Q\,(P_1, P_2)$.

Let $P_{i,h} = (w_{ih}, z_{ih}, s_{ih}) = (w_i, z_i, s_i) + h((W,Z)(w_i, z_i), S(w_i, z_i, s_i))$, $i=1,2$, and

$$\Delta w_h = w_{1,h} - w_{2,h}, \ldots, \Delta w = w_1 - w_2, \ldots,$$

then it is easy to deduce from (1.8) that

$$|\Delta w_h - \Delta w| \leq Kh\,(|\Delta w| + |\Delta z|)$$

$$|\Delta z_h - \Delta z| \leq Kh\,(|\Delta w| + |\Delta z|) \tag{3.3}$$

$$|\Delta s_h - \Delta s| \leq Kh\,(|\Delta w| + |\Delta z| + |\Delta s|)$$

Lemma 3.1 :

Suppose that two states P_1, P_2 are connected each other by either 1-wave or 2-wave then following inequality holds

$$Q\,(P_{1,h}, P_{2,h}) \leq Q\,(P_1, P_2) + 3Kh\,(|\Delta w| + |\Delta z|) \tag{3.4}$$

where $h < \dfrac{1}{2K}$ □

Proof :

For definiteness we may assume that P_2 is connected to P_1 by a 1-wave on the left. First suppose that the 1-wave is a 1-shock wave, which means that

$$\Delta s = -\,g(\Delta w),\; Q(\Delta w, \Delta z) = \Delta w, \Delta z = f(\Delta w) \tag{3.5}$$

Set $P_3 = (w_3, z_3, s_3)$ with

$$w_3 = w_1 - \Delta w_h\,, \qquad z_3 = z_1 - \Delta z_h\,, \qquad s_3 = s_1 - \Delta s_h \tag{3.6}$$

then $Q(P_1, P_3) = Q(w_h, z_h) = Q(P_{1,h}, P_{2,h})$ follows. Since $h < \frac{1}{2K}$ $|\Delta z| = |f'(\Theta\Delta w)| < |\Delta w|$, $|\Delta w_h - \Delta w| \leq Kh(|\Delta w| + |\Delta z|) < 2Kh|\Delta w| < \Delta w$, therefore

$$0 < \text{sign}\,\Delta w_h \,.\, \text{sign}\,\Delta w,\; |\Delta w_h| < 2|\Delta w| \tag{3.7}$$

We only consider the case involving a 1-shock wave in the solution of the Riemann problem (1.1)' with the initial value

$$P(x,0) = \begin{cases} P_1 & x < 0 \\ P_3 & x > 0 \end{cases} \tag{3.8}$$

The case involving a 1-rarefaction wave in the above solution can be handled similary. Set $P_6=(w_6,z_6,s_6)$ with $w_6=w_3$, $z_6=z_1-\Delta z'$, $\Delta z'=f(\Delta w_h)$, $s_6= s_1+g(\Delta w_h)$ (3.9)

Therefore $Q(P_1,P_6) = w_1-w_6=w_h>0$, here (3.7) is used, there are two cases :

Case 1 : $z_3<z_6$, i.e. the solution of Riemann problem (1.1)', (3.8) consists of four constant states P_1,P_4,P_5 and P_3 such that $P_4 \in S_1(P_1)$. $P_5 \in C(P_4)$, $P_5 \in S_2(P_4)$. We have

$$Q(P_{1,h},P_{2,h})=Q(P_1,P_3)=(z_5-z_3)+(w_1-w_4)=(z_5-z_6-\Delta z'+\Delta z_h)+(w_1-w_6)+(w_6-w_5)$$
$$=(w_1-w_6)+(\Delta z_h-\Delta z')+((z_5-z_6)-(w_5-w_6))$$
$$\leq (w_1-w_6)+|\Delta z_h-\Delta z|(w_1-w_6)+|\Delta z_h-\Delta z|+|\Delta z-\Delta z'|$$
$$\leq \Delta w_h+|\Delta z_h-\Delta z|+|f(w_h)-f(\Delta(w)| \leq Q(P_1,P_2)+3Kh(|\Delta w|+|\Delta z|) \quad (3.10)$$

There (3.9), (3.6), (3.5), (3.3) and $0<f'<1$, which implies $(z_5-z_6)-(w_5-w_6)<0$, are used.

Case 2 : $z_3 \geq z_6$, i.e. the solution of the Riemann problem (1.1) (3.8) consists four constant states P_1,P_4,P_5 and P_3 such that $P_4 \in S_1(P_1)$, $P_5 \in C(P_4)$ and $P_5 \in R_2(P_4)$. Then we have

$$Q(P_{1,h},P_{2,h}) = Q(P_1,P_3)=w_1-w_3=w_1-w_6=\Delta w+|\Delta w-\Delta w_h|\leq$$
$$\leq Q(P_1,P_2)+K_h(|\Delta w|+|\Delta z|) \quad (3.11)$$

Then suppose that the 1-wave, which connects the states P_1 and P_2, is a 1-rarefaction wave, then we have

$$Q(P_{1,h},P_{2,h})\leq Q(P_1,P_2)+|\Delta z_h-\Delta z|\leq Q(P_1,P_2)+K_h(|\Delta w|+|\Delta z|) \quad (3.12)$$

The inequalities (3.10), (3.11) and (3.12) completes the proof.

If follows from (1.8) that for any given $T > 0$, the initial value problem of the system of O.D.E

$$\frac{du}{dt} = U(u,v)$$
$$\frac{dv}{dt} = V(u,v) \quad (3.13)$$
$$(u,v)\Big|_{t=0}=(u_0(\mp L),v_0(\mp L))$$

has unique solution $(u_{\mp}(t),v_{\mp}(t))$ respectively such that $|u_{\mp}(t)|+|\log v_{\mp}(t)| \leq N$ for $t \in [0, T]$ (3.14)

here N is a constant depending only on $u_0(\mp L),v_0(\mp L)$, T and K.

Lemma 3.2 : For any given $T > 0$, $\eta < \frac{1}{2K}$ then the total variation of the first two components of the approximate solution $w^l(x,t)$ and $z^l(x,t)$ (resp. $u^l(x,t)$ and $v^l(x,t)$) are bounded uniformly for $l, \{\alpha_i, i \geq 0\}$ and $t \in [0,T]$, their upper and lower bounds and the positive lower bound of $v^l(x,t)$ are bounded uniformly for l, $\{\alpha_i, i \geq 0\}$ and $t \in [0,T] \times (-\infty,\infty)$. □

Proof : We set

$$F_1(nh) = \sum_{n+m=\text{even}} Q(P^l((m+2(\alpha_n-1))l, nh-0), P^l((m+2\alpha_n)l, nh-0) \tag{3.15}$$

for $n \geq 0$

$$F_1'(nh) = \sum_i Q(P^l(a_{i,n}l, nh-0), P^l(a_{i+1,n}l, nh-0) \tag{3.16}$$

for $n \geq 0$

where $\{a_{i,n} : i\text{-integer } a_{i,n} \leq a_{i+1,n}\} = \{m+2\alpha_n : m+n \text{ even}\} \cup I$

I is the set of all integers

and by (3.2) we have

$$F_1(nh) \leq F_1'(nh), \qquad \text{for } n \geq 0 \tag{3.17}$$

The lemma 3.1. yields

$$F_1'(nh) \leq F_1(n-1)h) + 3Kh\ TV(w^l, z^l)(.,(n-1)h+0),\ n \geq 1 \tag{3.18}$$

Here $TV(w^l, z^l)(.,(n-1)h+0)$ is the sum of the total variations of $w^l(x,(n-1)h+0)$ and $z^l(x,(n-1)h+0)$ over the x-interval $(-\infty, \infty)$. Thus

$$F_1'(nh) \leq F_1'((n-1)h)(1+12Kh)+12KhN, \qquad n \geq 0 \tag{3.19}$$

Here the inequality $TV(w^l, z^l)(.,(n-1)h+0) \leq 4F_1(n-1)h)+4N$ is used. By simple calculation we obtain

$$F_1'(0) \leq TV(w_0, z_0)(.) \tag{3.20}$$

$$F_1'(nh) \leq (F_1'(0) + 12KTN)e^{12KT}, \text{ for } nh \leq T \tag{3.21}$$

hence

$$TV(w^l, z^l)(.,t) \leq K \quad \text{for } t \in [0,T] \tag{3.22}$$

$$|w^l(x,t)| \leq K+N, |z^l(x,t)| \leq K+N \text{ for } x \in (-\infty,\infty) t \in [0,T] \tag{3.23}$$

$$|u^l(x,t)| \leq K+N,\ e^{-(K+N)} \leq v^l(x,t) \leq e^{(K+N)} \text{for } x \in (-\infty,\infty),\ t \in [0,T] \tag{3.24}$$

where $K = 4(TV(w_0, z_0)(.)+12KTN)e^{12KT}+4N$.

Since $0<f'(y)<1$, $0<f''(y)$, $0 < g'(y)$, $0 < g''(y)$, for $y>0$, $\lim_{y\to\infty} f'(y)=1$

and $TV(w^1,z^1)(.,t) \le K$, for $K \in [0,T]$, which implies that the strength, defined in (1.21) of any shock wave α or β in $(\hat{u}^1,\hat{v}^1,\hat{s}^1)(x,t)$ is not greater than K, we obtain that

(3.25) $$0 \le f'(2\|\alpha\|) \le f'(2K) < 1, \qquad 0 \le f'(2\|\beta\|) \le f'(2K) < 1$$

therefore

(3.26)
$$-D \le \frac{du}{dq} \le -1, \text{ for } (u,q,s) \text{ belongs to 1-shock curve}$$
$$1 \le \frac{du}{dq} \le D, \text{ for } (u,q,s) \text{ belong to 2-shock curve}$$

where

$$D = \frac{f'(2K)+1}{f'(2K)-1} \qquad \text{and}$$

$$0 \le \frac{ds}{dq} \le G, \text{ for } (u,q,s) \in \text{ 1-schock curve or 2-shock curve}$$

where $G=g'(2K)$. As an immediate consequence of (3.26) we have

$$|q_1-q_2| \le \frac{1}{\sqrt{2}}, |w_1-w_2| \le \frac{D+1}{2} |q_1-q_2| \text{ for any two states } P_i, i=1,2 \text{ in a 1-shock curve}$$

(3.28) $$|q_1-q_2| \le \frac{1}{\sqrt{2}}, |z_1-z_2| \le \frac{D+1}{2} |q_1-q_2| \text{ for any two states } P_i, i=1,2 \text{ in a 2-shock curve}$$

Lemma 3.3 : Suppose that two states P_1,P_2 are connected each other by either 1-wave or 2-wave, whose strength, defined in (1.21), is not greater than K, then the following inequality holds

(3.29) $$Q_1(P_{1,h},P_{2,h}) \le (1+3(D+1)h)\, Q_1(P_1,P_2)$$

where $h < \frac{1}{2K}$. □

Proof : For definiteness we may assume that P_2 is connected to P_1 by 1-wave on the left. First we suppose that the 1-wave is a 1-shock wave. The states $P_i=(u_i,q_i,s_i)$ i=3,4,5,6 are defined in the same way as in Lemma 3.1. We also consider the case there is a 1-shock wave in the solution of the Riemann problem (1.1), (3.8). The other cases can be handled similarly. It follows from the definition of Q_1 in (3.1) that

(3.30) $$|Q_1(P_1,P_6)-Q_1(P_1,P_2)| = \frac{1}{\sqrt{2}}|\Delta w_h - f(\Delta w_h) - \Delta w + f(\Delta w)| = \frac{1}{\sqrt{2}} |\Delta w_h - \Delta w|$$

and

$$Q_1(P_1,P_3)=Q(P_1,P_6)+\frac{1}{\sqrt{2}}\ |\Delta z_h - \Delta z'|,\ \text{for}\ |\Delta z_h|\le|\Delta z'|$$

(3.31)

$$Q(P_1,P_3<Q_1(P_1,P_6)+\frac{1}{\sqrt{2}}\ |\Delta z_h-\Delta z'|,\ \text{for}\ |\Delta z_h|>|\Delta z'|$$

Since $0<f'<1$, we know that

$$|\Delta z_h-\Delta z'|\le|\Delta z_h-\Delta z|+|\Delta z-\Delta z'|=|\Delta z_h-\Delta z|+|f(\Delta w)-f(\Delta w_h)|\le$$

(3.32) $$\le |\Delta z_h-\Delta z|+|\Delta w_h-\Delta w|$$

The inequalities (3.20), (3.31) and (3.32) yields

$$Q_1(P_{1,h},P_{2,h})=Q_1(P_1,P_3)\le Q_1(P_1,P_2)+\frac{1}{\sqrt{2}}|\Delta z_h-\Delta z|+\sqrt{2}|\Delta w_h-\Delta w|\le$$

(3.33) $$\le Q_1(P_1,P_2)+3\sqrt{2}kh|\Delta w|\le(1+3(D+1)Kh)Q_1(P_1,P_2)$$

Here (3.3) and (3.28) are used.

Then suppose that the 1-wave, which connected the states P_1 and P_2 is 1-rarefaction wave, similarly we have

(3.34) $$Q_1(P_{1,h},P_{2,h})\le Q_1(P_1,P_2)(1+(D+1)Kh)$$

The inequalities (3.33) and (3.34) imply Lemma 3.3.

Lemma 3.4 : Suppose that two states P_1,P_2 are connected each other by either 1-wave or 2-wave, whose strength defined in (1.21), is not greater than K, then the following inequality holds

(3.35) $$Q_2(P_{1h},P_{2h})\le Q_2(P_1,P_2)+(G+\sqrt{2}(D+1)+6G(D+1))Kh\,Q_1(P_1,P_2)$$

wnere $h<\frac{1}{2K}$. □

Proof : For definiteness we may assume that P_2 is connected to P_1 by 1-wave on the left. First we consider the case that the 1- wave is a shock wave. The state P_i, $i=3,4,5,6$ are defined in same way as in Lemma 3.1. We also consider the case there is a 1-shock wave in the solution of the Riemann problem (1.1)', (3.8). The other cases can be treated similarly. We introduce another auxiliary state $P_7=(w_7,z_7,s_7)$ where $w_7=w_6$ $z_7=z_6$ and $s_7=s_3$ and we have

(2.36) $$|(s_1-s_7)-\Delta s|=|(s_1-s_3)-\Delta s|=|\Delta s_h-\Delta s|\le Kh(|\Delta w|+|\Delta s|)$$

Thus we obtain

$$Q_2(P_1,P_7)-Q_2(P_1,P_2) = [|s_7-s_6|-(s_6-s_1)]-[-(s_2-s_1)]\le$$
$$\le |s_7-s_2|+|s_6-s_2|-(s_6-s_2)\le$$

(3.37) $\leq kh(|\Delta w|+|\Delta z|+|\Delta s|)+2|s_6-s_2|$

here the definition of Q_2 and (2.36) are used.

It follows from (3.27),(3.28),(3.6),(3.9) and (3.3) that

(3.38) $|s_6-s_2|\leq G|q_2-q_6|\leq \frac{G}{\sqrt{2}}\ |w_6-w_2|=\frac{G}{\sqrt{2}}\ |\Delta w_h-\Delta w|\leq \frac{G}{\sqrt{2}}Kh(|\Delta w|+|\Delta z|)\leq\sqrt{2}GKh\,|\Delta w|$

there are two cases ;

case 1, $z_3<z_6$. It is easy to known that

$Q_z(P_{1,h},P_{2h})-Q_z(P_1,P_7)=Q_z(P_1,P_3)-Q_z(P_1,P_7)=[|s_4-s_5|-(s_4-s_1)-(s_5-s_3)]$

(3.39) $-[|s_7-s_6|-s_6-s_1)]\leq z|s_6-s_4|.$

(3.40) $|s_6-s_4|\leq G|q_6-q_4|$

After introducing the state $P_8=(w_8,z_8,s_8)$ such that $P_8\in S_1(P_4)$ $\cap\ \{(w,z,s)_: w=w_3\}$ we obtain

(3.41) $|q_6-q_4|\leq|q_8-q_4|\leq(q_5-q_3)\leq\frac{1}{\sqrt{2}}\ |\Delta z_h-\Delta z'|\leq 2\sqrt{2}kh|\Delta w|$

The inequalities (3.37),(3.38),(3.39),(3.40) and (3.41) imply

$Q(P_{1,h},P_{2,h})\leq Q_2(P_1,P_2)+Kh(6\sqrt{2}G\Delta w+2\Delta w+GQ_1(P_1,P_2))\leq$

(3.42) $\leq Q_2(P_1,P_2)+(G+\sqrt{2}(D+1)+6G(D+1))KhQ(P_1,P_2)$

here (3.27) and (3.28) are used.

<u>Case 2</u> : $z_3\geq z_6$. Then $Q_2(P_{1,h},P_{2,h})=Q_2(P_1,P_3)=Q_2(P_1,P_7)$ follows, therefore the inequality (3.42) also holds.

Similarly for two states P_1,P_2 are connected by 1-rarefaction wave we obtain

(3.43) $Q_2(P_{1,h},P_{2,h})\leq Q_2(P_1,P_2)+\sqrt{2}\ Kh\ Q_1(P_1,P_2)$

The inequalities (3.42) and (3.43) imply the inequality (3.35). The proof is complete.

Similarly we can prove

<u>Lemma 3.5</u> : Suppose two states P_1 and P_2 are connected each other by a contact discontinuity, the following inequality holds

□

(3.44) $Q_2(P_{1,h},P_{2,h})\leq(1+Kh)Q_2(P_1,P_2)$

Now set

(3.45) $\bar{Q}(P_1,P_2)=Q_1(P_1,P_2)+MQ_2(P_1,P_2)$

where

(3.46) $M=\min\left(1,\ \frac{1}{2G}\right)$

We have

Lemma 3.6 : For any two states $P_i=(u^1,v^1,s^1)(x_i,nh+0)$ $i=1.2, x_1<x_2, nh < T$, the following inequality holds

(3.47) $$Q(P_{1,h},P_{2,h}) \leq Q(P_1,P_2)(1+|16Kh| (D+1))$$ □

Proof : Similar to Nishida [7] we have that

(3.48) $$Q(P_-,P_+)\leq Q(P_-,P_0)+ Q(P_0,P_+)$$

here $P_{\pm}=(u^1,q^1,s^1)(x_{\pm},nh+0), P_0=(u^1,q^1,s^1)(x_0,nh+0), x_-<x_0<x_+$ $nh<T$. Therefore without loss of generality we may assume that P_1,P_2 are connected each other by 1-wave, two wave or contact discontinuity. First we study the case that P_1,P_2 are connected each other by 1-wave or two wave. It follows from Lemma 3.3 and Lemma 3.4 that

$$Q_2(P_{1,h},P_{2,h})-Q_2(P_1,P_2)\leq(3(D+1)+M(G+\sqrt{2}(D+1)+6G(D+1))Kh.$$

$$Q_1(P_1,P_2)\leq(6(D+1)+\frac{1}{2}+\sqrt{2}(D+1))KhQ_1(P_1,P_2)\leq\delta(D+1)KhQ_1(P_1,P_2)\leq$$

(3.49) $$\leq 6(D+1)KhQ(P_1,P_2)$$

Here the inequality $Q_1(P_1,P_2)\leq 2Q(P_1,P_2)$ is used, which is deduced from following inequalities

$Q_2(P_1,P_2)=0$, for P_1,P_2 are connected each other by rarefaction wave.

$0<-MQ_2(P_1,P_2) \leq -\frac{1}{2G} Q_2(P_1,P_2) \leq \frac{1}{2} Q_1(P_1,P_2)$, for P_1,P_2 are connected each other by shock wave.

Then for the case P_1,P_2 are connected each other by a contact discontinuity, by Lemma 2.5 and $Q_1(P_1,P_2)=0$ we have

(3.50) $$Q(P_{1,h},P_{2,h}) \leq (1+Kh)\, Q(P_1,P_2)$$

The Lemma is proved from (3.49) and (3.50)

Lemma 3.7 : For any given $T >0$, $h <\frac{1}{2K}$ then the total variation of $s^1(x,t)$ one component of the approximate solution, is bounded uniformly for l, $\{\alpha_i, i\geq 0\}$ and $t \in [0,T]$ □

Proof : We set

$$F_2(nh) = \sum_{n+m=\text{even}} Q(P^1((m+2(\alpha_n-1))l,nh-0),P^1((m+2\alpha_n)l,nh-0)$$
for $n \geq 0$

$$F'2(nh) = \sum_i Q(P^1(a_{i,n}l, nh-0), P^1(a_{i+1,n}l,nh-0))$$

(3.51) for $n \geq 0$

where $a_{i,n}$ is defined as same as in (3.16)

the by (3.48) we have

(3.52) $F_2(nh) \le F_2'(nh)$, for $n \ge 0$

The Lemma 2.6 implies

(3.53) $F_2'(nh) \le F_2((n-1)h)(1+16Kh)$, $n \ge 1$

Therefore

$$F_2'(nh) \le F_2'((n-1)h)(1+16Kh) \le F_2'(0)(1+16Kh)^n) =$$

(3.54) $$=F'(0)\ (1+16Kh)^{\frac{T}{h}} \le F'(0)e^{16KT},\ n \ge 1$$

It is easy to known that

(3.55) $$F_2'(0) \le \frac{3}{2}\ (TVu_0(.)+\frac{1}{\delta}TVv_0(.))+MTVs_0(.)$$

But $TV\ s^l(.,nh-0) \le \sum_i Q_3(P^l(a_{i,n}l,nh-0),\ P^l(a_{i+1,n}l\ nh-0))$

$$\le \sum_i (Q_2+2GQ_1)(P^l(a_{i,n}l,nh-0),P^l(a_{i+1,n}l,nh-0)) \le$$

(3.56) $$\le \frac{1}{M}\ F_2'(nh)\ \text{for}\ 0 \le nh \le T$$

The Lemma is deduced from (3.54), (3.55) and (3.56).

Similar to Ying and Wang [9] by Lemma 3.2 and Lemma 3.7 we obtain that there is a subsequce $(u^{li},v^{li},s^{li})(x,t)$ such that

$$(u^{li},v^{li},s^{li})(x,t) \xrightarrow{a.e} (u,v,s)(x,t), l_i \to 0,\ \text{for}\ 0 \le t \le T$$

(3.57) $$-\infty<x<+\infty$$

4. Consistency of the difference scheme : In this section we will prove that the limit function obtained in (3.57) for almost random sequece $\{\alpha_i, i\ge 0\}$ equidistributed in (0,1) is the weak solution of the initial value problem (1.1) and (1.2), i.e. the following equality holds

$$\int_0^T\int_{-\infty}^{+\infty} (u\Phi_t+\frac{1}{v}\ \Phi_x+U(u,v)\Phi+v\Psi_t-u\Psi_x+V(u,v)\Psi+(e+\frac{u^2}{2})\chi_t+(pu)\chi_x\ +$$

$$+E(u,v,s)\chi)dxdt+\int_{-\infty}^{+\infty}(u_0(x)\Phi(x,0)+v_0(x)\Psi(x,0)+\frac{s_0(x)}{R}\ -\log v_0(x)+\frac{u_0^2(x)}{2})\chi(x,0))dx=0 \qquad (4.1)$$

where $(\Phi(x,t),\Psi(x,t),\chi(x,t)$ is any triplet of test functions for $-\infty<x<\infty$, $0\le t<T$.

Theorem 4.1. : If $u_0(x),v_0(x)$ and $s_0(x)$ are functions of bounded variation and $0<\delta\le v_0(x)$ for $-\infty<x<\infty$, for the nonhomogeneous terms in the system (1.1) the Lipschitz condition (1.8) holds, then for any $T>0$, the weak solution of the initial value problem (1.1), (1.2) exists in the strip $\{(x,t),-\infty<x<+\infty,\ 0 \le t \le T\}$ □

Proof : From (3.24) we may take time mesh length $h_i = \frac{1}{2^i}$ and space mesh length l_i such that $\frac{l_i}{h_i} = \exp(K+N)$, if i is large enough such that $h_i < \frac{1}{2K}$ then the approximate solution $(u^i, v^i, s^i)(x,t)$ can be constructed for the approximate solution obtained in section 3 hold.

Now we are going to show that for almost sequence $\{\alpha_i, i \geq 0\}$ equidistributed in (0,1) the limit function $(u,v,s)(x,t)$ obtained in (3.57) is the weak solution of the initial value problem (1.1), (1.2).

Noting that in each rectangle $Y_{m,n} = \{(x,t):(m-1)l \leq x<(m+1)l\, nh \leq t<(n+1)h$, m+n even$\}$ $(\hat{u}^l, \hat{v}^l, \hat{s}^l)(x,t)$ is a weak solution of system (1.1)' and approximate solution (u^l, v^l, s^l) is the solution of the system (1.1) in the domains of $Y_{m,n}$ in which $(\hat{u}^l, \hat{v}^l, \hat{s}^l)$ is a constant state. Then similar to Ying and Wang [10] or Wang [8] we obtain that

$$\Big| \iint_{Y_{m,n}} (u^l \Phi_t + \frac{1}{v^l} \Phi_x + U(u^l, v^l) \Phi + v^l \Psi_t - u^l \Psi_x + V(u^l, v^l)\Psi + (e^l + \frac{(u)^2}{2})\chi_t + (p^l u^l)\chi x +$$

$$+E(u^l, v^l, s^l)\chi)dxdt + \int_{(m-1)l}^{(m+1)l} (u^l(x, nh+0)\Phi(x,nh) - u^l(x,(n+1)h-0)\Phi(x,(n+1)h) +$$

$$+v^l(x,nh+0)\Psi(x,nh) - v^l(x,(n+1)h-0\Psi(x,(n+1)h-0) + (e^l + \frac{(u^l)^2}{2})(x,nh+0)\chi(x,nh) -$$

$$-(e^l + \frac{(u^l)^2}{2})(c,(n+1)h-0)\chi(x,(n+1)h)dx + \int_{nh}^{(n+1)h} \int \frac{\Phi}{v^l}((m-1)(l,t) + (u^l\Psi)((m-1)l,t) -$$

$$-(u^l\Psi)(m+1)l,t) + (p^l u^l \chi)((m-1)l,t) - (p^l u^l \chi)(m+1)l,t))dt \Big|$$

$$\leq C_1 l^3 + C_2 l^2 TV(a^l(.,nh+0) + v^l(.,nh+0) + s^l(.,nh+0)) \tag{4.2}$$

$$(m-1)l<x<(m+1)l$$

where C_1 and C_2 depend only on the upper and lower bounds of $u^l(x,t)$, $v^l(x,t)$, $s^l(x,t)$ and the constant K and the test functions Φ, Ψ and χ; $(m-1)l<x<(m+1)l$, TV $(u^l(.,nh+0)+v^l(.,nh+0)+v^l(.,nh+0)+s^l(.,nh+0))$ is the sum of the total variations of $u^l(x,nh+0)$, $v^l(x,nh+0)$ and $s^l(x,nh+0)$ over the interval $((m-1)l,(m+1)l)$. Here the integration by parts, Rankine-Hugoniot condition for shock wave in $(\hat{u}^l, \hat{v}^l, \hat{s}^l)$ and the relations between (u^l, v^l, s^l) and $(\hat{u}^l, \hat{v}^l, \hat{s}^l)$ defined and the relations between (u^l, v^l, s^l) and $(\hat{u}^l, \hat{v}^l, \hat{s}^l)$ defined by (2.1) and (2.2) are used.

By the fact that Φ, Ψ and χ has compact support in the strip $0 \leq t \leq T$, therefore the number of rectangles $Y_{m,n}$,m+n even in which at least one of Φ, Ψ, χ does not vanish identically is order of l^{-2} ; summing over all rectangles $\gamma_{m,n}$, m+n even we obtain

$$\Big| \int_0^T\int_{-\infty}^{+\infty} (u^l\Phi_t + \frac{1}{v^l}\Phi_x + U(u^l,v^l)\Phi + v^l\Psi_t - u^l\Psi_x + V(u^l,v^l)\Psi + (e^l + \frac{(u^l)^2}{2})\chi_t +$$

$$+ p^l u^l \chi_x + E(u^l,v^l,s^l,\chi)dxdt$$

$$+ \int_{-\infty}^{+\infty} (u_0^l(x)\Phi(x,0) + v_0^l(x)\Psi(x,0) + (\frac{s_0(x)}{R} - \log v_0(x) + \frac{u_0^2(x)}{2})\chi(x,0))dx \Big|$$

$$\leq \Big| \sum_{n=1}^{\frac{T}{h}} \int_{-\infty}^{+\infty} \Big[((u^l(x,nh-0) - u^l(x,nh+0))\Phi(x,nh)$$

$$+ (v^l(x,nh-0) - v^l(x,nh+0))\Psi(x,nh)$$

$$+ ((e^l(x,nh-0) + \frac{(u^l(x,nh-0))^2}{2}) - (e^l(x,nh+0) + \frac{(u^l(x,nh+0))^2}{2}))\chi(x,nh1\Big]dx\Big|$$

$$+ \Big| \int_{-\infty}^{+\infty} \Phi(x,0)(u_o(x) - u^l(x,0)) + \Psi(x,0)(v_0(x) - v^l(x,0))$$

$$+ \chi\ (x,0)((e_0(x) + \frac{u_0^2(x)}{2}) - (e^2(x,0)$$

$$+ \frac{(u^2(x,0))^2}{2}\ dx\ \Big| + 0(1)(C_1 + C_2 \sup_{t\in[0,T]} TV(u^l(x,t) + v^l(x,t) + s^l(x,t)) \qquad (4.3)$$

The second and third terms on the right-hand side of (3.3) tend to zero as $l_i \downarrow 0$. By Glimm's celebrated argument there is subsequence, denoted again by $\{l_i\}$ such that the first term in the right-hand side of (3.3) tends to zero for almost every selection of the sequence $\{\alpha_i\ i>0\}$, equildistributed in (0,1). This completes the proof.

REFERENCES

1. C.M. Dafermos & L. Hsiao, Hyperbolic systems of balance laws with inhomogeneity and dissipation, Indian U. Maht. J. 31 (1982), 471-491.

2. J. Glimm, Solutions in the large for nonlinear hyperbolic systems of equations, Comm. Pure. Appl. Math. 18 (1965), 697-715.

3. T.P. Liu, Quasilinear hyperbolic systems, Comm. Math. Phys. 68 (1979) 140-172.

4. T.P. Liu, Systems of quasilinear hyperbolic partial differential equations, in "Trends of Applications of Pure Mathematics to Machanics" Vol. III, (R.J. Knops,Ed), Pitmann, London, 1981.

5. M. Luskin & B. Temple, The existence of a global weak solution to the nonlinear waterhammer problem, Comm. Pure. Appl. Math. 35 (1982), 697-735.

6. T. Nishida, Global solutions for an initial boundary value problem of a quasilinear hyperbolic system, Proc. Japan. Acad. 44 (1968), 642-646.

7. T. Nishida, Nonlinear hyperbolic equations and related topics in fluid dynamics, Publications Mathématiques d'Orsay, 78.02.

8. J.H. Wang, An inhomogeneous quasilinear hyperbolic system. Acta Mathematica Scientia, 1 (1981), 403-421.

9. L.A. Ying & J.A. Wang, Global solutions of the Cauchy problem for a nonhomogeneous quasilinear hyperbolic system. Comm. Pure. Appl. Math. 33 (1980), 579-597.

10. L.A. Ying & J.A. Wang, Solutions in the large for nonhomogeneous quasilinear hyperbolic systems of equations J. Math. Anal. Appl. 78 (1980), 440-454.

FAR FIELD BOUNDARY CONDITIONS FOR STEADY STATE SOLUTIONS TO HYPERBOLIC SYSTEMS

Bertil Gustafsson
Lars Ferm
Uppsala University,
Uppsala and National Defense Research Center,
Stockholm

1. Introduction

In many applications, artificial boundaries are introduced in order to obtain a computational domain which is finite and such that the computation can be made efficiently. Usually there are no data available at these boundaries, or possibly data can be obtained but with poor accuracy. In such a case some kind of boundary conditions must be derived at the artificial boundary. A typical example is flow around an airplane, where the computational domain must be made rather small, but where only free stream data are available. Another example is seismological problems where waves are propagating through the earth, but where the computation must be limited to a small part near the surface. This latter problem gave rise to the first systematic attempt to derive accurate boundary conditions such that the artificial boundaries do not disturb the true solution. The result was the so called absorbing (or radiating) boundary conditions, which were developed by Engquist and Majda [2], [3]. Various versions of these conditions are in common use today, and we shall give a few comments about these in Section 2.

We shall consider hyperbolic problems in two space dimensions

$$u_t + Au_x + Bu_y = 0 \tag{1.1}$$

In particular we shall discuss the steady state problem

$$Au_x + Bu_y = 0 \tag{1.2}$$

which seems to be of primary interest in most applications. In [5] a new approach was introduced and applied to the isentropic Euler equations. The main idea is that the original system (1.2), which is usually nonlinear, is replaced outside the computational domain by the linearized system with constant coefficients. The exact form of the solution to this system can be derived, and this form is then used to derive conditions at the boundary. Hence, the only approximation introduced is the assumption that the system in the far field has constant coefficients.

We shall refer to these conditions as "fundamental boundary conditions", since they are based on the form of the fundamental solution to (the linearized form of) the system (1.2).

2. Absorbing (radiating) boundary conditions

Assume that we want to solve the system (1.1) in the domain $0\leq x\leq L$, $-\infty<y<\infty, 0\leq t$, and that boundary data are missing at x=L. The idea with absorbing boundary conditions is that waves should be allowed to pass through at x=L without any reflections. The form of the solution is assumed to be

$$u(x,y,t) = \hat{u}(x)e^{i(\omega y+\xi t)}$$

giving

$$A\hat{u}_x + i(\omega B + \xi I)\hat{u} = 0$$

If A is non-singular the equation is written in the form

$$\hat{u}_x = i(\xi\tilde{A}+\omega\tilde{B})\hat{u}$$

where

$$\tilde{A} = -A^{-1}, \quad \tilde{B} = -A^{-1}B$$

For convenience we assume that $\tilde{A}$ and $\tilde{B}$ are constant. If the matrix T is such that

$$T(\tilde{A} + \frac{\omega}{\xi}\tilde{B})T^{-1} = \begin{bmatrix} D_1 & 0 \\ D_{12} & D_2 \end{bmatrix}, \quad D_1 > 0, \; D_2 < 0$$

then with

$$\hat{v} = \begin{bmatrix} \hat{v}_1 \\ \hat{v}_2 \end{bmatrix} = T\hat{u}$$

the vector function

$$v_1(x,y,t) = \hat{v}_1(x)e^{i(\omega y+\xi t)} = e^{i\xi(x-L)D_1}\hat{v}_1(L)e^{i(\omega y+\xi t)}$$

represents the waves which are entering the domain at x=L. Hence the condition

$$(2.1) \qquad \hat{v}_1(L) = [T\hat{u}(L)]_1 = 0$$

is the "perfectly absorbing" boundary condition. This condition is imposed in Fourier space, and in order to obtain a condition in physical space, some approximations are made. If the matrix T has a series expansion

$$T = T_0 + \frac{\omega}{\xi}T_1 + \frac{\omega^2}{\xi^2}T_2 + \ldots, \quad |\omega| < |\xi|$$

the first approximation is

$$[T_0 \hat{u}(L)]_1 = 0 \tag{2.2}$$

This condition is exact if $\omega = 0$, i.e. for one-dimensional problems. In this case (2.2) is equivalent to prescribing zero values for the ingoing characteristic variables.

More accurate conditions are obtained by using higher order approximations of T. A second order approximation can be obtained from

$$[(T_0 + \frac{\omega}{\xi} T_1) \hat{u} (L)]_1 = 0 \tag{2.3}$$

which is rewritten as a polynomial in $i\omega, i\xi$ and then transformed to physical space by the substitutions

$$i\omega \to \partial/\partial y, \qquad i\xi \to \partial/\partial t.$$

Not all kinds of approximations of T lead to well posed problems, but a hierarchy of well posed ones were derived in [3].

It should be pointed out that the concept "more accurate" here is interpreted as "more absorbing", and the reflections are measured in terms of $|\omega/\xi|$. The conditions do not take into account waves that are entering the domain from the outside. However, these waves can of course be introduced by specifying non-zero data. For example, in (2.2), the values of the ingoing characteristic variables can be specified if they are available.

For large values of $|\omega/\xi|$ we must use other types of conditions. In particular, this is true for steady state problems corresponding to $\xi = 0$, which we shall consider in the next section.

Bayliss and Turkel [1] used a different approach, but their basic principle is the same : the boundary conditions are derived from the form of the solution which permits only outgoing waves.

3. Steady state problems and the fundamental boundary conditions

For vector functions U with m conponents, consider the non-linear system

$$A(U)U_x + B(U)U_y = 0 \tag{3.1}$$

in some unbounded domain. We wish to compute the solution in a bounded domain Ω, and in order to achieve this, we need conditions at the boundary $\partial\Omega$. Even if U is known at infinity, it is assumed that Ω is not large enough to give accurate solutions with these data given on $\partial\Omega$. The basic idea is now to approximate (3.1) outside Ω by the system

$$A(\bar{U})U_x + B(\bar{U})U_y = 0 \tag{3.2}$$

where $\bar{U}$ is independent of x,y. Under certain conditions on Ω, it is possible to derive the fundamental solution of (3.2), and from the form of this solution boundary conditions at $\partial\Omega$ can be derived.

As an example, consider the periodic channel $-\infty<x<\infty, 0\leq y\leq 2\Pi$ where $U(x,y) = U(x,y+2\Pi)$ with the computational domain defined as $\Omega=\{x,y/0\leq x \leq L, 0\leq y\leq 2\Pi\}$. The solution to (3.2) has the form

$$U(x,y) = \sum_{\omega} \hat{U}(x,\omega)e^{i\omega y}$$

where

$$A\hat{U}_x + i\omega B\hat{U}= 0$$

If $A^{-1}B$ has distinct eigenvalues, the solution $\hat{U}$ can be written in the form

$$\hat{U}(x,\omega) = \sum_{j=1}^{m} \alpha_j q_j e^{\lambda_j(x-L)} \tag{3.3}$$

where λ_j and the vectors q_j satisfy the eigenvalue problem

$$(A\lambda_j + i\omega B) q_j = 0, \quad j = 1,2,\ldots,m$$

Some of the eigenvalues will have a positive real part, and the condition of a bounded solution therefore implies that some of the coefficients α_j must be zero. We write the solution as

$$\hat{U}(x,\omega) = \sum_{\mathrm{Re}\,\lambda_j \leq 0} \alpha_j q_j e^{\lambda_j(x-L)}, \quad x \geq L \tag{3.4}$$

By setting x = L, relations can be derived between the components of q_j, and these are used as boundary conditions.

Considering for a moment the time-dependent equation

$$U_t + AU_x + BU_y = 0$$

we can Fourier-transform it with respect to y and Laplace-transform with respect to t, obtaining

$$s\hat{U}+ A\hat{U}_x + i\omega B\hat{U}= 0$$

The solution has again the form (3.4), and it is well known (see e.g. [8]) that for Re s > 0 there are no eigenvalues λ_j on the imaginary axis. In fact, the number of eigenvalues with positive real part is the same as the number of characteristics entering the domain from the right at x = L. We denote these eigenvalues by $\lambda_1,\ldots,\lambda_r$, when letting s approach zero, some of the eigenvalues may approach the imaginary axis, but they cannot cross over it (for $\omega=0$ we always get $\lambda_j = 0$).

An example of a periodic channel problem is given in [7], where flow in a water turbine is computed.

Consider next flow in a channel with solid walls, see Fig. 3.1. (see also [6]. It is assumed that the flow is subsonic at both boundaries.

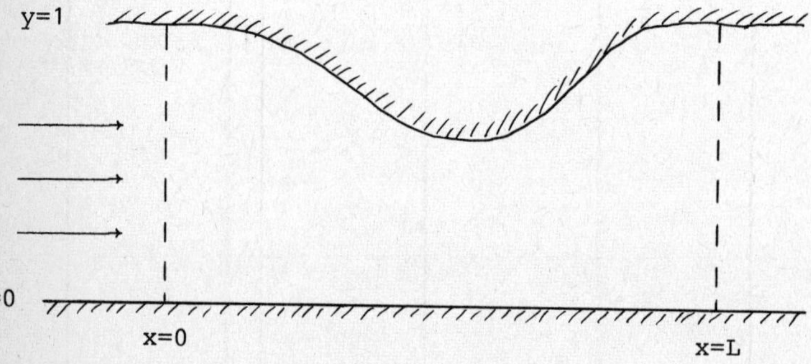

Fig. 3.1. FLow in a channel.

The Euler equations are (3.1) with

$$U=\begin{bmatrix}\rho\\ u\\ v\\ p\end{bmatrix},\qquad A(U)=\begin{bmatrix}u & \rho & 0 & 0\\ 0 & u & 0 & 1/\rho\\ 0 & 0 & u & 0\\ 0 & c^2\rho & 0 & u\end{bmatrix},\quad B(U)=\begin{bmatrix}v & 0 & \rho & 0\\ 0 & v & 0 & 0\\ 0 & 0 & v & 1/\rho\\ 0 & 0 & c^2\rho & v\end{bmatrix}$$

where c is the speed of sound.

In the down-stream region $x \geq L$ the system (3.2) is used with $\bar{v} = 0$. From this system u and ρ can be eliminated, and we obtain

$$\begin{bmatrix}v\\ p\end{bmatrix}_x + \begin{bmatrix}0 & \frac{1}{\bar{\rho}\bar{u}}\\ -\frac{\bar{c}^2\bar{\rho}\bar{u}}{\bar{r}} & 0\end{bmatrix}\begin{bmatrix}v\\ p\end{bmatrix}_y = 0,\quad \bar{r} = \sqrt{\bar{c}^2 - \bar{u}^2} \tag{3.5}$$

By expanding v and p in trigonometric series

$$v = \sum_{\omega=1}^{\infty} \hat{v}\,(x,\omega)\,\sin \Pi\, \omega y$$

$$p = \sum_{\omega=0}^{\infty} \hat{p}\,(x,\omega)\,\cos \Pi\omega y$$

we obtain from (3.5)

$$\frac{\partial}{\partial x}\begin{bmatrix}\hat{v}\\ \\ \hat{p}\end{bmatrix} + \Pi\omega\begin{bmatrix}0 & -\dfrac{1}{\bar{\rho}\bar{u}}\\ \\ -\dfrac{\bar{c}^2\bar{\rho}\bar{u}}{\bar{r}} & 0\end{bmatrix}\begin{bmatrix}\hat{v}\\ \\ \hat{p}\end{bmatrix} = 0 \tag{3.6}$$

The solution is

$$\begin{bmatrix}\hat{v}\\ \\ \hat{p}\end{bmatrix} = \alpha_1\, q_1\, e^{\lambda_1(x-L)} + \alpha_2\, q_2\, e^{\lambda_2(x-L)}$$

where

$$\lambda_1 = -\lambda_2 = \frac{\bar{c}\omega\Pi}{\bar{r}}, \quad q_1 = \begin{bmatrix}\bar{r}\\ \\ \bar{c}\bar{\rho}\bar{u}\end{bmatrix}, \quad q_2 = \begin{bmatrix}\bar{r}\\ \\ -\bar{c}\bar{\rho}\bar{u}\end{bmatrix}$$

For $\omega > 0$ we must have $\alpha_1 = 0$, and the form of q_2 implies the condition

$$\hat{p}\,(L,\omega) = -\frac{\bar{c}\bar{\rho}\bar{u}}{\bar{r}}\,\hat{v}\,(L,\omega), \quad \omega = 1,2,\ldots \tag{3.7}$$

For $\omega = 0$ in (3.6) we get $\partial\hat{p}/\partial x=0$, showing that the value of $\hat{p}(L,0)$ can be taken as $\hat{p}\,(\infty,0) = \int_o^i p(\infty,y)dy$.

By multiplying (3.7) with $\cos \Pi\omega y$, we get the final down-stream condition

$$p(L,y) - \int_o^1 p(\infty,y)dy = -\frac{\bar{c}\bar{\rho}\bar{u}}{\bar{r}} \sum_{\omega=1}^{\infty} \hat{v}\,(L,\omega)\,\cos \Pi\omega y \tag{3.8}$$

At the inflow boundary the corresponding condition can be derived from (3.6). Furthermore, we get from the linearized system (3.2)

$$(3.9)\qquad \begin{cases} u_x + \dfrac{1}{\bar{\rho}\bar{u}}\, p_x = 0 \\[2ex] \rho_x - \dfrac{1}{\bar{c}^{\,2}}\, p_x = 0 \end{cases}$$

Hence, the complete set of conditions at the inflow boundary x=0 can be given as

$$(3.10)\qquad \begin{cases} u(0,y) + \dfrac{1}{\bar{\rho}\bar{u}}\, p(0,y) = u(-\infty,y) + \dfrac{1}{\bar{\rho}\bar{u}}\, p(-\infty,y) \\[2ex] \rho(0,y) - \dfrac{1}{\bar{c}^2}\, p(0,y) = \rho\,(-\infty,y) - \dfrac{1}{\bar{c}^2}\, p(-\infty,y) \\[2ex] p(0,y) - \displaystyle\int_0^1 p(-\infty,y)\, dy = \dfrac{\bar{c}\bar{\rho}\bar{u}}{\bar{r}} \displaystyle\sum_{\omega=1}^{\infty} \hat{v}\,(0,\omega)\cos \Pi\omega y \end{cases}$$

(3.8), (3.10) are called <u>the fundamental boundary conditions</u>.

For the computational implementation, the FFT can be used to compute $\hat{v}\,(x,\omega)$, $x = 0,L$ and the cos-sums occuring in (3.8) and (3.10). Even if the FFT is not used, the extra work at the boundaries is negligable compared to the work at inner points.

4. External flow problems

For external problems the principles described in the previous section can still be applied. Consider again the Euler equations and flow around an airfoil, see Fig. 4.1.

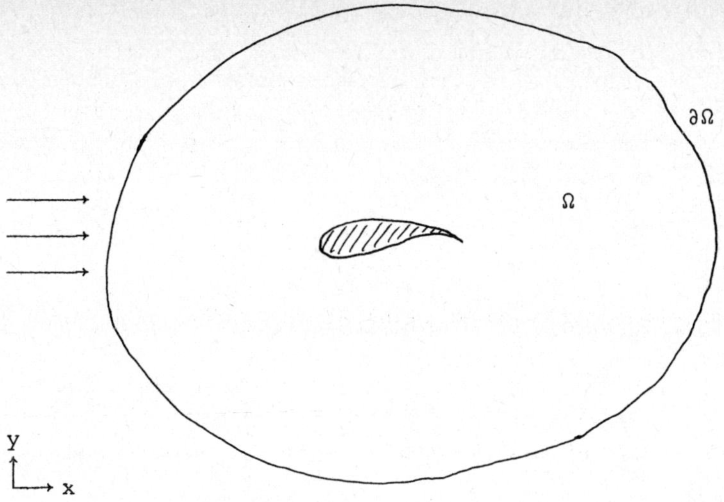

Fig. 4.1 Flow around an airfoil

It is assumed that the flow is subsonic everywhere on $\partial\Omega$.

The 2×2 - system (3.5) was derived independent of the geometry, and under the only assumption that $\bar{v} = 0$ for the state which we have linearized around. It is therefore valid also in the present case. Assuming that $\partial\Omega$ is an ellipse with its main axis parallel to the free stream, we introduce the new coordinates r, Θ by

$$\begin{cases} x = r \cos \Theta \\ y = a\, r \sin \Theta \end{cases}$$

such that $r = r_0$ represents $\partial\Omega$.

The system (3.5) written as

$$\frac{\partial V}{\partial x} + Q \frac{\partial V}{\partial y} = 0, \quad V = \begin{bmatrix} v \\ p \end{bmatrix}$$

takes the form

$$(4.1) \qquad \left(\cos \Theta I + \frac{\sin\Theta}{a} Q\right) \frac{\partial V}{\partial r} - \frac{1}{r} \left(\sin \Theta I - \frac{\cos \Theta}{a} Q\right) \frac{\partial V}{\partial \Theta} = 0$$

$$r_0 \leq r < \infty, 0 \leq \Theta \leq 2\Pi$$

Since the solution is periodic in Θ, it can be written in the form

$$V(r,\Theta) = \sum_{\omega=0}^{\infty} \hat{V}_c\ (r,\omega)\ \cos\ \omega\Theta + \sum_{\omega=1}^{\infty} \hat{V}_s\ (r,\omega)\sin\omega\Theta, \hat{V} = \begin{bmatrix} \hat{v} \\ \hat{p} \end{bmatrix}$$

The differential equation (4.1) has variable coefficients, therefore the form of the solution cannot be derived in such a direct way as in Section 3. However, one can prove

Theorem The Fourier coefficients of the solution V to (4.1) satisfy

$$(4.2)\qquad \begin{aligned} &\sqrt{r_2}\ \hat{p}_s\ (r_o,\omega) - \sqrt{r_1}\hat{v}_c\ (r_o,\omega) = 0 \\ &\sqrt{r_2}\ \hat{p}_c\ (r_o,\omega) + \sqrt{r_1}\hat{v}_s\ (r_o,\omega) = 0 \end{aligned} \qquad ,\ \omega = 1,\ 2,...$$

where $r_1 = \dfrac{\bar{c}^2\bar{\rho}\,\bar{u}}{\bar{c}^2 - \bar{u}^2}$, $r_2 = \dfrac{1}{\bar{\rho}\ \bar{u}}$

□

For $\omega = 0$ we have

$$(4.3)\qquad \frac{\partial}{\partial r}\ \hat{p}_c\ (r,0) = \frac{\partial}{\partial r}\ \hat{v}_c\ (r,0) = 0$$

The last condition shows that the constant free stream values p_∞ and $v_\infty = 0$ can be used for $\hat{p}_c\ (r_0,0)$ and $\hat{v}_c\ (r_0,0)$. Note that the conditions (4.2), (4.3) can be used everywhere at the boundary $\partial\Omega$, regardless of in- or outflow. At inflow points we use also the conditions

$$(4.4)\qquad \left\{ \begin{aligned} &u\ (r_0,\Theta) + \frac{1}{\bar{\rho}\ \bar{u}}\ p(r_0,\Theta) = u_\infty + \frac{1}{\bar{\rho}\ \bar{u}}\ p_\infty \\ &\rho\ (r_0,\Theta) - \frac{1}{\bar{c}^2}\ p\ (r_0,\Theta) = \rho_\infty - \frac{1}{\bar{c}^2}\ p_\infty \end{aligned} \right.$$

derived from (3.9)

5. The numerical algorithm for steady state problems

The boundary conditions discussed in this paper are derived without any reference to the numerical method used for the actual computation. We shall briefly discuss the implementation when explicit marching procedures are used to obtain the steady state for the external problem.

It is assumed that the boundary $\partial\Omega$ is a grid line, and that the grid is of 0-type with M points in the r-direction and N points in the Θ-direction, see Fig. 5.1. For each time-step all inner points (including points on solid surfaces in Ω) are first computed. At all boundary points on

$\partial\Omega$ the values of the outgoing characteristics are extrapolated, giving three conditions at outflow points and one condition at inflow points. The equations (4.4) give two more conditions at inflow points, and the complete set is obtained everywhere by using (4.2), (4.3) (for the N first coefficients). It is possible to reduce these four conditions along $\partial\Omega$ analytically to one condition, i.e. we can define N variables w_j related to each other by a linear system

$$Sw = b$$

The inverse of S can be computed once and for all. The procedure at the boundary is therefore at each time step :

i) Compute b (which depend on the values of U near the boundary

ii) Compute $w = S^{-1}b$

iii) Compute the remaining 3N variables at $\partial\Omega$.

Compared to standard procedures, the only additional work is the computation of $S^{-1}b$, which takes $2N^2$ arithmetic operations. This should be compared to γMN for the work at inner points, where γ depends on the method. Usually γ is greater than 50, and the extra work is therefore negligable.

The type of boundary conditions presented here can be worked out for other systems. This has been done for incompressible flow governed by the system

$$(5.1) \qquad \begin{cases} p_x + uu_x + vu_y = 0 \\ uv_x + p_y + vv_y = 0 \\ u_x + v_y = 0 \end{cases}$$

This particular problem could of course be solved by using the potential equation

$$(5.2) \qquad \Delta\Phi = 0, \qquad u = \Phi_x, \qquad v = \Phi_y,$$

but we use the form (5.1) for illustration of the application to first order systems. The equation (5.2) is satisfied also by the solution u,v to the constant coefficient equation (3.2), but since p does not in general satisfy $p_x + \bar{u}u_x + \bar{v}u_y = 0$, only an approximation is obtained by our procedure.

We used the marching procedure and mesh developed by Eriksson and Rizzi [4]. (Actually their original program was used, but with the boundary modification). It uses the finite volume for the "hyperbolized" system with an extra term p_t/c^2 added to the last equation. The solid lines in Fig. 5.2, 5.3 show the accurate solution (isobars) when a large computational domain Ω_1 is used. The dashed lines in Fig. 5.2 show the solution on a smaller domain Ω_2 where the free stream values of the ingoing characteristic variables have been specified, i.e. the condition (2.2) with non-zero right hand side. The accuracy is completely lost. Fig. 5.3 shows the solution with the fundamental boundary conditions. As can be seen, the accuracy is very good.

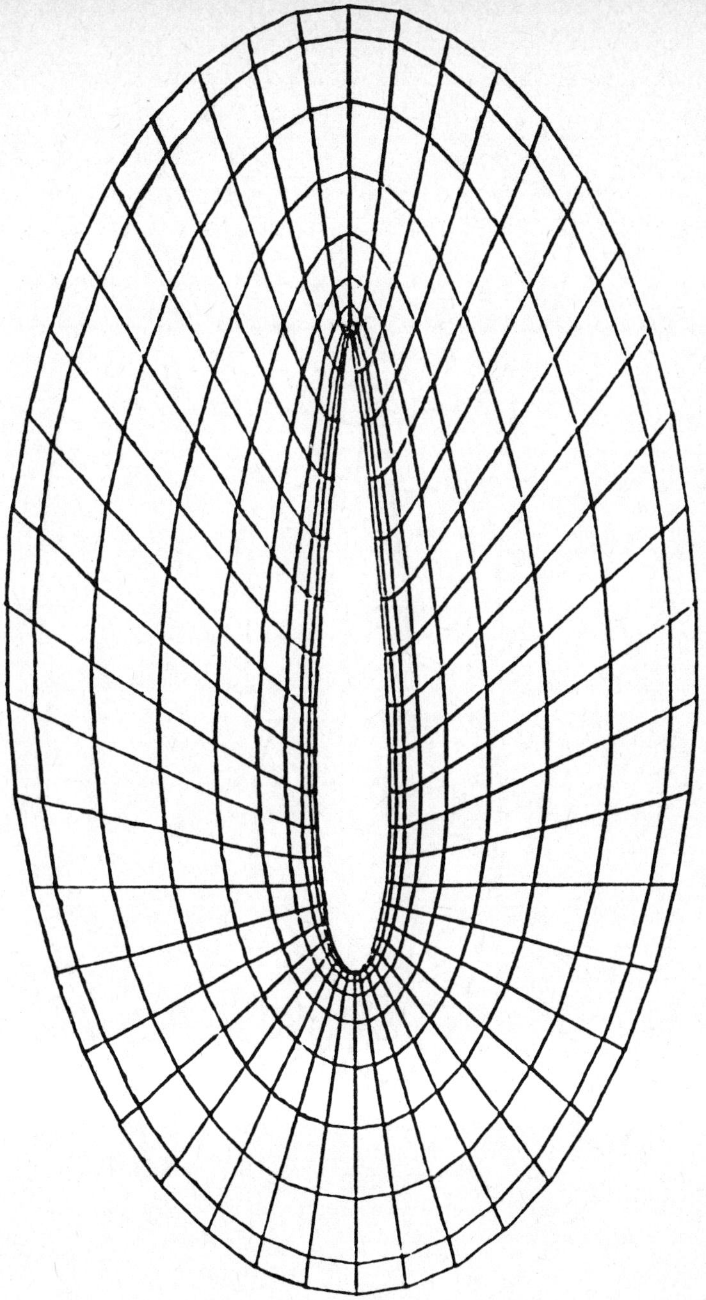

Fig. 5.1 The grid

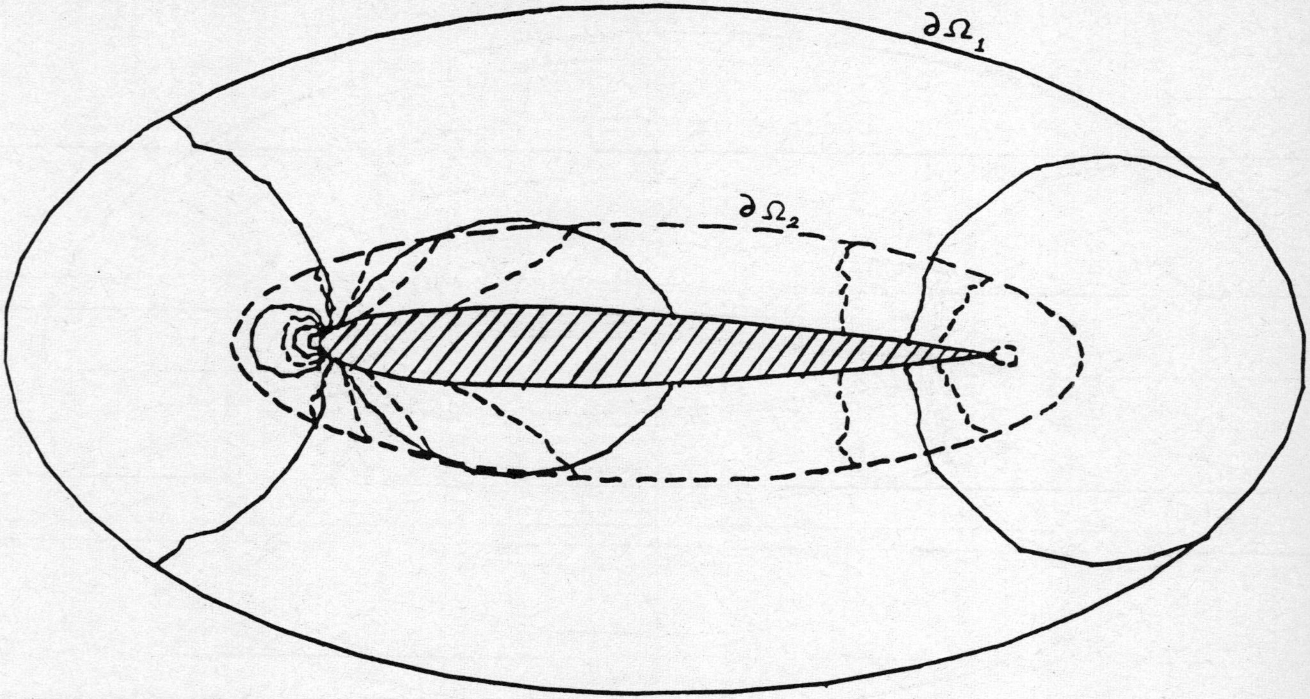

Fig. 5.2 Isobars with characteristic variables specified at the boundary. Solid and dashed lines for large and small domain calculation resp.

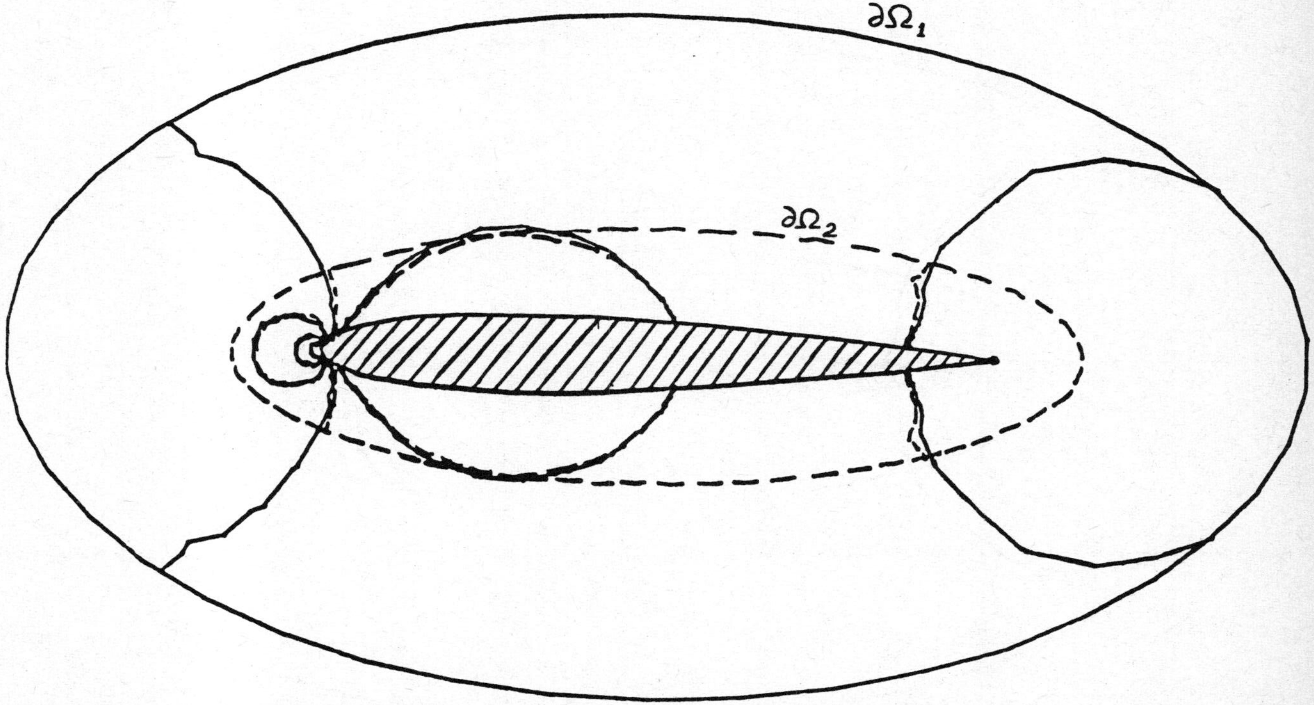

<u>Fig. 5.3</u> Isobars with fundamental boundary conditions. Solid and dashed lines for large and small domain calculation resp.

REFERENCES

1. Bayliss A., Turkel E. : Far field boundary conditions for compressible flow. J. Comp. Phys., Vol 48, pp. 182-199 (1982).

2. Engquist B., Majda A; : Absorbing boundary conditions for the numerical simulation of waves. Math. Comp. Vol 31, No. 139, pp. 629-651 (1977).

3. Engquist B., Majda A; : Radiation boundary conditions for acoustic and elastic wave calculation. Comm. Pure and Appl. Math., Vol XXXII, pp. 313-357 (1979).

4. Eriksson L.E., Rizzi A., Therre J.P. : Numerical solutions of the steady incompressible Euler equations applied to water turbines. AIAA 2nd Applied aerodynamics conference 1984, Seattle.

5. Ferm L., Gustafsson B. : A down-stream boundary procedure for the Euler equations. Comp. and Fluids, Vol 10, No. 4, pp. 261-276 (1982).

6. Ferm L. : Open boundary procedures for stationary inviscid flow problems. Uppsala University, Dept. of Comp. Sciences, Report No. 94 (1984).

7. Ferm L. : Open boundary conditions for the flow around water turbines. Uppsala University, Dept. of Comp. Sciences, Report No. 102 (1985).

8. Kreiss H.-O. : Initial boundary value problems for hyperbolic systems. Comm. Pure and Appl. Math., Vol XXIII, pp. 277-298 (1970).

CAN HYPERBOLIC SYSTEMS OF CONSERVATION LAWS BE WELL-POSED IN BV($\mathbf{R};\mathbf{R}^N$) ?

Michelle Schatzman

Abstract: The solution of a hyperbolic system of conservation laws is bounded in BV in terms of the BV norms of its initial data. It depends continuously on its initial data, if an appropriate distance in BV is defined. These results hold on piecewise smooth solutions satisfying the entropy condition.

1. Introduction and Notations

Let

(1) $u_t + f(u)_x = 0$

be a hyperbolic system of conservation laws. Here, f is a mapping from an open subset $\mathcal{U}$ of $\mathbf{R}^N$ to $\mathbf{R}^N$, and the eigenvalues of Df(u), the differential of f, are real and distinct ; they are denoted :

(2) $\lambda_1(u) < \lambda_2(u) < \ldots < \lambda_N(u)$,

and the corresponding eigenvectors are denoted $r_i(u)$; they satisfy the following relation :

(3) $Df(u)\, r_i(u) = \lambda_i(u)\, r_i(u)$, $\forall\, i = 1, 2, \ldots, N$.

We shall need also the left eigenvectors of Df(u) ; it can be convenient to think of them as eigenlinear forms ; they are denoted $\ell_i(u)$, and they satisfy the relations

(4) $\ell_i(u)\, Df(u) = \ell_i(u)\, \lambda_i(u)$,

(5) $\ell_i(u)\, r_j(u) = \delta_{ij}$, $\forall\, i,j = 1, 2, \ldots, N$.

Finally the space BV($\mathbf{R},\mathbf{R}^N$), which is denoted more shortly as BV or BV($\mathbf{R}$), is the space of locally integrable functions from $\mathbf{R}$ to $\mathbf{R}^N$ whose gradient is a bounded measure . There is a similar definition for BV($\mathbf{R}\times\mathbf{R}^+,\mathbf{R}^N$)

The question we are concerned with is to know whether the Cauchy problem for (1) is well posed; this means, that admissible solutions should exist, be unique, bounded when the initial data are bounded, and continuous with respect to these initial data .

Such a question depends crucially on the choice of the functional space in which we work ; let us recall that the existence theory for scalar equations started with O.A. Oleinik [9], and was rendered definitive by S.N. Kruzkov [6], after the work of A.I. Vol'pert

Oleinik [9], and was rendered definitive by S.N. Kruzkov [6], after the work of A.I. Vol'pert [15] ; it is known that the scalar equation is well posed in the three spaces $L^1(\mathbf{R})$, $L^\infty(\mathbf{R})$ and BV(**R**) . After B. Keyfitz (B. Quinn) [10] remarked that the semigroup of solutions of (1) is a semigroup of contractions in $L^1(\mathbf{R})$, the well-posedness in the above spaces was made clear by P. Benilan [1], M. Crandall [2] ...

The existence theory for systems was started by J. Glimm [4], and was studied in a BV frame ; B. Temple is explaining his recent results on L^1 stability in one of the talks of this conference [14] ; and last, but certainly not least, the compensated compactness results of R. DiPerna [3] take place in a L^∞ frame . Of this choice of functional spaces, I shall prefer to look at BV(**R**), because L^1 has so little compactness, that it seems very hard to obtain something, and L^∞ is an enormously complicated space .

In this talk, I shall present some ideas on stability, and of continuity of solutions with respect to the initial data ; the stability will relie on estimates of continuous Glimm functionals, and the continuity involves a discussion of what is, for our purposes a good topology in BV(**R**) . All the results which are presented here are concerned only with a subclass of piecewise smooth solutions of (1), which it would be very interesting to extend .

2 . A priori estimates in BV : the statement .

We assume here that the characteristic fields are genuinely linear, or linearly degenerate ; namely, we must have all index i :

either

(6) $D\ell_i(u)\, r_i(u) \neq 0$ for all u in $\mathcal{U}$, (genuine nonlinearity)

or

(7) $D\ell_i(u)\, r_i(u) = 0$ for all u in $\mathcal{U}$, (linear degeneracy) .

The set of indices i for which (6) holds will be denoted GNL .The notions of genuine nonlinearity and linear degeneracy have been introduced by Lax [7], and are described in a very attractive fashion in Lax [8] .The main result is as follows :

Theorem. Let v be a constant state belonging to $\mathcal{U}$. If u is an admissible piecewise Lipschitz solution of (1), which belongs to $BV(\mathbf{R}\times\mathbf{R}^+)^N$, if the L^∞ norm of u - v is smaller than some constant depending on f and v, and the BV norm of the initial data u_0 is smaller than a certain constant depending on f and v, then $|u(.,t)|_{BV}$ is bounded independantly of t by an expression which tends to zero as $|u_0|_{BV}$ tends to zero.

In order to describe the idea of the proof, it is necessary to introduce some technical tools. Nevertheless, the main idea is to use Glimm estimates in a continuous setting. Recall that Glimm introduced a pair of functionals, a linear one which is equivalent to the BV norm, but does not decrease (in general) on the approximate solutions constructed by the random choice method, and the other one, a quadratic functional, which estimates the potential of wave interaction; this functional, applied to approximate solutions, decreases as a function of time, because after having interacted, two waves do not interact any more. When the initial data are small enough, in BV, and close enough to a constant state in L^∞, the decrease in the quadratic functional estimates the increase in the linear functional, .

The idea of the present proof is therefore to construct a continuous analog of the Glimm functionals, to show that they behave as in the discrete case, when u belongs to a certain functional class, and to deduce the stability result.

It is of course obvious that this kind of stability holds for solutions constructed by the random choice method; but, as uniqueness is still an open problem, one must be interested in obtaining stability by an a priori argument, not depending on the construction of the solutions.

3. Glimm functionals.

Let us describe briefly the tools which are needed now. The self-similar solution of Riemann problem, with data u^ℓ and u^r, is given by a sequence of simple centered waves, which join constant states denoted u_j, $1 \leq j \leq N-1$; By convention, $u^0 = u^\ell$, and $u^N = u^r$. Between the state u^{j-1} and the state u^j, there is a simple wave of the j-th family, and it is possible to define a real parameter, called the strength of the wave. If we normalize the eigenvectors r_j so that condition (6) becomes

(6bis) $D\ell_j(u)\, r_j(u) > 0, \ \forall u \text{ in } \mathcal{U},$

then, the usual convention is that for a genuinely nonlinear characteristic field, the strength of a wave is negative if the wave is a shock, and positive if it is a rarefaction wave . When the characteristic field is linearly degenerate, the waves are all contact discontinuities, irrespective of the sign of any parameter .

The strengths of the waves are obtained by solving an implicit equation, with u^{ℓ} and u^r as data ; we shall write

(8) $\varepsilon_j = E_j(u^{\ell},u^r)$.

The functions E_j satisfy the expansion

(9) $E_j(u^l,u^r) \simeq \ell_j(u^{\ell})\,(u^r - u^{\ell})$ + higher order terms

in the neighborhood of u^{ℓ} .

For U a function from **Z** to $\mathcal{U}$, the discrete linear Glimm functional is defined as

(10) $$L(U) = \sum_{i=1}^{N} L_i(U)\ ;\ L_i(U) = \sum_{k=-\infty}^{+\infty} |E_i(U_k,U_{k+1})| .$$

and the quadratic functional is defined from an auxiliary function Δ, defined on couples of vectors :

(11) $$\Delta(\varepsilon,\zeta) = \sum_{i>j} |\varepsilon_i|\,|\zeta_j| + \sum_{i\in GNL} (|\varepsilon_i|\,|\zeta_i| - \varepsilon_i^+ \zeta_i^+) .$$

the formula for the quadratic Q is then given by

(12) $$Q(U) = \sum_{k<n} \Delta(E(U_k,U_{k+1}), E(U_n, U_{n+1})) .$$

In order to define analogous functionals in the continuous case, we consider functions u in $BV(\mathbf{R};\mathbf{R}^N)$, with values in $\mathcal{U}$; for such functions, du/dx is a measure with values in $\mathbf{R}^N$, which admits a decomposition in the sum of its diffuse and atomic parts :

(13) $$\frac{du}{dx} = \mu + \sum_k [u](\xi_k)\,\delta(. - \xi_k) .$$

Here,

(14) $$\mu = \text{diffuse part of } \frac{du}{dx} ;$$

(15) $$[u](\xi_j) = u(\xi_j + 0) - u(\xi_j - 0)) = \text{jump of } u \text{ at } \xi_j ,$$

$$(16)\qquad \sum_k [u](\xi_k)\,\delta(.-\xi_k) = \text{atomic part of } \frac{du}{dx}\,.$$

With the help of these notations, we can define the measure-strength of j-waves in u by

$$(17)\qquad M_j(u) = \text{measure} = \ell_j(u)\,\mu + \sum_n E_j(u(\xi_n-0),u(\xi_n+0))\,\delta(.-\xi_n)\,.$$

If u is a smooth function, then, (9) shows that

$$(18)\qquad M_j(u) = \ell_j(u)\,u_x\,dx.$$

We define now the continuous analogues of L and Q ; let

$$(19)\qquad \mathcal{L}_k(u) = \int_{\mathbb{R}} |M_k(u)| \;;\quad \mathcal{L}(u) = \sum_{k=1}^{N} \mathcal{L}_k(u)\,.$$

It is suitable to define the quadratic functional in several steps ; for i>k, let

$$(20)\qquad \mathcal{Q}_{ik}(u) = \int_{x<y} |M_i(u)(x)|\,|M_k(u)(y)|\,.$$

If i is the index of a genuinely nonlinear field, (i ∈ GNL), we let

$$(21)\qquad \mathcal{Q}_{ii}(u) = \int_{x<y} \{|M_i(u)(x)|\,|M_i(u)(x) - M_i(u)(x)^+\,M_i(u)(y)^+\}$$

but

$$(22)\qquad \mathcal{Q}_{ii}(u) = 0 \text{ otherwise}\,.$$

Then, the quadratic is given by

$$(23)\qquad \mathcal{Q}(u) = \sum_{i\geq k} \mathcal{Q}_{ik}(u)\,.$$

The functionals $\mathcal{L}$ and $\mathcal{Q}$ are clearly the continuous analogues of L and Q defined at (10) and (12).

It is possible to prove that there exists a real number K such that

$$(24)\qquad \frac{d}{dt}(\mathcal{L} + K\mathcal{Q}) \leq 0\,,$$

provided that $\mathcal{L}(u_0)$ is small enough, and $|u - u_0|_\infty$ remains suitably bounded.

4. Conservation laws for wave strength.

Relation (24) is obtained in several steps which involve working in different regions of the half-plane : smooth regions, regions involving a shock, and regions involving a meeting point of shocks . From the technical point of view, two different approaches have been given by the author ; in [12], one considers $\mathfrak{L}$ and $\mathfrak{Q}$ defined not only on lines t = constant, but on space-like lines ; this enables one to work in regions involving, for instance, only one shock . In [13], the proof is sketched with the help of characteristic tubes . Both these approaches relies on approximate conservation laws satisfied by the wave strengths, and the potential of wave interaction . The complete description, in the case of a system, is quite intricate, but it can be carried out easily in the case of an equation . Thus, we consider a solution u of the conservation law

(25) $u_t + f(u)_x = 0,$

and we assume that u is piecewise continuously differentiable ; such a situation is known to be generic, thanks to the work of D. Schaeffer [11] . Let s_j parameterize the shock curve numbered by j . We make the following convention :

$$\{u_z\} = \begin{cases} u_z \text{ where u is differentiable, i.e. } x \neq s_j(t) \,; \\ 0 \text{ elsewhere .} \end{cases}$$

Then, we can write rules for differentiating discontinuous functions :

$$u_x = \{u_x\} + [u](s,.)\, \delta(x-s) \,.$$

$$u_t = \{u_t\} - [u](s,.)\, \delta(x-s)\, (ds/dt) \,.$$

With the help of these rules, it is possible to write |M(u)| in the general case ; But the computation will be more legible, if we assume that there is only one shock, which will be parameterized by s . Then, according to the previous rules, and rules of differentiation of a Dirac mass, we have :

(26) $|M(u)|_t = \{|\{u_x|_t\} - [|\{u_x\}|](s)\, \delta(x-s)\, (ds/dt) + (d|[u]|/dt)\, \delta(x-s) -$

$|[u]|(s)\delta'(x-s)\, (ds/dt) \,.$

The aim is to find an N(u) such that

$$\frac{\partial}{\partial t}|M(u)| + \frac{\partial}{\partial x} N(u) = R(u),$$

and the rest R(u) is a measure which behaves nicely . If we did not have a shock, R(u) would be readily defined : it is enough to differentiate the equation once :

(27) $u_{xt} + (f'(u)_x)_x = 0,$

and to multiply it by the sign of u_x, and after simple manipulations, we obtain

(28) $|u_x|_t + (f'(u)|u_x|)_x = 0 \,.$

It is necessary to remark that no masses are created by this process : when u_x changes sign, $f'(u)|u_x|$ vanishes, and (27) can be justified simply .

Thus, intuitively, we can try for N the following choice :

(29) $N(u) = f'(u)|[u_x]| + |[u]|(s)\ \delta(x-s)\ (ds/dt)$.

We compute the space derivative of the measure N :

(30) $N(u)_x = \{(f'(u)|[u_x]|)_x\} + [f'(u)|[u_x]|](s)\ \delta(x-s) + |[u]|(s)\ \delta'(x-s)\ (ds/dt)$.

Adding (26) and (30), we obtain

(31) $|M(u)|_t + N(u)_x = ([(f'(u) - (ds/dt))|[u_x]|] + (d|[u]|/dt))(s)\ \delta(x-s)$.

Observe that identity (28) in smooth regions caused the absolutely continuous term to disappear . We can see that the rest R(u), defined by the above expression is a measure indeed, and that the wave strength satisfies an approximate conservation law . In order to estimate $(d|[u]|/dt)$, we can write for u_1, u_2, u_3 and u_4 as in figure below

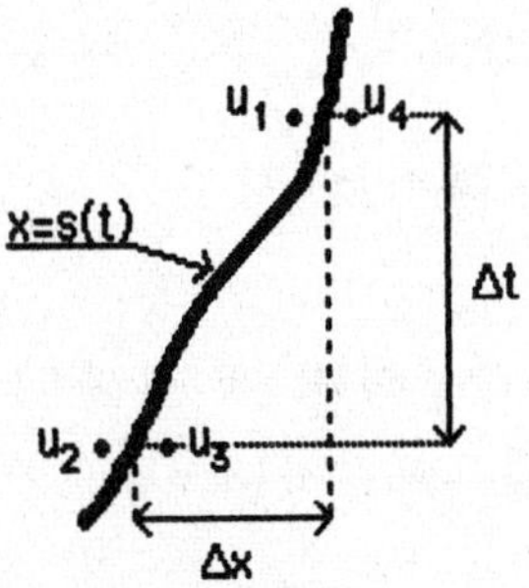

$$u_4 - u_1 = [u](s(t+\Delta t),t)$$

$$u_3 - u_2 = [u](s(t),t),$$

so that

$$(u_4 - u_1) - (u_3 - u_2) = (u_4 - u_3) - (u_1 - u_2) = [((ds/dt) - f'(u))u_x]\ \Delta t +$$

higher order terms .

Therefore,

(32) $$\frac{d[u](s)}{dt} = [(\frac{ds}{dt} - f'(u))\{u_x\}] .$$

This relation enables us to rewrite (31) as

(33) $|M(u)|_t + N(u)_x = ([(f'(u) - (ds/dt))(|[u_x]| - \{u_x\}\ \mathrm{sgn}([u]))]$.

If the geometric entropy condition holds,

$$f'(u^r) - (ds/dt) < 0 \text{ and } f'(u^l) - (ds/dt) > 0,$$

Therefore, the rest R(u) is a negative measure .

The conclusion generalizes obviously to the case of N shocks ; in the case of a system, the situation is more complicated . By analogy to the scalar case, we write

(34) $|M_i(u)|_t + N_i(u)|_x = R_i$.

Here,

$$N_i(u) = \lambda_i(u)\,|\ell_i(u)(u_x)| + \sum_j |E_i(u^\ell, u^r)(s_j(t),t)| \frac{ds_j}{dt}\delta(x - s_j(t)),$$

and R_i admits the decomposition

$$R_i = R_{i,ac} + R_{i,s} + R_{i,m},$$

with

$$R_{i,ac} = \mathrm{sgn}(\ell_i(u)(u_x)) \sum_{k<n} D\ell_i(u)\,(r_k \bullet r_n - r_n \bullet r_k)(\lambda_n - \lambda_k)(\ell_k u_x)(\ell_n u_x).$$

The term $R_{i,s}$ is carried by shock curves, and is given by

$$R_{i,s} = \sum_{s_j \text{ is an i-shock}} \frac{d|E_i(u)|}{dt} d(x-s_j) + \sum_j [|\ell_i u_x|(l_i - \frac{ds}{dt})]\,\delta(x - s_j).$$

The term $R_{i,m}$ is a made out of Dirac masses, and is carried by the set of meeting points of shocks, and of points where shocks appear or disappear . If u were self-similar around a point (x_q,t_q), it would be immediate that $R_{i,m}$ is given by

$$R_{i,m} = (\mathfrak{T}_i(u)(t_q+1) - \mathfrak{T}_i(u)(t_q-1))\delta(x - x_q)\delta(t - t_q).$$

In the general situation, we assume that u is locally self-similar, that is, the sequence $u_{\varepsilon,q}$ defined by

$$u_{\varepsilon,q}(X,T) = u(x_q + \varepsilon X,\, t_q + \varepsilon T)$$

has a limit u_q in the sense of distributions as the positive real parameter ε tends to zero . Then, $R_{i,m}$ is given by

$$R_{i,m} = \sum_q (\mathfrak{T}_i(u_q)(1) - \mathfrak{T}_i(u_q)(-1))\delta(x-x_q)\delta(t-t_q).$$

Define a local potential of wave interaction by

$$Q_{ik} = |M_i(u)(x)|\,|M_k(u)(y)|Y(y-x), \text{ if } i > k$$

$$Q_{ii} = (|M_i(u)(x)|\,|M_i(u)(y)| - (M_i(u)(x))^+ (M_k(u)(y))^+)Y(y-x), \text{ if } i \in GNL,$$

where Y is the Heaviside function . We can write similar conservation laws for the Q_{ik} . There is a positive measure Λ such that

$$(35)\quad \frac{\partial Q_{ik}}{\partial t} + \frac{\partial(N_i(x)\,|M_k(y)|\,Y(y-x))}{\partial x} + \frac{\partial(N_k(y)\,|M_i(x)|\,Y(y-x))}{\partial y} = -\Lambda\delta(x-y) + S_{ik}\,.$$

Here S_{ik} is a measure depending on x,y and t, and it can be proved that

$$|S_{ik}|(x,y) \leqslant \text{constant}\,(|M_k(x)|\,\Lambda(y) + |M_i(y)|\,\Lambda(x))\,.$$

Similar, but more complicated conservation laws hold for Q_{ii}, when $i \in GNL$.

It is possible to perform the previous computations on functions which are Lipschitz continuous except on a locally finite set of curves and points . The exceptional curves must be Lipschitz continuous ; the function must have limits from either side of the exceptional curves . At the exceptional points, the function must be locally self-similar, in the above-defined sense ; It will be a consequence of the analysis that u_x has limits on either side of a shock curve, in a suitable sense ; finally, the solution must satisfy the geometric entropy condition .

5 . COMPARISON BETWEEN BV FUNCTIONS .

With an adequate notion of continuity, shocks which have close locations and close left and right limits should be considered as close, as they look intuitively close to the human eye ; the distance defined by the BV norm, that is

$$\|u\|_{BV} = \int |u_x|\,dx + |u(-\infty)|$$

does not fulfill this condition : if the shocks are at very close, but different locations, their distance is finite . On the other hand, the weak topology considers as close shocks which are close in the previous sense, but allows for a shock and an overshooting shock to be close . This is not precise enough . thus, I propose to consider a topology on BV defined by a distance ; in the scalar case, the idea is to deform the variable x so that the largest shocks of v deformed have the same location as the largest ones of u, and to compute the BV distance between u and the deformed v ; then, one has to pay for the deformation by adding to this number the distance between the deformation and the identity . The final result is obtained by minimizing over all the deformations . In mathematical terms,

$$\Phi = \{\text{increasing homeomorphisms from } \mathbf{R} \text{ to itself}\}$$

$$d(u,v) = \inf_{\varphi\in\Phi} \{\|u - v\circ\varphi\|_{BV} + \|\varphi - Id\,\|_{BV}\}\,.$$

For this distance, which is not translation invariant, BV is a complete space . Other distances could be imagined . One nice property is that the previously defined continuous Glimm functionals are continuous with respect to this distance . This distance has to be generalized properly to the higher dimensional case ; in the case of a system of two equations, one uses Riemann invariants .

6 . CONTINUITY WITH RESPECT TO THE INITIAL DATA IN THE SCALAR CASE .

The following result can be proved :

Theorem . Let u be a piecewise smooth solution of the scalar equation

$u_t + f(u)_x = 0$.

Then, for almost every time t, and for all positive ε, there exists an α such that, if $d(u^\circ, v^\circ) < \alpha$, and if the solution v is piecewise smooth, then the distance between u(.,t) and v(.,t) in BV is smaller than ε.

The idea of the proof is close to the idea of the proof of the BV estimates : one has to consider the problem in different regions, smooth regions, regions with one shock, and regions with a meeting point of shocks . It is because of meeting points of shocks that the result holds only for almost every t ; it cannot be expected that shocks of close solutions will meet at the same time ; they will only meet at close times . Let us prove the result in the smooth case : u and v are two smooth solutions of a scalar conservation law ; denote by X and Y respectively the characteristic of u and v, namely the solutions of

$$\partial X(x,t)/\partial t = f'(u(X(x,t),t)) \; ; \; X(x,0) = x .$$

$$\partial Y(x,t)/\partial t = f'(v(Y(x,t),t)) \; ; \; Y(x,0) = x .$$

Then, we use the characteristics to compare u and v : the norm $\|u(.,t) - v(.,t)\circ Y(t)\circ X(t)^{-1}\|_{BV}$ is equal to $\|u(.,t)\circ X(t) - v(.,t)\circ Y(t)\|_{BV}$, which is the constant $\|u_0 - v_0\|_{BV}$. The cost of the deformation is $\|Y(t)\circ X(t)^{-1} - I\|_{BV} = \|Y(t) - X(t)\|_{BV}$, which is estimated by writing :

$$\frac{\partial Y}{\partial x} - \frac{\partial X}{\partial x} = t\,(f(v_0)_x - f(u_0)_x)$$

with the help of the properties of the characteristics of a scalar conservation law . From this relation, it is clear that the cost of the deformation is small when the distance between the initial data is small .

In the case when we have a shock, we let r and s denote parame-terizations of the shock in u and of the shock in v . Let

$$z^{\ell} = \min(Y^{-1}(s^{\ell}), X^{-1}(r^{\ell})) \; ; \; z^{r} = \max(Y^{-1}(s^{r}), X^{-1}(r^{r})) .$$

Outside of the interval $[z^{\ell}, z^{r}]$, the comparison is as before . Inside, there is a difficulty due to the fact that an interval of initial data disappears into the shock, so to say, and that this interval is not the same for u and v . But, an argument of continuity shows that for close initial data, the strength and the location of the shocks are close ; the size of the interval which cannot be accounted for depends continuously on the difference r - s ; we use then the following lemma :

Lemma . Let u and v be two BV functions . Let φ be a homeormophism from $\mathbf{R} \setminus [a,b]$ to $\mathbf{R} \setminus [c,d]$. Denote

$$\varepsilon = TV(u|_{\mathbf{R} \setminus [a,b]}) + TV(v|_{\mathbf{R} \setminus [c,d]}) + b - a + c - d .$$

Then,

$$\lim_{\varepsilon \to 0} d(u,v) \leqslant \|u \circ \varphi - v\|_{BV(\mathbf{R} \setminus [a,b])} + \|\varphi - I\|_{BV(\mathbf{R} \setminus [a,b])} .$$

The question adressed here is close to the results of Golubitsky and Schaeffer [5] .

In the case of meeting points of shocks, one has to use the continuity of shocks with respect to initial data, and the continuity of the meeting time ; the result is proved individually for each shock .

REFERENCES

[1] P. Benilan, Equations d'évolution dans un espace de Banach quelconque et applications, Thèse de doctorat d'Etat, Orsay 1971 ;

[2] M.G. Crandall, The semi-group approach to first order quasilinear equations in several space variables, Israel J. Math. 12(1972)108-132 .

[3] R.J. DiPerna, Convergence of approximate solutions to conservation laws, Arch. Rat. Mech. Anal. 82(1983)27-70 .

[4] J. Glimm, Solutions in the large for nonlinear systems of conservation laws, Comm. Pure Appl. Math. 18(1965)685-715 .

[5] M. Golubitsky, D.G. Schaeffer, Stability of shock waves for a single conservation law, Advances in Math. 16(1975)65-71 .

[6] S.N. Kruzhkov, first order quasilinear equations with several independant variables, Mat. Sb. (NS) 81(123)(1970)228-255 = Math. USSR Sb 10(1970) 217-243 .

[7] P.D. Lax, Hyperbolic systems of conservation laws, II, Comm. Pure Appl. Math. 10(1957)537-556 .

[8] P.D. Lax, Hyperbolic systems of conservation laws and the mathematical theory of shock waves, C.B.M.S. S.I.A.M. Philadelphie, 1973.

[9] O.A. Oleinik, Discontinuous solutions of nonlinear differential equations, Uspehi Mat. Nauk, 12(1957)3(75)3-73 = AMS Transl. (2)26(1963)95-172.

[10] B. Quinn, solutions with shocks : An example of L^1 contractive semi-groups, Comm. Pure Appl. Math. 24(1971)125-132.

[11] D. Schaeffer, A regularity theorem for conservation laws, Advances in Math. 11(1973)368 - 386.

[12] M. Schatzman, Continuous Glimm functionals and uniqueness of solutions of Riemann problem, Indiana Univ. Math. J., 34(1985)533-589.

[13] M. Schatzman, The geometry of continuous Glimm functionals, A.M.S. Lectures Appl. Math.,23(1986)417-439.

[14] B. Temple, this conference.

[15] A.I. Vol'pert, The space BV and quasilinear equations, Math. USSR Sb. 21(1967)225-267

Mathématiques,
Université Claude-Bernard,
69622 Villeurbanne Cedex, France.

PROPAGATION DES OSCILLATIONS
DANS LES SYSTEMES HYPERBOLIQUES
NON LINEAIRES

Denis SERRE
Université de Saint-Etienne
23 rue du Dr. Paul Michelon
42023 Saint-Etienne cedex 2 FRANCE

Abstract :

We study in this note the solutions of hyperbolic 2 × 2 systems of conservation laws with oscillating data, e.g. $u^{\varepsilon}(x,0) = a(x,\frac{x}{\varepsilon})$, $a(x,.)$ being periodic. The analysis is done using the compensated compactness theory, following an idea of L. Tartar. The initial oscillations can be killed in a small time if the system is fully nonlinear. But they propagate in linearly degenerate systems. A general analysis gives rise to a relaxed problem, for which the unknown is a field $U(x,t,y)$, $y \in]0,1[$. The resulting system is differential in (x,t), and y is a coupling parameter. When the initial problem is a system of balanced laws, then the relaxed problem is also integrodifferential in y.

I. Introduction

1- Cadre général

Nous étudions ici des systèmes strictement hyperboliques de lois de conservations, sous la forme

$$(1.1) \qquad u_t + f(u)_x = 0, \quad x \in \mathbb{R}, \quad t \in]0,T[.$$

Le domaine des phases, dans lequel $u(x,t)$ prend ses valeurs, est un convexe de $\mathbb{R}^n$. La fonction de flux $f : \mathbb{R}^n \longrightarrow \mathbb{R}^n$ est régulière.

De tels systèmes apparaissent en mécanique et en physique, et expriment les principes fondamentaux de la dynamique et de la thermodynamique, ainsi que des conditions de compatibilité cinématiques. Dans tous les systèmes réalistes, il existe une entropie strictement convexe de (1.1). Rappelons qu'une fonction réelle Φ, régulière sur l'espace des phases, est une entropie s'il existe une autre fonction Ψ (appelée flux de Φ) telle que

(1.2) $$\frac{\partial \Psi}{\partial u_i} = \sum_{j=1}^{n} \frac{\partial f_j}{\partial u_i} \frac{\partial \Phi}{\partial u_j}, \qquad 1 \leq i \leq n.$$

Nous supposerons donc qu'il existe une entropie $E(u)$ de classe $\mathcal{C}^2$, telle que $E''(u)$ soit définie positive en tout point. Nous notons F le flux de E.

Une solution faible de (1.1) dans $Q_T = \mathbb{R} \times]0,T[$ est alors un élément u de $L^\infty(Q_T)$ vérifiant pour toute fonction test $\Theta \in \mathcal{D}(Q_T)$

(1.3) $$\iint_{Q_T} \{\Theta_t\, u + \Theta_x f(u)\}\, dx\, dt = 0,$$

et lorsque $\Theta \geq 0$:

(1.4) $$\iint_{Q_T} \{\Theta_t\, E(u) + \Theta_x\, F(u)\}\, dxdt \geq 0.$$

L'inégalité (1.4), appelée inégalité entropique permet de sélectionner parmi toutes les solutions de (1.3), solutions aux sens des distributions de (1.1), celles qui ont un sens physique. On espère en particulier que (1.4) assure l'unicité pour le problème de Cauchy, mais celà n'a jamais été prouvé pour les systèmes ($n \geq 2$), excepté dans le cas linéaire.

En ce qui concerne l'existence d'une solution du problème de Cauchy

(1.1) et $u(x,0) = a(x)$,

outre les méthodes numériques on peut envisager d'approcher (1.1) par un système parabolique, muni d'une "viscosité artificielle" dont le coefficient $\varepsilon > 0$ est destiné à tendre vers zéro :

(1.1.ε) $$u_t + f(u)_x = \varepsilon u_{xx} \text{ dans } Q_T.$$

Puisque E est convexe, les solutions de (1.1.ε) vérifient un analogue de (1.4), soit

(1.4.ε) $$E(u)_t + F(u)_x \leq \varepsilon\, E(u)_{xx}.$$

2. La compacité par compensation appliquée à (1.1)

Lorsque ε tend vers zéro, on ne parvient pas à trouver une estimation à priori des dérivées u_x^ε et u_t^ε de la suite des solutions de (1.1.ε). On ne peut donc pas utiliser de théorème de Sobolev pour s'assurer que

$$\lim_{\varepsilon\to 0} f(u^\varepsilon) = f(\lim_{\varepsilon\to 0} u^\varepsilon)$$

et passer à la limite dans (1.1.ε) et (1.4.ε) pour obtenir (1.3) et (1.4). Pour remédier à ce manque d'information, L. Tartar [15] a proposé d'appliquer la théorie de la compacité par compensation, due à lui-même et à F. Murat [7]. Comme R. Di Perna [2] a montré que cette méthode pouvait s'appliquer aussi à une suite u^ε de solutions faibles du système (1.1), nous allons l'exposer dans un cadre général.

Dans les deux cas, il s'agit d'étudier une suite u^ε de fonctions mesurables bornées sur Q_T, vérifiant

$$(1.5)\quad \begin{cases} u^\varepsilon_t + f(u^\varepsilon)_x = h^\varepsilon \\ \\ E(u^\varepsilon)_t + F(u^\varepsilon)_x \leq E'(u^\varepsilon).\, h^\varepsilon \end{cases}$$

Pour les solutions faibles de (1.1), $h^\varepsilon=0$, et pour celles de (1.1.ε), $h^\varepsilon = \varepsilon u^\varepsilon_{xx}$. On suppose par ailleurs que la suite u^ε est bornée dans $L^\infty(Q_T)$, ce qui est démontré dans certains cas modèles (Chueh-Conley-Smoller [1], Serre [13], [14]) mais reste une question ouverte en général. La première étape consiste alors à prouver que pour toute entropie Φ, de classe $\mathcal{C}^2$ et de flux Ψ, la suite $\Phi(u^\varepsilon)_t + \Psi(u^\varepsilon)_x$ appartient à la somme d'un compact de H^{-1} et d'un borné de l'espace des mesures, dans toute partie bornée de Q_T. Comme cette suite est également bornée dans $W^{-1,\infty}(Q_T)$, une application d'un lemme dû à F. Murat [8] montre qu'elle est relativement compacte dans $H^{-1}(\omega)$, si ω est compact.

On applique alors le lemme divergence-rotationnel, qui est un cas particulier des résultats de compactié par compensation, pour obtenir l'égalité suivante

$$(1.6)\quad \begin{aligned} &\mathrm{Lim}\, \{\Phi_1(u^\varepsilon)\, \Psi_2(u^\varepsilon) - \Phi_2(u^\varepsilon)\, \Psi_1(u^\varepsilon)\} \\ &= \lim \Phi_1(u^\varepsilon) \lim \Psi_2(u^\varepsilon) - \lim \Phi_2(u^\varepsilon) \lim \Psi_1(u^\varepsilon) \end{aligned}$$

Dans cette égalité, Φ_1 et Φ_2 sont deux entropies régulières quelconques de (1.1), et Ψ_1, Ψ_2 sont leurs flux. Les limites sont comprises pour la topologie faible-étoile de $L^\infty(Q_T)$, et peuvent toujours être obtenues par extraction d'une sous-suite convenable, puisque $\{u^\varepsilon\}$ est bornée dans $L^\infty(Q_T)$.

3. Mesures de Young . Oscillations

Etant donnée une suite bornée u^ε dans $L^\infty(Q_T)$, par exemple celle de l'alinéa précédent, on peut toujours en extraire une sous-suite qui converge faiblement-étoile. Mais si G est une fonction numérique régulière sur $\mathbb{R}^n$, la suite $G(u^\varepsilon)$ est également bornée, et peut subir le même sort. Comme $\mathcal{C}(\mathbb{R}^n)$ est séparable, on peut utiliser le procédé diagonal pour extraire de $\{u^\varepsilon\}$ une sous-suite, qui sera encore notée ainsi, telle que la suite $G(u^\varepsilon)$ converge dans $L^\infty(Q_T)$ faible-étoile pour chaque fonction continue G ;

On n'a pas en général

$$\lim G(u^\varepsilon) = G(\lim u^\varepsilon),$$

mais pour presque tout (x,t) de Q_T, l'application

$$G \longrightarrow (\lim G(u^\varepsilon))(x,t)$$

est une mesure positive de masse unité, donc une probabilité, notée $\nu_{x,t}$, clairement mesurable sur Q_T. La mesure $\nu_{x,t}$ traduit la façon dont la suite u^ε oscille au point (x,t) ou au voisinage de celui-ci, lorsque $\varepsilon \longrightarrow 0$. Par exemple, si $\bar{u} : \mathbb{R} \longrightarrow \mathbb{R}^n$ est périodique de période 1, avec

$$\bar{u}(\sigma) = \begin{cases} \alpha \text{ sur } [0,p[, \\ \beta \text{ sur } [p,1[, \end{cases} \qquad u^\varepsilon(x,t) = \bar{u}\left(\frac{x}{\varepsilon}\right),$$

alors $\nu_{x,t} \equiv \nu = p\delta_\alpha + (1-p)\delta_\beta$.

De telles mesures, appelées mesures de Young de la suite u^ε, apparaissent chaque fois que la résolution d'un problème posé dans $L^\infty(Q_T)$ risque de ne pas avoir de solution parce qu'une suite de solutions approchées ne converge pas vers une bonne solution. Les mesures de Young peuvent donc être considérées comme les inconnues d'un problème relaxé. C'est ainsi qu'il faut voir, à mon avis, les solutions à valeurs dans les mesures (measure valued solutions) introduites par R. Di Perna [3], et les solutions oscillantes qu'on va étudier ici. Ajoutons qu'il n'est pas suffisant de comprendre quelle est la forme des variables relaxées, encore faut-il sav oir écrire le problème relaxé. C'est ce que nous ferons dans les paragraphes suivants, notamment au IV.

4. L'équation de Tartar

En terme de mesures d'Young, l'équation (1.6) se traduit par

$$<\nu,\Phi_1\ \Psi_2 - \Phi_2\ \psi_1> =<\nu,\ \Phi_1><\nu,\ \Psi_2>-<\nu,\Phi_2><\nu,\ \Psi_1>,$$

pour tous les couples entropies-flux $(\Phi_i,\ \Psi_i)$ réguliers, et pour presque tout (x,t) de Q_T, en notant $\nu=\nu_{x,t}$.

Cette équation, proposée par L. Tartar pour l'étude de la convergence des solutions approchées [15], montre que l'inconnue relaxée n'est pas une probabilité quelconque. Lorsque le nombre d'entropie est suffisamment grand (essentiellement si n=2), (1.7) entraîne que ν ne dépend que de probabilités définies sur des pavés de $\mathbb{R}^m$, $m < n$. Plus précisément, l'auteur a conjecturé en 1983 [12] que les solutions de (1.7) pour n=2 sont, à un poids explicite près, des produits tensoriels de mesures définies sur des intervalles réels. Cette formule a été démontrée récemment par M. Rascle [10] par une méthode très élégante. Son application la plus spectaculaire concerne le système de l'élasticité (Rascle [9])

$$(1.8)\qquad \begin{cases} v_t + w_x = 0, \\ w_t + \sigma(v)_x = 0. \end{cases}$$

Dans ce système, $\sigma(v)$ est la tension d'une corde, fonction de son allongement v, et vérifie $\sigma'(v) > 0$. On a montré [11] que (1.8) est bien posé dans $L^\infty\ (Q_T)$, pourvu que l'on dispose d'une estimation

$$(1.9)\qquad \|v,w\|_\infty \leq L(\|v_o,w_o\|_\infty).$$

Ce résultat avait été obtenu auparavant par Di Perna [2] dans le cas particulier où sgn σ'' = sgn v, sans avoir recours au théorème de Rascle (dans ce cas, (1.9) est également démontré).

5. Notations

Nous précisons ici quelques points techniques, bien connus depuis le travail de Lax [5].

Nous supposerons tout au long, de cet article que n=2, afin d'assurer un ensemble assez vaste d'entropie. Le système (1.1) sera strictement hyperbolique, c'est-à-dire que la matrice Jacobienne f'(u) aura deux valeurs propres réelles et distinctes $\lambda_1(u) < \lambda_2\ (u)$. On peut définir des champs de vecteurs propres associés $r^1(u)$ et $r^2(u)$, qui ne s'annulent pas.

Il existe alors des fonctions $w_1(u)$, $w_2(u)$ aussi régulières que f, dont les gradients ne s'annulent pas, et qui vérifient

$$(1.10) \qquad r^1 . \nabla_u w_2 = 0, \qquad r^2 . \nabla_u w_1 = 0.$$

Ces fonctions sont les "invariants de Riemann" du système (1.1). L'application $u \longrightarrow (w_1, w_2)$ est un changement de variables, et le système (1.1) peut s'écrire à l'aide des nouvelles variables, pour des solutions différentiables, sous la forme

$$\frac{\partial w_k}{\partial t} + \lambda_k(w) \frac{\partial w_k}{\partial x} = 0, \quad k=1,2.$$

II. Oscillations : Propriétés communes à tous les systèmes 2 x 2

1 - Position du problème

On suppose donc n=2, et on s'intéresse au problème de Cauchy pour (1.1) :

$$(2.1) \qquad \begin{cases} u_t + f(u)_x = 0 \text{ dans } Q_T \\ E(u)_t + F(u)_x \leq 0 \text{ dans } Q_T \\ u(x,0) = a(x) \text{ sur } \mathbb{R} \end{cases}$$

Etudier la propagation des oscillations consiste à prendre une suite bornée de conditions initiales $a^\varepsilon(.)$ qui converge dans $L^\infty(Q_T)$ faible-étoile ainsi que chacune des suites $G \circ a^\varepsilon$ (G continue sur $\mathbb{R}^2$) lorsque ε tend vers zéro. On introduit bien sûr la famille mesurable $\{\nu_x^o; x \in \mathbb{R}\}$ des mesures d'Young définies par cette suite :

$$\langle \nu_x^o, G \rangle = (\lim_{\varepsilon \to 0} G \circ a^\varepsilon)(x),$$

pour presque tout x de $\mathbb{R}$, et toute fonction continue G sur $\mathbb{R}^2$.

On suppose ensuite qu'à chaque condition initiale a^ε correspond (au moins) une solution u^ε de (2.1), et que la suite u^ε est bornée dans $L^\infty(Q_T)$. Celà a été prouvé dans les cas modèles (par exemple R. Di Perna [2], Serre [14]). Quitte à extraire une sous-suite convenable, on peut également supposer que $G \circ u^\varepsilon$ converge dans $L^\infty(Q_T)$ faible-étoile pour toute fonction continue G, et introduire les mesures d'Young correspondantes par

$$(2.3) \qquad \langle \nu_{x,t}, G\rangle = (\lim_{\varepsilon\to 0} G \circ u^\varepsilon)(x,t),$$

presque partout dans Q_T.

Puisque les familles ν_x^o et $\nu_{x,t}$ caractérisent les oscillations respectives des suites a^ε et u^ε, l'objectif de ce travail est de trouver un système d'évolution dont les mesures $\nu_{x,t}$ soient solution, et dont les mesures ν_x^o soient la condition initale.

2- Première réduction

La première étape est commune à tous les systèmes 2×2. Comme on l'a rappelé dans l'introduction, les mesures $\nu_{x,t}$ sont des solutions de l'équation de Tartar, pour presque tout (x,t) :

$$(2.4) \qquad \langle \nu, \Phi_1 \Psi_2 - \Phi_2 \Psi_1\rangle = \langle \nu, \Phi_1\rangle\langle \nu, \Psi_2\rangle - \langle \nu, \Phi_2\rangle\langle \nu, \Psi_1\rangle,$$

pour tous couples entropies-flux (Φ_i, Ψ_i). D'après le théorème de Rascle, conjecturé par l'auteur, les solutions de (1.6) sont nécessairement de la forme

$$(2.5) \qquad \nu = e^{K(w)}\, \nu_1 \otimes \nu_2,$$

où les mesures positives ν_1 et ν_2 opèrent sur les variables w_1 et w_2 respectivement. Le poids $K(w)$ est donné par la formule

$$(2.6) \qquad \frac{\partial^2 K}{\partial w_1 \partial w_2} = -(\lambda_2 - \lambda_1)^{-2} \frac{\partial(\lambda_1,\lambda_2)}{\partial(w_1,w_2)}.$$

Il est faux en général que la réciproque soit vraie, c'est-à-dire que (2.5) implique (2.4). C'est cependant le cas lorsque les deux champs caractéristiques sont linéairement dégénérés, comme on le verra au paragraphe IV. Par ailleurs, on prendra garde que les probabilités ν_x^o, qui sont tout à fait arbitraires, ne vérifie pas (2.5) en général. Par contre, en appliquant cette formule à $\nu_{x,t}$, on en déduit l'existence de deux familles mesurables de mesures positives $\nu_{x,t}^1$ et $\nu_{x,t}^2$, définies chacune sur un intervalle de $\mathbb{R}$, et telles que

$$\nu_{x,t} = e^{K(w)}\, \nu_{x,t}^1 \otimes \nu_{x,t}^2.$$

On a donc simplifié la caractérisation des inconnues relaxées, qui sont en fait deux mesures à une seule variable, et paramétrées par x et t. Grosso modo, si on connait l'oscillation des suites w_1^ε et w_2^ε, alors on connait l'oscillation de la suite $(w_1^\varepsilon , w_2^\varepsilon)$: au terme de couplage près (exp K), les variables aléatoires w_1^ε et w_2^ε sont indépendantes. Celà sera précisé à l'alinéa suivant.

3- Fonctions de répartition

Bien que notre problème ait été simplifié, car deux mesures d'une variables forment une inconnue moins générale qu'une mesure de deux variables, il reste délicat à traiter. En effet, les mesures d'Young ne sont pas pratiques à manipuler dans des équations aux dérivées partielles. Suivant une idée de L. Tartar, sous-jacente à son étude des oscillations dans le système de Carleman [6], nous allons les remplacer par leurs fonctions de répartition.

Etant donnée une mesure positive non nulle σ, définie sur un intervalle réel $[a,b]$. La fonction de répartition de σ est l'unique fonction croissante et continue à gauche

$$S :]0,1[\longrightarrow [a,b]$$

qui vérifie pour toute fonction g continue sur $[a,b]$:

$$\langle\sigma,g\rangle = \langle\sigma, 1\rangle \int_0^1 g \circ S(y)\, dy.$$

Pour chaque (x,t) dans Q_T, nous introduisons donc les fonctions de répartition $A_i(x,t,.)$ de $\nu^i_{x,t}$, i = 1,2. La formule (2.5) signifie alors que pour toute fonction G continue sur $\mathbb{R}^2$,

$$\langle\nu,G\rangle = R^{-1} \int_0^1 \int_0^1 (G \exp K)\ (A_1(y),A_2(z))dy\, dz$$

où

$$R = \int_0^1 \int_0^1 (\exp K)(A_1(y),A_2(z))dydz.$$

Appliqué à la suite u^ε, celà donne

$$\text{(2.7)} \qquad (\text{Lim } G(u^\varepsilon))(x,t) = R(x,t)^{-1} \int_0^1\int_0^1 (G \exp K)(A_1(x,t,y), A_2(x,t,z))\ dydz$$

avec

$$R(x,t) = \int_0^1\int_0^1 (\exp K)(A_1(x,t,y),A_2(x,t,z))dydz. \tag{2.8}$$

Il est équivalent de traiter $\nu^i_{x,t}$ ou A_i, mais les fonctions sont plus maniables dans les équations aux dérivées partielles. C'est donc à l'aide de A_1 et A_2 que nous traduirons (souvent de façon formelle) les résultats obtenus. Ceux-ci dépendant fortement de la nature du système, nous distinguerons plusieurs cas.

III- Cas des systèmes vraiment non-linéaires

1- Champs caractéristiques vraiment non-linéaires

Suivant Lax [5], on dit que le i-ème champ caractéristique est vraiment non-linéaire si

$$\frac{\partial\lambda_i}{\partial w_i}$$

ne s'annule pas (et est donc de signe constant). Le principal effet de la non linéarité est de faire se rencontrer les caractéristiques $(t,x_i(t))$ définies par $\dot{x}_i = \lambda_i(w(x_i))$ lorsque w_i oscille. Plus précisément, soit ν une solution positive de l'équation de Tartar (2.4), et soit $[a_1,b_1]\times[a_2,b_2]$ le plus petit rectangle du plan $\mathbb{R}^2_w$ qui contient le support de ν. Dans [12], nous avons généralisé une formule de R. Di Perna [2] :

$$\text{Si } a_i < b_i, \qquad < \nu,\ h(w_i)\ \rho_i \frac{\partial\lambda_i}{\partial w_i} > = 0,$$

pour toute fonction h continue sur $[a_i,\ b_i]$. La fonction $\rho_i(w)$, calculable explicitement, est continue et strictement positive. Il s'ensuit que si

$$\frac{\partial\lambda_i}{\partial w_i} > 0$$

dans une bande $\alpha_i<w_i<\beta_i$, alors la restriction de ν à cette bande est nulle. Comme $\nu_{x,t}$ n'est pas nulle, on en déduit

Proposition 3.1 (R. Di Perna) : Si le i-ème champ caractéristique est vraiment non linéaire alors $a_i = b_i$.

Celà signifie que $\nu^i_{x,t} = \delta(w_i = w_i\ (x,t))$ est une masse de Dirac, ce qui équivaut à

$$\text{Lim } h\ (w_i^\varepsilon) = h\ (\text{Lim } w_i^\varepsilon).$$

En d'autres termes, la suite w_i^ε n'est pas oscillante.

Dans le cas le plus simple où les deux champs caractéristiques sont vraient non-linéaires, $\nu_{x,t}$ est une masse de Dirac pour presque tout (x,t) de Q_T, et on a pour toute fonction G continue sur $\mathbb{R}^2$

$$\text{Lim } G(u^\varepsilon) = G(\text{Lim } u^\varepsilon).$$

C'est vrai en particulier pour le flux f du système (1.5), ce qui permet de passer à la limite dans l'espace des distributions sur Q_T. La fonction $u = \text{Lim } u^\varepsilon$ vérifie donc l'équation (1.1). De même, en prenant G=E puis G=F, on obtient l'inégalité entropique (1.4). Finalement :

Théorème 3.2 (R. Di Perna) : Supposons que les deux champs caractéristiques soient vraiment non linéaires. Soit $a^\varepsilon(.)$ une suite bornée dans $L^\infty(\mathbb{R})$, qui converge vers $a(.)$ pour la topologie faible-étoile. Soit u^ε une suite de solutions faibles du problème de Cauchy (2.1) vérifiant $u^\varepsilon(x,0) = a^\varepsilon(x)$. On suppose que la suite u^ε converge dans $L^\infty\ (Q_T)$ faible-étoile, ainsi que toutes les suites $G \circ u^\varepsilon$, G continue. Alors sa limite $u(x,t)$ est une solution faible du problème de Cauchy (2.1) pour la condition initiale $u(x,0) = a(x)$.

□

En conclusion, le problème de Cauchy (2.1) est bien posé dans $L^\infty(Q_T)$, du point de vue de l'existence, pourvu qu'on dispose d'une estimation à priori de la forme

$$\|u\|_{\infty,Q_T} \leq L\ (\|a\|_\infty).$$

Il n'y a donc pas lieu d'introduire ici un problème relaxé. Les oscillations à hautes fréquences de la suite a^ε sont immédiatement amorties par le développement de chocs dissipatifs.

2- Un système plus réaliste : l'élasticité

En élasticité non linéaire, le mouvement d'une corde sur la droite réelle peut être modélisé par le système

$$(3.2) \qquad \begin{cases} w_t = v_x \\ v_t = \sigma(w)_x \end{cases}$$

où v est la vitesse et w l'allongement au point matériel $y(x,t)=y_o(x)+\int_o^t v(x,s)\,ds$.

La fonction de tension vérifie $\sigma' > 0$, et un calcul facile montre que

$$(3.3) \qquad \frac{\partial\lambda_2}{\partial w_2} = \frac{\partial\lambda_1}{\partial w_1} = \frac{\sigma''}{2\sigma'}(w)$$

L'expérience montrant que pour les matériaux courants, σ n'est ni convexe ni concave, les quantités

$$\frac{\partial\lambda_i}{\partial w_i}$$

peuvent changer de signe pour une ou plusieurs valeurs de w. On ne peut donc appliquer le théorème 3.2 directement. En utilisant une forme particulière de l'équation (3.1), R. Di Perna a cependant généralisé son théorème au cas où σ'' s'annule une seule fois. Ce résultat a ensuite été amélioré à l'aide de la formule (2.4) (Rascle-Serre [11], Rascle [9]).

Théorème 3.3 (Rascle) : Les conclusions du théorème 3.2 sont valables pour le système de l'élasticité, sans condition sur le signe de σ''.

□

IV- Cas d'un système linéairement dégénéré

1- Généralités

L'égalité (3.1) est triviale lorsque le i-ème champ caractéristique est linéairement dégénéré, c'est-à-dire satisfait à l'égalité

$$\frac{\partial\lambda_i}{\partial w_i} \equiv 0.$$

Pour simplifier, nous considérons dans ce paragraphe un système dont les deux champs caractéristiques son linéairement dégénérés. Pour éviter le cas linéaire, on supposera toutefois que

$$\frac{\partial\lambda_1}{\partial w_2} \quad \text{et} \quad \frac{\partial\lambda_2}{\partial w_1}$$

ne s'annulent pas. On peut alors choisir comme invariants de Riemann $w = \lambda_2$ et $z = \lambda_1$, ce qui conduit au système suivant

$$(4.1) \qquad w_t + zw_x = 0, \quad z_t + wz_x = 0$$

Le cas d'un système dont un seul champ caractéristique est linéairement dégénéré, a été étudié à Saint-Etienne par M. Bonnefille, avec quelques exemples particulièrement dégénérés de systèmes 3×3.

A priori, le système (4.1) n'a aucun sens lorsque w ou z est discontinu à travers une courbe $(x(t),t)$ du domaine Q_T. Mais on vérifie aisément que toutes les formes conservatives $u_t + f(u)_x = 0$ de (4.1) conduisent à la même condition de Rankine-Hugoniot :

$$(4.2) \qquad \begin{cases} w\,(x(t)+0,\ t) = w(x(t)-0,\ t) = \dot{x}\,(t) \\ \text{ou bien} \\ z\,(x(t)+0,\ t) = z(x(t)-0,\ t) = \dot{x}\,(t) \end{cases}$$

La résolution du problème de Riemann est donc très simple. Si $a(x) = (w^{\pm}, z^{\pm})$ selon que x est positif ou négatif, la solution autosimilaire de (4.1) est, sous l'hypothèse Max $(w^{\pm})$ < Min $(z^{\pm})$:

$$(4.3) \qquad u(x,t) = \begin{cases} (w^-, z^-) & \text{si } x \leq tw^- \\ (w^-, z^+) & \text{si } tw^- < x < tz^+ \\ (w^+, z^+) & \text{si } tz^+ < x \end{cases}$$

Celà permet de mettre en oeuvre le schéma de Glimm [4] lorsque la condition initiale satisfait à l'hypothèse

$$(4.4) \qquad \gamma \leq w_o\,(x) \leq \alpha \leq \beta \leq z_o(x) < \delta \quad \text{sur } \mathbb{R}$$

Puisque dans le problème de Riemann, on a le principe du maximum et la conservation de la variation totale des invariants de Riemann, ce schéma converge pour chaque forme conservative de (4.1) :

<u>Théorème 4.1</u> : Soit w^o, z^o deux fonctions à variation bornée sur $\mathbb{R}$, vérifiant (4.4). Alors il existe deux fonctions w,z définies sur Q_T et à variation bornée, telles que

i) $\gamma \leq w(x,t) \leq \alpha$, $\beta \leq z(x,t) \leq \delta$ pour tout (x,t) de Q_T.

ii) Pour toute entropie Φ du système (4.1), de flux Ψ, on a au sens des mesures

(4.5) $$\Phi\,(w,z)_t + \Psi\,(w,z)_x = 0.$$

iii) $\text{Lim}\,(w(.,t) - w^o) = \text{Lim}\,(z(.,t) - z^o) = 0$ dans $L^1(\mathbb{R})$.

□

Le problème de Cauchy pour (4.1) est donc bien posé dans la classe des fonctions à variation bornée. Remarquons que l'inégalité entropique est ici remplacée par les égalités (4.5) qui sont toutes compatibles entre elles. Les entropies du système sont les fonctions $\Phi(w,z)=(a(w)+b(z))/(w-z)$ où a(.) et b(.) sont des fonctions arbitraires d'une variable. Le flux associé à Φ est alors $\Psi(w,z) = (za(w) + wb(z))/(w-z)$.

2- Les inconnues relaxées

Contrairement au cas vraiment non linéaire, la formule (2.5) est équivalente à l'équation (2.4), et on a $K(w,z) = \text{Log}\,(z-w)$. Les mesures $\nu^i_{x,t}$, ou leurs fonctions de répartition W et Z, n'obéissent donc à aucune contrainte de structure. Les oscillations des suites w^ε et z^ε peuvent se propager comme dans le système de Carleman [6]. Nous allons établir un système qui gouverne l'evolution de W et Z, analogue à celui que Tartar construisit dans cet article.

Nous supposerons que pour tout $\varepsilon > 0$, les inégalités (4.4) sont satisfaites par les conditions initiales w^ε_o, z^ε_o, avec des constantes α, β, γ, δ indépendantes de ε. Grâce au théorème 4.1, les solutions w^ε, z^ε ont lesm êmes bornes dans le cas des solutions BV obtenues par le schéma de Glimm. Plus généralement, nous traiterons une suite $(w^\varepsilon, z^\varepsilon)$ qui satisfait à ces inégalités. Par passage à la limite, il est immédiat que

(4.6) $$\gamma \leq W(x,t,y) \leq \alpha \quad , \quad \beta \leq Z(x,t,y) \leq \delta \text{ sur } Q_T \times]0,1[.$$

Réécrivant (2.7)et (2.8) pour une entropie de la forme $\Phi = a(w)/(z-w)$, il vient

(4.7) $$\text{Lim}\,\Phi\,(w^\varepsilon,z^\varepsilon) = u(x,t)\int_0^1 a(W(x,t,y))dy,$$

où

(4.8) $$\text{Lim}\,(z^\varepsilon-w^\varepsilon)^{-1} = u = \{\int_0^1 (Z(x,t,y) - W(x,t,y))dy\}^{-1}$$

De même pour le flux

$$\Psi = za(w) \,/\, (z-w) :$$

(4.9) $$\text{Lim}\,\Psi(w^\varepsilon, z^\varepsilon) = v(x,t)\int_0^1 a(W(x,t,y))dy,$$

où

$$\text{(4.10)} \qquad \mathrm{Lim}\ z^{\varepsilon}\ (z^{\varepsilon}-w^{\varepsilon})^{-1} = v = u \int_0^1 Z(x,t,y)dy.$$

3- Le système d'évolution relaxé

Pour obtenir ce système, il suffit de passer à la limite dans la collection d'égalités entropiques (4.5). On suppose ces égalités vérifiées par chaque $(w^{\varepsilon},z^{\varepsilon})$, ce qui est vrai pour les solutions obtenues par le schéma de Glimm. Lorsque l'entropie et le flux sont $a(w)/(z-w)$ et $za(w)/(z-w)$, on obtient grâce à (4.7) et (4.9) :

$$\text{(4.11)} \qquad \frac{\partial}{\partial t}\{u \int_0^1 a \circ W\, dy\} + \frac{\partial}{\partial x}\{v \int_0^1 a \circ W\, dy\} = 0,$$

pour toute fonction continue a (.) sur $[\gamma,\alpha]$.

En prenant une entropie de la forme $b(z)/(z-w)$, on obtiendra aussi

$$\text{(4.12)} \qquad \frac{\partial}{\partial t}\{ u \int_0^1 b \circ Z\, dy\} + \frac{\partial}{\partial x}\{(1+v) \int_0^1 b \circ Z\, dy\} = 0,$$

pour toute fonction continue b (.) sur $[\beta,\delta]$.

Avant d'interpréter plus clairement ces deux collectionsd'équations, déterminons les conditions initiales pour W et Z. Etant donnée une fonction test $\Theta \in \mathcal{D}(\mathbb{R} \times [0,T[)$, on écrit l'égalité (4.5), sous la forme :

$$\iint_{Q_T} \{\Phi(w^{\varepsilon},z^{\varepsilon})\Theta_t + \Psi(w^{\varepsilon},z^{\varepsilon})\Theta_x\}dxdt + \int_{\mathbb{R}} \Theta(x,0)\Phi(w_o^{\varepsilon},z_o^{\varepsilon})dx = 0.$$

Passant à la limite pour $\Phi = a(w)/(z-w)$, on obtient

$$\iint_{Q_T} (\Theta_t u + \Theta_x v)(\int_0^1 a \circ W\, dy)\, dxdt$$

$$+ \int_{\mathbb{R}} \Theta(x,0) < \nu_x^o , \frac{a(w)}{z-w} > dx = 0.$$

Et compte tenu de 4.11, il reste

$$\int_{\mathbb{R}} \Theta(x,0)\ \{< \nu_x^o , \frac{a(w)}{z-w}> - (u\int_0^1 a \circ W\, dy)(x,0)\}\ dx = 0,$$

pour toute fonction test θ. C'est-à-dire

$$(4.13)\qquad u\int_0^1 a \circ W\,dy = \langle \nu_x^o , \frac{a(w)}{z-w} \rangle \qquad \text{à } t = 0,$$

au sens des traces, et pour toute fonction a(.) continue sur $[\gamma,\alpha]$. Celà peut s'écrire encore

$$(4.14)\qquad \int_0^1 a\,(W(.,0,y))dy = \text{Lim } p_o^\varepsilon\, a\,(w_o^\varepsilon) \,/\, \text{Lim } p_o^\varepsilon,$$

avec $p_o^\varepsilon = (w_o^\varepsilon - z_o^\varepsilon)^{-1}$. Ces égalités, vérifiées pour toute fonction a (.), déterminent de manière unique W (x,0,.) comme fonction croissante et continue à gauche, et celà pour presque tout x. De même, on obtient Z (x,0,y) par les égalités

$$(4.15)\qquad \int_0^1 b(Z(.,0,y))\,dy = \text{Lim } p_o^\varepsilon\, b(z_o^\varepsilon)/\text{Lim } p_o^\varepsilon,$$

pour toute fonction continue b.

Il est clair que le système relaxé qui decrit les solutions oscillantes de (4.1) est constitué des équations d'évolution (4.11) et (4.12), et des conditions initiales (4.14) et (4.15). La définition un peu bizarre de celles-ci montre que le système (4.1), tel qu'il est écrit, est mal posé dans $L^\infty\,(Q_T)$. Par contre, il existe au moins uneforme de ce système sous laquelle il est bien posé. La voici :

$$(4.16)\qquad \bar u_t + \bar v_x = 0, \qquad \bar v_t + ((1+\bar v)\,\bar v/\bar u)_x = 0,$$

où

$$\bar u = (z-w)^{-1},\ \bar v = z\bar u,$$

et

$$(1+\bar v)\,\bar v/\bar u = wz/(z-w).$$

En effet $(\bar u^\varepsilon, \bar v^\varepsilon)$ satisfont ce système, et si on pose $u = \text{Lim } \bar u^\varepsilon$, $v = \text{Lim } \bar v^\varepsilon$ on retrouve les définitions (4.8) et (4.10). En prenant $a \equiv 1$ dans (4.11), on retrouve bien $u_t + v_x = 0$, et en prenant $a \equiv w$, on obtient

$$v_t + (\text{Lim } w^\varepsilon z^\varepsilon/(z^\varepsilon - w^\varepsilon)^{-1})_x = 0.$$

Or, d'après (2.7)

$$\text{Lim}\ w^\varepsilon z^\varepsilon/(z^\varepsilon - w^\varepsilon)^{-1} = u \int_0^1\int_0^1 W(x,t,y)Z(x,t,y')dydy'$$

$$= u \int_0^1 W\,dy. \int_0^1 Z\,dy' = u(v/u)((1+v)/u)$$

$$= v\ (1+v)/u.$$

Finalement $(u,v) = (\text{Lim}\ u^\varepsilon, \text{Lim}\ v^\varepsilon)$ satisfont bien le système (4.16). Ce raisonnement fournit le résultat suivant :

Théorème 4.2 : Soit (w_o, z_o) deux fonctions de $L^\infty(\mathbb{R})$, satisfaisant l'inégalité (4.4), et posons $u_o = (z_o - w_o)^{-1}$, $v_o = z_o u_o$. Alors le système (4.16) possède une solution (u,v) de classe $L^\infty(Q_T)$, dont la condition initiale est (u_o, v_o).

□

Démonstration :

Construire une suite (w_o, z_o) de fonctions à variation bornée sur $\mathbb{R}$, vérifiant (4.4), et telles que $u_o = \text{Lim}\ (z_o^\varepsilon - w_o^\varepsilon)^{-1}$ et $v_o \triangleq \text{Lim}\ z_o^\varepsilon(z_o^\varepsilon - w_o^\varepsilon)^-$ Appliquer le théorème 4.1, puis passer à la limite.

4 - Interprétation

Nous effectuons maintenant des calculs formels dans le but d'aboutir à un système d'équations aux dérivées partielles en (x,t), dans lequel y joue le rôle d'un paramètre de couplage. Développons d'abord (4.11), en supposant que la dérivation dans le produit et celle sous le signe d'intégration sont licites. En vertu de la relation $u_t + v_x = 0$, il vient :

$$\int_0^1 a'(W)\ (W_t + \lambda W_x)\ dy = 0, \tag{4.17}$$

où $\lambda(x,t) = v/u = \int_0^1 Z\ dy$.

Puisque celà doit être vérifié pour toute fonction régulière $a(.)$, et que W est croissante, il s'ensuit que $W_t + \lambda W_x$ doit être nul, sauf peut-être sur les paliers de W. Mais si sur un ouvert ω de $Q_T \times]0,1[$, W ne dépend que de x,t, alors on doit avoir $(W_t + \lambda W_x)_y = W_{yt} + \lambda W_{yx} \equiv 0$. De sorte que $W_t + \lambda W_x$ est constant sur les paliers de W qui ne sont pas isolés. Enfin, (4.17) implique que la moyenne de $W_t + \lambda W_x$ soit nulle sur un tel palier. Finalement il vient

(4.18) $$W_t + W_x \int_0^1 Z\,(y')\,dy' = 0,$$

et de même

(4.19) $$Z_t + Z_x \int_0^1 W\,(y')\,dy' = 0,$$

dans $Q_T \times \,]0,1[$.

Ce système exprime le plus clairement possible l'évolution des oscillations de la suite $(w^\varepsilon, z^\varepsilon)$. Remarquons que si nous en prenons la moyenne sur $]0,1[$, on obtient un système analogue à (4.1) pour λ et $\mu = \int_0^1 W(y')\,dy'$.

5 - Vitesses de groupe

En principe, la valeur de w(x,t) se propage avec la vitesse (variable) z(x,t) car (4.1) apparaît comme un système de deux équations de transport. De même z est constant sur les courbes (x(t),t) de pente w(x(t),t). Mais lorsqu'on met des oscillations importantes dans les conditions initiales, on observe des vitesses de groupe, qui ne sont pas Lim w^ε et Lim z^ε , mais μ et λ. Il est remarquable qu'on puisse les comparer. Le résultat est conforme à ce qui se passe pour les schémas aux différences pour une équation de transport : les vitesses de groupe sont encadrées par les vitesses des structures individuelles. Plus précisement :

Proposition : On a

$$\text{Lim } w^\varepsilon \leq \lambda \leq \alpha < \beta \leq \mu \leq \text{Lim } z^\varepsilon$$

□

Démonstration : On a d'après (2.7), et puisque $u \geq 0$:

$$\text{Lim } w^\varepsilon = u \int_0^1\!\!\int_0^1 W(y)\,(Z(y') - W(y))dy$$

$$\leq u \int_0^1 W(y)\,dy \,.\, \int_0^1 (Z(y') - W(y'))dy',$$

d'après l'inégalité de Cauchy-Schwartz. C'est à dire, d'après (4.8), Lim $w^\varepsilon \leq \lambda$. On montre de même que $\mu \geq$ Lim z^ε.

CQFD.

V- Un système avec second membre

Considérons plus généralement un système analogue à (4.1), avec un second membre d'ordre zéro :

(5.1) $$w_t + zw_x = F(w,z)\ ,\quad z_t + wz_x = G(w,z),$$

où F et G sont régulières. En supposant que F et G permettent un principe du maximum, on a un résultat analogue au théorème 4.1 dans la classe BV ($\mathbb{R}$). On peut donc procéder comme au paragraphe précédent, pour obtenir un système nettement plus compliqué que (4.18), (4.19). En effet, d'après (2.7) on obtient pour toute fonction continue a :

$$\text{(5.2)} \qquad \frac{\partial}{\partial t} u \int_0^1 a \circ W \, dy + \frac{\partial}{\partial x} v \int_0^1 a \circ W \, dy$$

$$= \text{Lim} \{a' (w^\varepsilon) F (z^\varepsilon - w^\varepsilon)^{-1} + a (w^\varepsilon)(G^\varepsilon - F^\varepsilon)(z^\varepsilon - w^\varepsilon)^{-2}\}$$

$$= u \int_0^1 \int_0^1 \{a'(W) F(W,Z) + a (W) H(W,Z)\} dy \, dy',$$

où $H = (G-F)/(z-w)$. Faisant d'abord $a \equiv 1$, on a

$$\text{(5.3)} \qquad u_t + v_x = u \int_0^1 \int_0^1 H(W,Z) \, dy \, dy'.$$

Développant alors (5.2) et utilisant (5.3), on obtient formellement

$$\text{(5.4)} \qquad \int_0^1 a'(W) \{W_t + \lambda W_x\} dy = \int_0^1 \int_0^1 \{a'(W)F(W,Z)$$

$$+ a(W) H(W,Z)\} dy dy' - \int_0^1 a(W) dy \, . \int_0^1 \int_0^1 H(W,Z) dy \, dy'.$$

Mais $H(W(y),Z(y')) - \int_0^1 H(W(\sigma), Z(y')) d\sigma$

est la dérivée par rapport à y d'une fonction Φ (y,y'), nulle pour y=0 ou 1. En intégrant par partie, on trouve

$$\int_0^1 \int_0^1 a(W) H(W,Z) dy \, dy' - \int_0^1 a(W) \, dy \, . \int_0^1 \int_0^1 H(W,Z) \, dy \, dy'$$

$$= - \int_0^1 \int_0^1 a'(W) \, \Phi \, \frac{\partial W}{\partial y} \, dy \, dy'.$$

Le même raisonnement qu'au paragraphe précédent conduit donc à l'équation

$$W_t + \lambda W_x = \int_0^1 F(W,Z) \, dy' - \int_0^1 \frac{\partial W}{\partial y} \Phi(y,y') \, dy',$$

c'est-à-dire :

(5.5)
$$W_t + \lambda W_x = \int_0^1 F(W,Z)\,dy' + \frac{\partial W}{\partial y}\left\{ y\int_0^1\int_0^1 H(W,Z)\,d\sigma dy' - \int_0^y\int_0^1 H(W,Z)\,d\sigma dy'\right\}.$$

Et de même,

(5.6)
$$Z_t + \mu Z_x = \int_0^1 G(W,Z)\,dy + \frac{\partial Z}{\partial y'}\left\{y'\int_0^1\int_0^1 H(W,Z)\,dy d\sigma' - \int_0^1\int_0^{y'} H(W,Z)\,dy\, d\sigma'\right\}.$$

Par contre, les conditions initiales restent inchangées. Il est clair que pour des fonctions F et G arbitraires, il est impossible d'extraire de (5.5) et (5.6) un système d'équations aux dérivées partielles pour un nombre fini de fonctions de x et t, contrairement au cas homogène où l'on avait (4.16).

Bibliographie

[1] K. Chueh, C. Conley, J. Smoller : Positively invariant regions for systems of non linear diffusion equations. Ind. J. Maths. 26 (1977) 373-392.

[2] R. Di Perna : Convergence of approximate solutions to conservation laws. Arch. Rat. Mech. Anal. 82 (1983) 27-70.

[3] R. Di Perna : Measure - valued solutions to conservation laws. Arch. Rat. Mech. Anal. 88 (1985) 223-270.

[4] J. Glimm : Solutions in the large for nonlinear hyperbolic systems of equations. Comm. Pure Appl. Maths. 18 (1965) 697-715.

[5] P.D. Lax : Hyperbolic systems of conservation laws, II. Comm. Pure Appl. Maths. 10 (1957) 537-566.

[6] Mc Laughlin, G. Papanicolaou, L. Tartar : Lecture Notes in Physics no. 230, Macroscopic modelling of turbulent flows. Frisch, Keller, Papanicolaou, Pironneau eds. Springer-Verlag.

[7] F. Murat : Compacité par compensation. Ann. Scuola Norm. Sup. Pisa 5 (1978) 489-507.

[8] F. Murat : L'injection du cône positif de H^{-1} dans $W^{-1,q}$ est compacte pour tout $q < 2$.

[9] M. Rascle : Thèse, Lyon 1983.

[10] M. Rascle : Un résultat de "compacité par compensation" à coefficients variables. Application à l'élasticité non linéaire. C.R.A.S. 302 (1986) 311-314.

[11] M. Rascle, D. Serre : Compacité par compensation et systèmes hyperboliques de lois de conservation. Applications. Comptes Rendus Acad. Sciences. 299 (1984) 673-676.

[12] D. Serre : Compacité par compensation et systèmes hyperboliques de lois de conservation C.R.A.S. 299 (1984), 555-558.

[13] D. Serre : Domaines invariants pour les systèmes hyperboliques de lois de conservation. Publication de l'UA 740 Lyon-Saint-Etienne, no. 53 (1986).

[14] D. Serre : Solutions à variation bornée pour certains systèmes hyperboliques de lois de conservation. J. Diff. Equ. A paraître

[15] L. Tartar : Compensated Compactness and applications to PDE, in Research Notes in Maths : Nonlinear Analysis and Mechanics, Heriot-Watt symposium, 4 (1979). R.J. Knops ed. Pitman Press.

STABILITY AND DECAY IN SYSTEMS OF CONSERVATION LAWS

Blake Temple
Department of Mathematics
University of Wisconsin-Madison
Madison, Wisconsin 53706

§1. INTRODUCTION

I am going to discuss the issue of the continuous dependence of solutions on the initial data for systems of hyperbolic conservation laws. Systems of conservation laws arise in many applications - a prime example being Eulers equations for compressible, inviscid flow in gas dynamics - and although these equations have been studied extensively during the last century, it is still not known that the initial value problem is well-posed. It is believed that the initial value problem in one space dimension is well-posed in the norm L^1, but even in this setting the stability and uniqueness of solutions has remained an unsolved problem since the time of Euler. In [24] the author obtained a decay result which implies that the constant state is stable with respect to perturbations of the initial data in L^1_{loc}. This provides the first stability result for systems in a norm in which it is reasonable to believe that continuous dependence holds for the general initial value problem. Here I am going to discuss the issue of continuous dependence with the view that the problem is difficult in systems due to the spontaneous formation and decay of "spikes" in solutions. This phenomenon is a consequence of the nonlinear coupling between the characteristic fields. We introduce the idea that these spikes give rise to singularities in the stability problem. We indicate this from two perspectives. First, the time evolution of a solution from initial data given by a single spike involves an infinite family of reflected waves which effect its decay. In the continuous dependence problem, spikes form in solutions, rescaling occurs, and stability becomes a consequence of local decay. Thus, in this setting a spike represents a singularity in the sense that an infinite family of reflected waves will appear in a neighborhood of the point where the local decay occurs. From a second point of view, the results in [22] indicate that the formation of a spike in a continuous dependence process also leads to a singularity in the function $C(t)$ which relates the derivative of the L^1 difference between solutions at time t to the L^1-difference itself: $C(t)$ becomes infinite at times where local decay occurs. This implies, among other things, the failure of simple Grönwall inequalities in the norm L^1. In short, the formation and decay of spikes in a continuous dependence process leads to singularities in the wave structure of solutions as well as singularities in the function which relates L^1-differences to their time derivatives. The author believes that the above singularities are fundamental to the continuous dependence and uniqueness problem for systems of conservation laws; and that the central difficulty in obtaining continuous dependence is one of obtaining estimates which link local decay to stability in a general setting. The proof given in [22] can be viewed as the first method of analysis sufficient for linking local decay to stability in the simplest setting of the stability problem for the constant state.

In Section 1 we state Glimm's theorem [6]. In Section 2 we summarize the decay results in [24]. In Section 3 we indicate that L^1 is the most reasonable of the classical norms in which to expect continuous dependence to hold for systems. In Section 4 we describe the formation and decay of spikes in simple continuous dependence problems and interpret these as singularities; and in Section 5 we outline the proof of the decay result, and interpret the method as providing a link between local decay and stability in the stability problem for the constant state.

§1. We consider the initial value problem for a system of conservation laws

$$u_t + f(u)_x = 0 , \tag{1}$$

$$u(x,0) = u_0(x) . \tag{2}$$

Here $u = (u_1,\ldots,u_n)$, $x \in R$, $t > 0$ and $u_0(x)$ denotes the initial data. We look for solutions $u(x,t)$ of (1) which satisfy (2). In principle, solutions of (1) propagate in waves along characteristics something like the linear equation $u_t + u_x = 0$, but for (1) f is nonlinear and so wave speeds depend on the solution u. The nonlinearity leads to the focusing of characteristics and the formation of shock waves in the presence of smooth solutions, as well as the smoothing of discontinuities by rarefaction waves in the presence of unstable discontinuities. Thus the meaning of "wave" and "characteristic" must be taken in a nonlinear sense, and we must look for generalized (weak) solutions of (1), (2).

In 1965, Glimm gave a beautiful proof that discontinuous solutions of (1), (2) exist for all time [6]. Moreover, he gave an explicit and original method for constructing these discontinuous solutions. It is believed that these solutions are unique and depend continuously on the initial data, but it is not known in what sense these statements hold. Let $\| \ \|_S$ denote the supnorm and TV the total variation. Glimm's theorem in the case $n = 2$ can be stated as follows [23]:

THEOREM (Glimm). If $\|u_0(\cdot) - C\|_S << 1$, and $TV\{u_0(\cdot)\} < \infty$, then there exists a solution $u(x,t)$ of (1) and (2) defined for all time. Moreover,

$$\|u(\cdot,t) - C\|_S \leq \text{Const}\|u_0(\cdot) - C\|_S , \tag{3}$$

$$\|u(\cdot,t_2) - u(\cdot,t_1)\|_{L^1} \leq \text{Const}|t_2 - t_1| . \tag{4}$$

The assumptions on f are that the matrix df have real and distinct eigenvalues, and $\nabla\lambda_i \cdot R_i > 0$ for (λ_i, R_i) eigenpairs of df. Conditions on f involve λ_i because $\lambda_i(u)$ are the characteristic speeds for the state u. The above conditions are called strict hyperbolicity and genuine nonlinearity, respectively.

§2. DECAY RESULT

The author recently obtained the following decay result which holds for solutions $u(x,t)$ obtained by Glimm's method, $n = 2$ (see [24] for details)

THEOREM 2: Assume $u_0(\pm \infty) = C$. Then

$$\|u(\cdot,t) - C\|_S \leq F\left(\frac{t}{\|u_0(\cdot) - C\|_{L^1}}\right) , \tag{5}$$

where $F(\xi) \to 0$ as $\xi \to \infty$. In fact,

$$F(\xi) = \{\log(\xi)\}^{-1/2} , \tag{6}$$

(constants have been omitted).

Fixing $\|u_0(\cdot) - C\|_{L^1}$ and letting $t \to \infty$ in (5) gives that the solution decays to zero in the supnorm. Thus solutions that are close to the constant state in L^1 decay to the constant state in the supnorm at a rate given by (6). Previous decay results of Glimm/Lax, DiPerna and Liu [4,7,11-13] estimate decay in the total variation. Simple examples show that the total variation does not decay at a rate independent of the support of the initial data. Thus it is crucial that in [5] we estimate decay in the supnorm.

Because (5) gives decay with a rate independent of the support of the initial data, it is also a continuous dependence result for the constant state C; i.e., let u_0^ε be any sequence of initial data satisfying

$$\|u_0^\varepsilon(\cdot) - C\|_{L^1} \to 0 \quad \text{as} \quad \varepsilon \to \infty .$$

Then (5) implies that $\|u^\varepsilon(\cdot,t) - C\|_\infty \to 0$ for all $t > 0$. This is the first continuous dependence result in the norm L^1 for systems of conservation laws. It is an open problem in what sense continuous dependence holds for initial data more general than the constant state, but below we indicate that Theorem 2 is the first stability result in a norm in which it is reasonable to believe that the initial value problem is well-posed.

The methods also give directly that periodic solutions decay to the average value C in each period at a rate given by (5), which is independent of the period; and an analysis which uses the period P to estimate the L^1-norm gives

$$\|u(\cdot,t) - C\|_S < \left(\frac{P}{t}\right)^{1/4} .$$

(Again constants have been omitted.) This is algebraic in t, but P/t is obtained in [7] for decay in the total variation, and this is sharp.

§3. THE SCALAR CASE

We next consider the case when (1) is a single equation in order to determine in what sense one expects continuous dependence to hold for systems of conservation laws. We first make some definitions. We say that solutions of (1), (2) are <u>strongly stable</u> in a norm $\|\ \|$ if

$$\|u(\cdot,t) - v(\cdot,t)\| < C\|u(\cdot,0) - v(\cdot,0)\|$$

for any solutions u, v of (1). We say that solutions are weakly stable in $\|\ \|$ if

$$\|u(\cdot,t) - v(\cdot,t)\| < F(\|u(\cdot,0) - v(\cdot,0)\|)$$

where $F(\xi) \to 0$ as $\xi \to 0$. Because of finite speed of propagation, we need only consider strong and weak stability in $\|\ \|$ locally, by which we mean that

$$\|u(\cdot,t) - v(\cdot,t)\|_{[-M,M]} \to 0 \quad \text{as} \quad \|u(\cdot,0) - v(\cdot,0)\|_{[-N,N]} \to 0$$

in the above senses, where $[-N,N]$ contains the domain of dependence of $[-M,M]$ at time t. In this language, Theorem 2 implies that the constant state is weakly stable in L^1_{loc}. (For systems, it is an open problem whether the constant state is strongly stable in L^1.) In the case (1) is a scalar equation, solutions are L^1-contractive [8]; i.e.

$$\int_{-\infty}^{\infty} |u(x,t) - v(x,t)|dx \leq \int_{-\infty}^{\infty} |u(x,0) - v(x,0)|dx \tag{8}$$

for all solutions u and v. Thus in the scalar case, solutions are strongly stable in L^1 with constant $C = 1$. We now give an example which shows that solutions of the scalar equation are not weakly stable in L^∞ or in the total variation norm, and are not strongly stable in any L^p, $p > 1$. This leaves L^1 as a leading candidate for stability in systems of conservation laws. (In fact in [22] it is proved that in general, solutions to systems of conservation laws are not L^1-contractive when the absolute value in (8) is replaced by any nondegenerate metric. We interpret this in the context of the stability issue in the next section.)

For our example, consider the simplest nonlinear scalar conservation law (the inviscid Burger's equation)

$$u_t + \left(\tfrac{1}{2} u^2\right)_x = 0 \, . \tag{9}$$

Rewriting (9) gives

$$u_t + uu_x = D_{\vec{V}} u = 0 \, , \tag{10}$$

where $D_{\vec{V}}$ is the derivative in the direction $(u,1)$ in xt-space. Thus for smooth solutions, (10) says that if $u(\bar{x},0) = \bar{u}$, then $u(x,t) = \bar{u}$ along the line

$$x - \bar{x} = \bar{u}t \, . \tag{11}$$

Thus discontinuities form when values of $u_0(x)$ are chosen so that two such lines (11) intersect (when $\frac{du_0}{dx} < 0$). To analyze discontinuous solutions, we first find solutions that evolve from a simple discontinuity,

$$u_0(x) = \begin{cases} u_L & x < 0 \\ u_R & x > 0 \, . \end{cases} \tag{12}$$

The problem (1), (12) is called the Riemann problem. When $u_R > u_L$, we can use (10) and (11) to solve (12); i.e., we obtain a solution by taking $u(x,t) = \bar{u}$ on the lines satisfying

$$\frac{dx}{dt} = \frac{d}{du}\left(\tfrac{1}{2} u^2\right)\Big|_{u=\bar{u}} = \bar{u} \, .$$

The solution looks like a fan (see Figure 1). When $u_R < u_L$ this procedure fails, since then the slopes $\bar{u}$ decrease from u_L to u_R. In this case we appeal to the Rankine Hugoniot condition (which derives from an integral form of (1))

$$s(u_R - u_L) = f(u_R) - f(u_L) \, . \tag{13}$$

For (9), (13) states that any discontinuity must propagate with speed s satisfying

$$s = \frac{u_L + u_R}{2} \, .$$

Thus the solution is a "shock wave" or discontinuity that moves with speed $\frac{dx}{dt} = s$, and separates the two constant states u_L and u_R (see Figure 2).

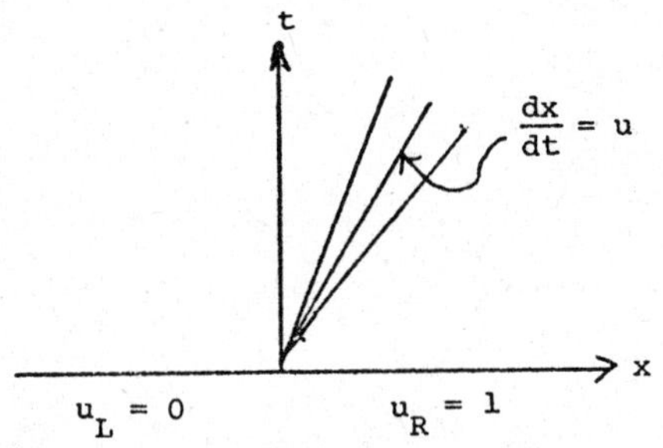

Figure 1. Rarefaction wave solution.

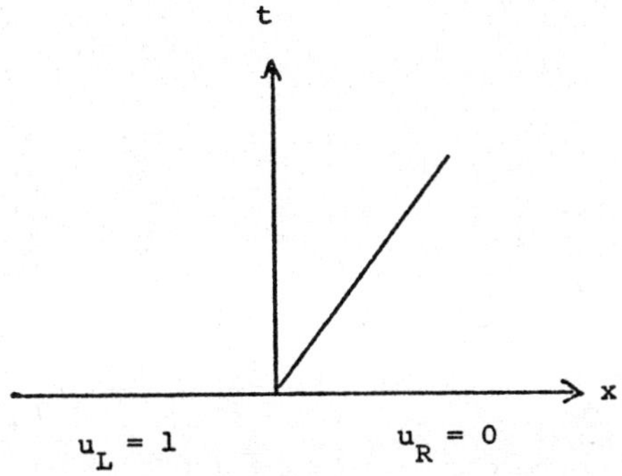

Figure 2. Shock wave solution.

Now consider the shock wave solution $u(x,t)$ that evolves from initial data

$$u_0(x) = \begin{cases} 1, & x < 0 \, , \\ 0, & x > 0 \, , \end{cases}$$

as depicted in Figure 2. Let $u^\varepsilon(x,t)$ be the solution obtained by perturbing the initial data to

$$u_0^\varepsilon(x) = \begin{cases} 1 & x < 0 , \\ -\varepsilon & 0 < x < 1 , \\ 0 & 1 < x . \end{cases}$$

The corresponding solution is diagrammed in Figure 3 for $t < 1$, ε sufficiently small. One can verify that

$$\begin{aligned} &TV\{u_0^\varepsilon(\cdot) - u_0(\cdot)\} = 2\varepsilon , \\ &TV\{u^\varepsilon(\cdot,1) - u(\cdot,1)\} > 1 , \\ &\| u_0^\varepsilon(\cdot) - u_0(\cdot)\|_{L^\infty} = \varepsilon , \\ &\| u^\varepsilon(\cdot,1) - u(\cdot,1)\|_{L^\infty} > 1 , \\ &\| u_0^\varepsilon(\cdot) - u_0(\cdot)\|_{L^p} = \varepsilon , \\ &\| u^\varepsilon(\cdot,1) - u(\cdot,1)\|_{L^p} = O(1)\varepsilon^{1/p} . \end{aligned} \tag{14}$$

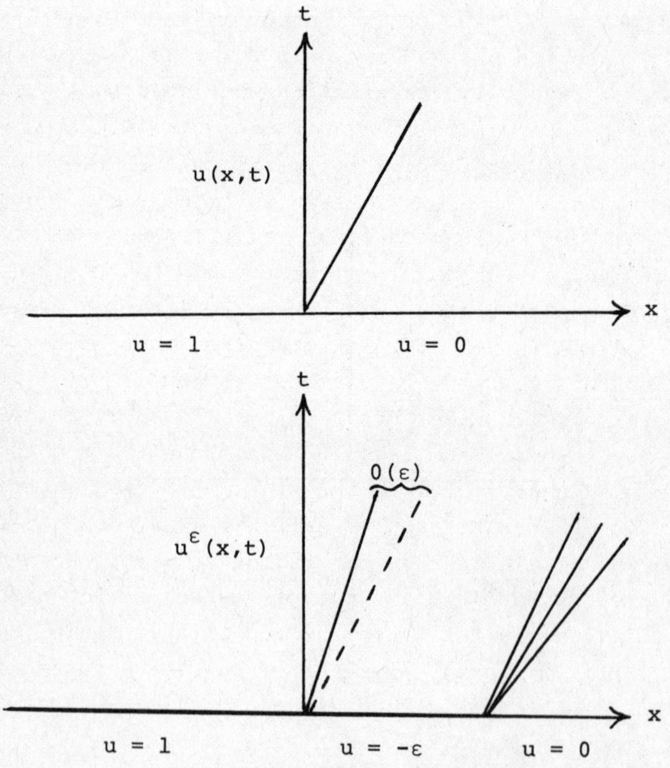

Figure 3. Perturbation of a shock wave.

Statement (14) verifies the claim that solutions are not stable in TV or L^∞, and not strongly stable in L^p for $p > 1$. Thus we expect L^1 to be the leading candidate for a norm in which to prove continuous dependence for systems.

§4. FORMATION AND DECAY OF SPIKES

A major complication in systems is that "spikes" spontaneously form and decay in solutions due to the fact that more than one family of waves is present (see [20] for examples of coupled systems in which this phenomenon does not occur). In a continuous dependence problem, spikes can form, rescaling occurs, and stability becomes a consequence of local decay. We interpret the formation of a spike in this setting as a singularity in the problem.

In order to describe such singularities in examples, we review the solution of the Riemann problem (1), (2), (12) for systems of conservation laws [9] (we take $n = 2$). In this case there are in general two waves which appear in the solution of a Riemann problem. Indeed, for each state u_L we can construct a coordinate system that indicates the solution of the Riemann problem as follows (we assume $u_L \approx u_R$): For fixed u_L, construct the curve $S_1(\underset{\sim}{u}_L) \cup R_1(\underset{\sim}{u}_L)$; and for each $u_M \in S_1(u_L) \cup R_1(u_L)$ construct $S_2(u_M) \cup R_2(u_M)$, where $R_i(v)$ denotes the i-rarefaction curve through v (the integral curve of R_i in the direction $\lambda_i(u) > \lambda_i(v)$) and $S_i(v)$ denotes the i-shock curve through v (the component of the Hugoniot locus of v that completes $R_i(v)$ to a C^2 curve through v [9]). The curves $R_i(u_L)$ give the right states u_R of the rarefaction waves with left state u_L, and the curves $S_i(u_L)$ give the right states of the shock waves with left state u_L. The Riemann problem (1), (2), (12) is solved by a 1-wave with right state u_M followed by a 2-wave with right state u_R, where the state u_M is obtained from the constructed coordinate system as the state with the 1-coordinate of u_L and the 2-coordinate of u_R (see Figure 4). Now generically, the shock curves do not lie on the integral curves R_i, and this leads to the formation of spikes in solutions of systems of conservation laws. (We assume this case in the examples, and this is equivalent to assuming $\nabla R_i \cdot R_i \neq 0$ in the presence of genuine nonlinearity [20].) We now describe the formation and decay of spikes in two simple continuous dependence problems. In these examples, $u_0^\varepsilon(x)$ denotes a sequence of initial data tending to $u_0(x)$ in L^1 as $\varepsilon \to 0$, and

$u\ (x,t)$ and $u(x,t)$ denote the corresponding solutions of (1), (2) generated by Glimm's method.

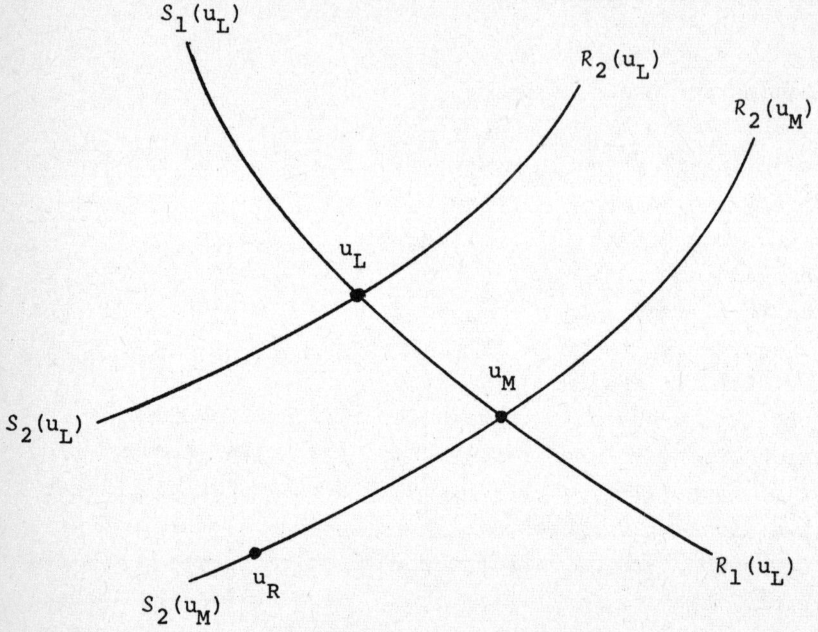

Figure 4. The coordinate system of shock and rarefaction curves determined by u_L.

The first example is relevant to the stability problem for the constant state. Here

$$u_0^\varepsilon(x) = \begin{cases} \bar{v} & 0 < x < \varepsilon \\ u & \text{otherwise} \end{cases} \tag{15}$$

where $\bar{v} \in S_1(u)$, so that $u_0^\varepsilon \to u$ in L^1 as $\varepsilon \to 0$. In Figure 5A the configuration of states is sketched in Riemann invariant coordinates r and s (so that integral curves $R_1(w)$ lie on $r =$ const. and $R_2(w)$ lie on $s =$ const. for all states w). The initial data $u_0^\varepsilon(x)$ is sketched in Figure 5B, and is a spike of height $|\bar{v} - u|$ and width ε. Under the time evolution governed by (1), this spike decays to u schematically in a manner diagrammed in Figure 5C. Here all waves are drawn as sharp discontinuities. In principle waves of different families cross with essentially only a change in wave speed, but waves in the same family intersect, and transmit and reflect waves of diminished strength. It is this diminishing of strength, or "cancellation" that leads to the decay of the solution to u in the supnorm [7,15]; i.e., the sequence of states $\{v_n\}$ in Figure 5C tends to the value u as $n \to \infty$. Now consider the continuous dependence problem $\varepsilon \to 0$, $u_0^{\varepsilon} \to u$ in L^1. By the rescaling properties of (1),

$$u^{\varepsilon}(x,t) = u^1(\frac{1}{\varepsilon} x, \frac{1}{\varepsilon} t) .$$

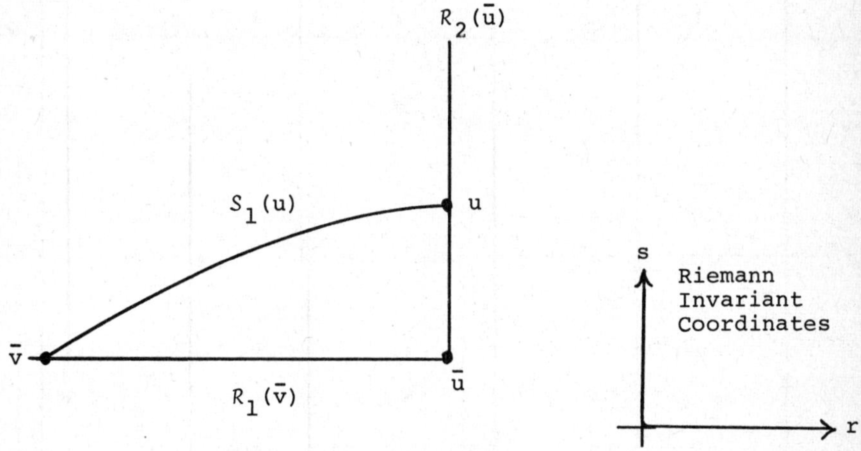

Figure 5A. States in $u_0^{\varepsilon}(x)$ for Example 1.

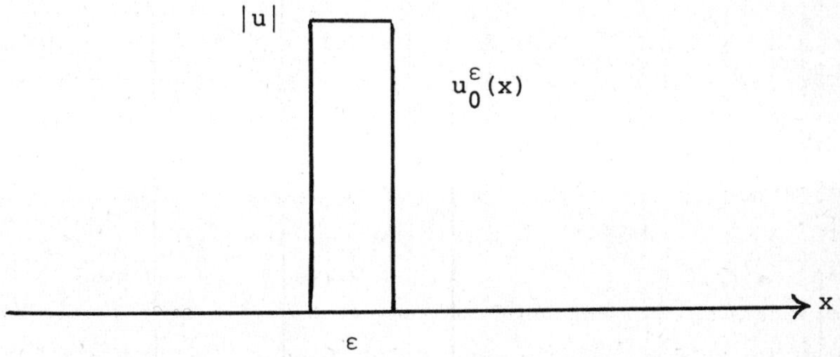

Figure 5B. The spike in $u_0^{\varepsilon}(\cdot)$ for Example 1.

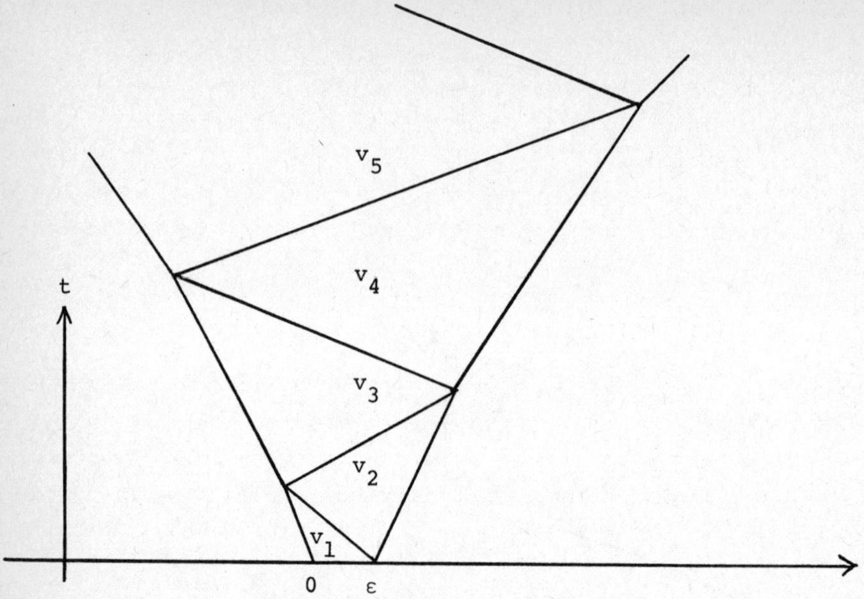

Figure 5C. Decay of $u^{\varepsilon}(x,t)$ for Example 1.

Thus as $\varepsilon \to 0$, the diagram of waves and states $\{v_n\}$ in Figure 5C get rescaled into the origin - and stability in L^1 is preserved because the solution $u^{\varepsilon}(\cdot,t)$ tends to u in the supnorm at times $t > 0$ as $\varepsilon \to 0$. Thus stability is a consequence of local decay. In fact, every reflected wave must be accounted for in order to obtain that $u^{\varepsilon}(\cdot,t) \to u$ in the supnorm for $t > 0$: and hence in order to get L^1-stability of the constant state for positive times. We say that there is a singularity in this problem at time $t = 0$ because an infinite family of reflected waves appears in a neighborhood of the point $x = 0$, $t = 0$ as $\varepsilon \to 0$. But moreover, in [22] we showed that given any neighborhood U of u-space (where our other assumptions hold) and any metric D on U which is compatible with the Euclidean norm, there exists states u and $\bar{v}$ in U, $\bar{v} \in S_1(u)$, such that

$$\frac{d}{dt}\int_{-\infty}^{\infty} D(u^1(x,t),u)dx = \mu > 0$$

at time $t = 0$. This implies that

$$\frac{d}{dt}\int_{-\infty}^{\infty} D(u^{\varepsilon}(x,t),u)dx = \mu \tag{16}$$

for all $\varepsilon > 0$. Let

$$I_\varepsilon(t) = \int_{-\infty}^{\infty} D(u^\varepsilon(x,t),u)dx ,$$

so that (16) gives

$$\frac{d}{dt} I_\varepsilon(t) = \mu ,$$

and let $C_\varepsilon(t)$ be the constant that equates the derivative of the L^1-difference between u^ε and u at time t to the L^1 difference itself; i.e.,

$$\frac{d}{dt} I_\varepsilon(t) = C_\varepsilon(t) I_\varepsilon(t) .$$

Then since $I_\varepsilon(0) = D(\bar{v},u)\varepsilon \to 0$ as $\varepsilon \to 0$, we have that $C_\varepsilon(0) \to \infty$ as $\varepsilon \to 0$. Thus the continuous dependence problem $u_0^\varepsilon \to u$ in L^1 also has a singularity at time $t = 0$ in the sense that, in general, the constant that relates the L^1 difference to its time derivative has a singularity at $t = 0$ as $\varepsilon \to 0$. This implies that the simple Grönwall inequality

$$\frac{d}{dt} \| u(\cdot,t) - v(\cdot,t)\|_{L^1} \le \text{Const}\| u(\cdot,t) - v(\cdot,t)\|_{L^1}$$

fails for solutions of (1) in any metric compatible with the Euclidean norm; and thus for systems, there is no such metric for which solutions with shocks form an L^1-contractive semigroup. This indicates the sense in which "spikes" represent singularities in the continuous dependence problem. In the next example, we indicate that analogous spikes form spontaneously in the general stability problem.

For the second example, the solution $u(x,t)$ consists of two slow shock waves which intersect a third fast shock wave at $x = x_0$, $t = t_0$. The interaction at time $t = t_0$ produces one slow shock and one fast shock which are noninteracting thereafter (see Figure 6). Specifically,

$$u_0(x) = \begin{cases} u_L & \text{for} \quad x < x_1 , \\ u_1 & \text{for} \quad x_1 < x < x_2 , \\ u_2 & \text{for} \quad x_2 < x < x_3 , \\ u_R & \text{for} \quad x > x_3 , \end{cases}$$

and $u_1 \in S_1(u_L)$, $u_2 \in S_1(u_1)$, $u_R \in S_2(u_2)$, and x_1, x_2, x_3 are chosen so that all waves interact simultaneously at $t = t_0$. The solution $u^\varepsilon(x,t)$ evolves from initial data $u_0^\varepsilon(x)$ which is obtained from $u_0(x)$ by perturbing the position of the middle shock at time $t = 0$ by an amount $\varepsilon << 1$ in the positive x-direction, so that

$$\| \acute{u}_0(\cdot) - u_0^\varepsilon(\cdot)\|_{L^1} = O(\varepsilon) ,$$

(see Figure 7). This has the effect of producing a spike in the

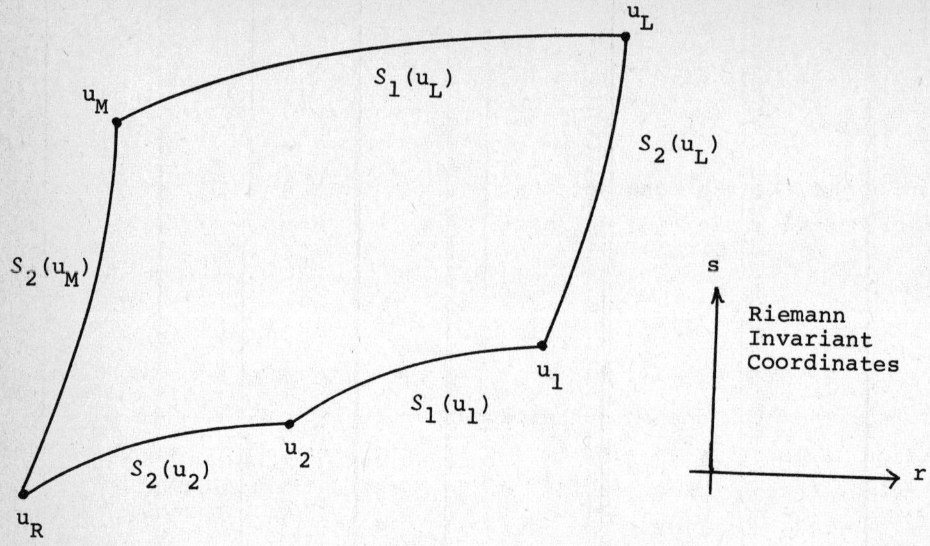

Figure 6A. States in $u_0(\cdot)$ for Example 2.

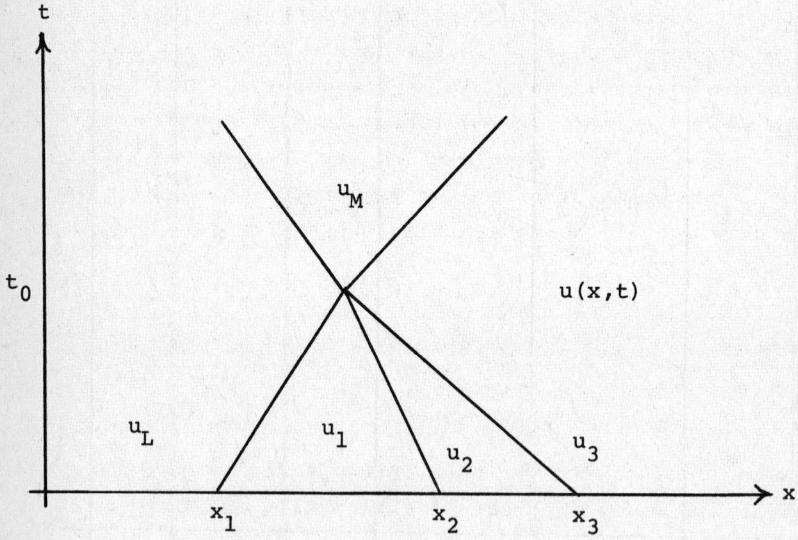

Figure 6B. Time evolution of $u(x,t)$ for Example 2.

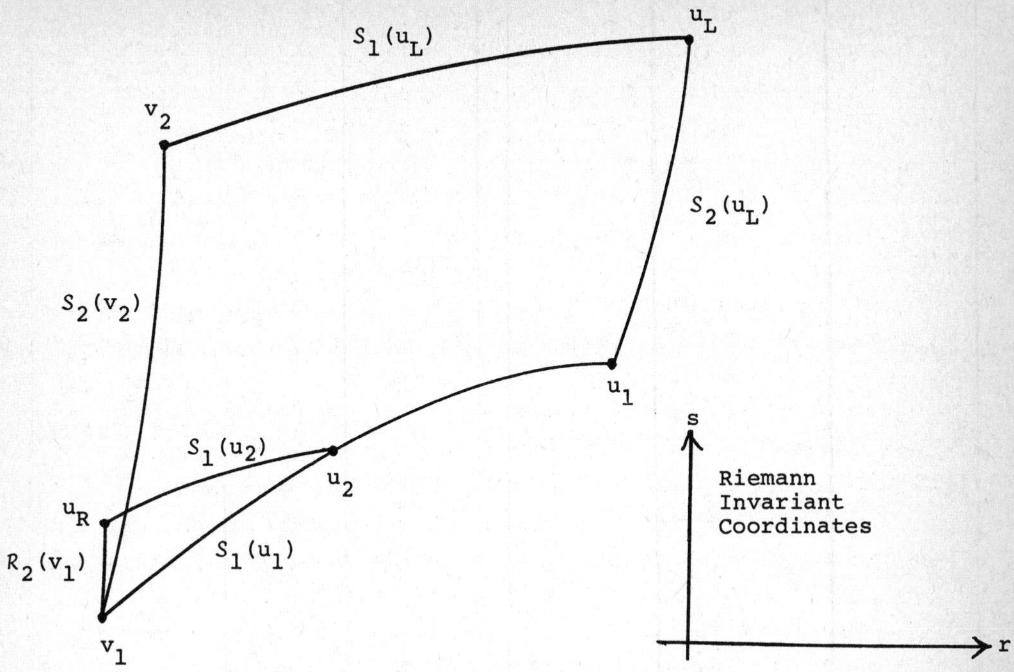

Figure 7A. States in $u^{\varepsilon}(x,t)$ for Example 2.

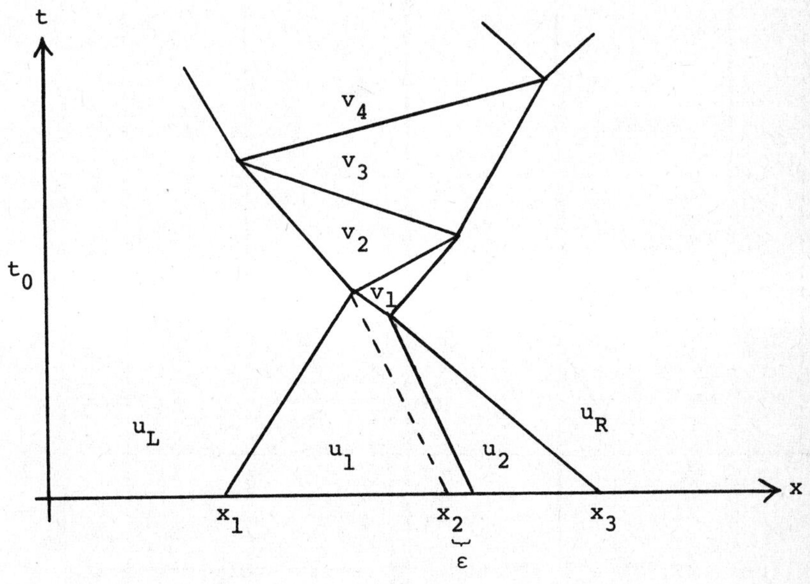

Figure 7B. Decay of $u^{\varepsilon}(x,t)$ for Example 2.

solution $u^\varepsilon(x,t)$ at time $t \approx t_0$, and this spike has no analogue in the solution u. The width of this spike is on the order of ε, but the height is independent of ε, and in fact is on the order of S^3, where S is the strength of the 1-shocks in the problem at time zero. (This follows from the fact that $S_1(u_L)$ and $R_1(u_L)$ have no more than C^3 contact at $u = u_L$ [9].) In Figures 6A and 7A the configuration of states defining $u(x,t)$ and $u^\varepsilon(x,t)$ are diagrammed in Riemann invariant coordinates. The time evolution of the solution $u(x,t)$ is diagrammed in Figure 6B, and the time evolution of $u^\varepsilon(x,t)$ is drawn schematically in Figure 7B. In the latter case all waves are drawn as sharp discontinuities. The s-profiles of the solutions at t_0+ are given in Figures 8A and 8B, and the spike which appears in $u^\varepsilon(\cdot,t_0+)$ is seen to have a height on the order of $|u_R - v_1|$. For fixed ε, the spike decays schematically in a manner diagrammed in Figure 7B. Again, in principle, waves of different families cross with essentially only a change in wave speeds, but waves in the same family intersect, and transmit and reflect waves of diminished strength. It is this diminishing of strength, or cancellation, that causes the solution $u^\varepsilon(x,t)$ to decay to a solution near $u(x,t)$ as $t \to \infty$. Specifically, the sequence of constants $\{v_n\}$ in Figure 7B tend to the constant u_M as $n \to \infty$. Now consider the continuous dependence problem $\varepsilon \to 0$, $u_0^\varepsilon \to u_0$ in L^1. Again by rescaling

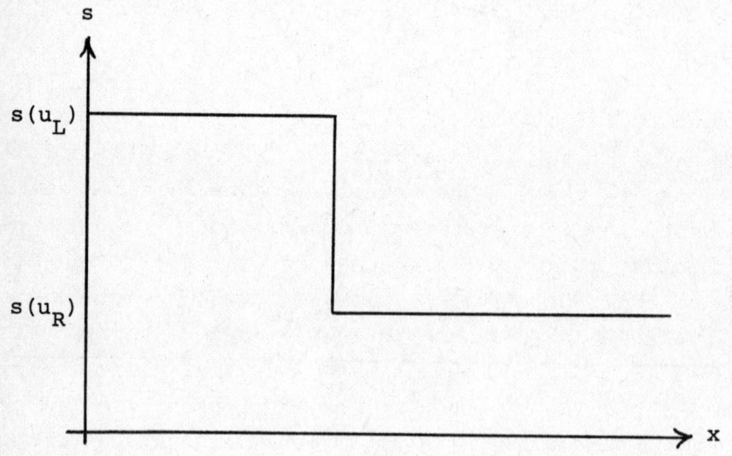

Figure 8A. S-Profile of $u(\cdot,t)$ at $t = t_0+$.

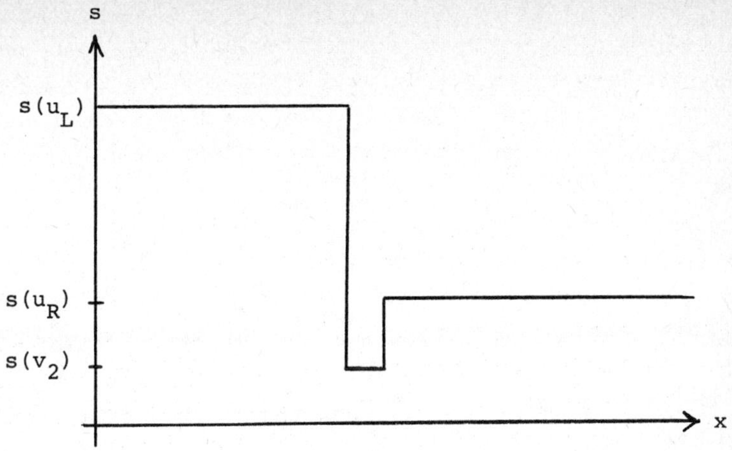

Figure 8B. S-Profile of $u^{\varepsilon}(\cdot,t)$ at $t = t_0+$.

properties of (1), the diagram of waves and states $\{v_n\}$ in Figure 7B get rescaled into the point (x_0,t_0) as $\varepsilon \to 0$ – and stability in L^1 for times $t > t_0$ is preserved because the solution $u^{\varepsilon}(x,t)$ decays in the supnorm to the state u_M near the point (x_0,t_0). Thus again, stability is a consequence of local decay. We say that there is a singularity in the problem at time $t = t_0$ because an infinite family of reflected waves appear in a neighborhood of (x_0,t_0) as $\varepsilon \to 0$. One cannot help but notice the analogy with the first example, and in fact the two spikes are essentially equivalent. Thus, by [22] we expect that in general, the function $C_{\varepsilon}(t)$ defined by

$$\frac{d}{dt}\|u^{\varepsilon}(\cdot,t) - u(\cdot,t)\|_{L^1} = C_{\varepsilon}(t)\|u^{\varepsilon}(\cdot,t) - u(\cdot,t)\|_{L^1}$$

will also tend to infinity at $t = t_0$ as $\varepsilon \to 0$. Thus we see that spikes form at time $t > 0$ in continuous dependence problems even when none exist at $t = 0$. The author believes that the central problem in stability is one of obtaining estimates that link local decay to stability in general. The method of analysis in [24] does this in the case of the stability problem for the constant state, and we indicate this in the next section. The idea is to isolate the time intervals where local decay occurs, and show that these intervals have small measure. The idea is then to use the Lipschitz bound (4) to estimate L^1-differences over the small time intervals which contain the singularities. Note that the measure of the time intervals that contain the singularities must be small in order for solutions to be strongly stable in L^1, i.e., by (16), if for any T

$$\int_0^T C_\varepsilon(t)dt \to \infty \quad \text{as} \quad \varepsilon \to 0 ,$$

then solutions are not strongly stable in L^1.

§5. STABILITY OF THE CONSTANT STATE

We now illustrate the technique used in the proof of (5) in the setting of the decay of a spike (Figure 2). In fact, the techniques rely only on the bounded L^1-norm, not the support of the initial data, and thus local decay is estimated without appealing to global mechanisms (cf. [7]). For simplicity, we assume $u = 0$ in (15). (For details see [24]). The main step in Glimm's Theorem is to show that a certain quadratic function is positive and decreasing. The idea is that since waves in a solution intersect, cross, and never intersect again, it is reasonable to define a function $Q(t)$ that sums the product of the strength of waves in the solution at a given time t that will intersect at a later time;

$$Q(t) = \Sigma |\gamma_i||\gamma_j| . \tag{17}$$

Roughly Q decreases because a term is lost from the sum in (17) whenever two waves γ_i and γ_j cross. For the decay of a spike, we define times $t_j(\varepsilon)$ between which Q decreases by some small number δ, and we view these times as functions of the L^1 norm ε. Let t_{m+1} be the first time that does not tend to zero as $\varepsilon \to 0$. The L^1 norm of the solution at time t_m+ tends to zero by (4) which gives the L^1-Lipschitz continuity of the solutions of Glimm; and so we have reduced the problem to the case that

$$\|u(\cdot,t_m+)\|_{L^1} \to 0 ,$$

and Q does not decrease by more than δ over a time interval (t_m,t_{m+1}) that does not tend to zero with ε. We then conclude that $\|u(\cdot,t_m+)\|_\infty << 1$ by arguing that if not, then there is a spike at time t_m+ which contains strong waves whose first interactions would occur before time t_{m+1}, contradicting the assumption that $Q(t_m) - Q(t_{m+1}) < \delta$.

In the context above, as $\varepsilon \to 0$ the times $\{t_k(\varepsilon)\}$ squeeze together at times where singularities are present, since these are times where many waves cross in a short amount of time. In the case of stability for the constant state, the singularity is at $t = 0$.

Thus the Lipschitz bound (4) is used to estimate the L^1-differences over the small time interval $[0,t_m]$ containing the singularity. The inequality that quantifies the time and strength of first interactions as a function of the L^1-norm and supnorm of the initial data is the main estimate (statement (18) below), and its proof involves an analysis of approximate characteristics. The proof of (18) requires estimates for the time evolution of left and right states on approximate characteristics and this is along the lines of the wave tracing techniques developed by Liu in [15]. Previous decay results have relied on estimates for the time evolution of wave strength and speeds on approximate characteristics, but not left and right states.

For the analysis, a formalism is developed to organize the information contained in the analysis of approximate characteristics. This is an extension of the ideas in [16]. The author believes that the techniques outlined above have applications in the general continuous dependence problem. We now indicate how the main estimate (18) is used to obtain Theorem 2. For simplicity, we assume $C = 0$ in Theorems 1 and 2.

<u>MAIN ESTIMATE</u>. Assume that $u_0(\pm\infty) = 0$, and that u_0 satisfies the conditions of Theorem 1. Then (constants have been omitted)

$$\left| Q\left(\frac{\|u_0(\cdot)\|_{L^1}}{\|u_0(\cdot)\|_S^2}\right) - Q(0)\right| \geq \|u_0(\cdot)\|_S^2 , \tag{18}$$

where Q is the Glimm-quadratic functional associated with any given solution $u(x,t)$ obtained by Glimm's method from initial data u_0. In words, (18) states that Q will decrease on the order of the supnorm squared in a time on the order of the L^1-norm divided by the supnorm squared. Choose $\delta << 1$, and define times $\{t_k\}$ between which Q decreases by no more than δ; specifically,

$$t_{k+1} = \mathrm{Inf}\{t \geq t_k : |Q(t) - Q(t_k+)| \geq \delta\} ,$$

so that

$$|Q(t_{k+1}+) - Q(t_k+)| \geq \delta ,$$

$$|Q(t_{k+1}-) - Q(t_k+)| \leq \delta .$$

It follows from the fact that Q is bounded positive and decreasing, that $t_k = +\infty$ for k larger than a number on the order of $1/\delta$; i.e. for $t > 1/\delta$ where we ignore constants. Let S_k and ε_k denote the supnorm and L^1-norm of $u(x,t)$ at time t_k+:

$$\varepsilon_k = \| u(\cdot, t_k+) \|_{L^1} ,$$

$$S_k = \| u(\cdot, t_k+) \|_S .$$

Since (18) holds when u_0 is replaced by $u(\cdot, t)$, a direct consequence of (18) is that if

$$|t_{k+1} - t_k| \geq \frac{\varepsilon_k}{S_k^2} , \tag{19}$$

then

$$S_k^2 \leq \delta . \tag{20}$$

By (3), if the supnorm is small at time t_k, then it is small for all $t > t_k$. Thus, from the point of view of (19) and (20), the longest time it takes for $S_k^2 \leq \delta$ occurs when

$$S_k^2 \geq \delta , \tag{21}$$

and

$$t_{k+1} - t_k \leq \frac{\varepsilon_k}{S_k^2} , \tag{22}$$

for all $k \leq \frac{1}{\delta}$. But $t_k = \infty$ for $k > \frac{1}{\delta}$, thus (19) must hold for $k = \frac{1}{\delta}$, in which case (20) gives $S_k^2 \leq \delta$ for $k = \frac{1}{\delta}$. It remains only to estimate $t_{1/\delta}$ under the assumptions (21) and (22). From (21) and (22),

$$t_{k+1} \leq t_k + \frac{\varepsilon_k}{S_k^2} \leq t_k + \frac{\varepsilon_k}{\delta} . \tag{23}$$

Now by (4)

$$\varepsilon_k \leq \text{Const.}(\varepsilon_0 + t_k) \equiv \varepsilon_0 + t_k ,$$

so without loss of generality, assume

$$\varepsilon_k \leq t_k, \quad k \geq 1 ,$$

$$t_1 \leq \frac{\varepsilon_0}{\delta} .$$

Then (23) reads

$$t_{k+1} \leq t_k + \frac{t_k}{\delta} = \left(1 + \frac{1}{\delta}\right) t_k , \quad k \geq 1 ,$$

$$t_1 \leq \frac{\varepsilon_0}{\delta} \leq \left(1 + \frac{1}{\delta}\right) \varepsilon_0 .$$

Thus

$$t_{1/\delta} \leq \left(1 + \frac{1}{\delta}\right)^{1/\delta} \varepsilon_0 .$$

Hence $\| u(\cdot, t) \|_S^2 \leq \delta$ for $t \geq \left(1 + \frac{1}{\delta}\right)^{1/\delta} \varepsilon_0$. Now for fixed t, choose δ so

$$t = \left(1 + \frac{1}{\delta}\right)^{1/\delta} \varepsilon_0 , \tag{24}$$

and hence

$$\|u(\cdot,t)\|_S^2 < \delta .$$

Solve for δ in (24) and obtain

$$\log \frac{t}{\varepsilon_0} = \frac{1}{\delta} \log\left(1 + \frac{1}{\delta}\right) < \frac{1}{\delta^{1+\sigma}} , \quad \sigma << 1 .$$

Thus we conclude

$$\|u(\cdot,t)\|_S^2 < \delta = \left\{\log\left(\frac{t}{\|u_0(\cdot)\|_{L^1}}\right)\right\}^{-1/1+\sigma} ,$$

which gives

$$\|u(\cdot,t)\|_S < \left\{\log\left(\frac{t}{\|u_0(\cdot)\|_{L^1}}\right)\right\}^{-1/2+\sigma} , \quad \sigma << 1 .$$

This is (5), (6) except that in (6) we neglect the small number σ.

As a final comment, we note that the main estimate (18) in this analysis follows via estimates which are embedded in the formalism of wave tracing which was developed by Liu in [15]. This formalism is a way of locating the influence at time t of a piece of the solution at time $t = 0$ by means of approximate characteristics. The program of Liu is to use this formalism to construct approximations to solutions at time t from initial data at time $t = 0$, and then to use this construction to obtain continuous dependence. In this direction, a complex system for keeping track of "first order" changes in state values along approximate characteristics is developed in [15]. In principle, errors are estimated in terms of Q, so that the sum of all first order errors are bounded by the total change in Q. Because of the appearance of an infinite family of approximate characteristics near singular points, one expects that (in a sense) estimates of all orders are needed for the stability problem. The techniques outlined above reduce the problem of stability for the constant state to an estimate which can be proved within the framework developed by Liu because it follows from an analysis of first order changes in state values on approximate characteristics. We look for such a reduction in the general stability problem.

ACKNOWLEDGMENT: The author thanks James Glimm, Ronald J. DiPerna, Tai Ping Liu and I. Liang Chern for many helpful discussions involving this work. Sponsored by the United States Army under Contract No. DAAG29-80-C-0041. This material is based upon work supported by the National Science Foundation under Grant No. DMS-8210950, Mod. 1.

REFERENCES

[1] Chern, I-Liang, On the decay of a strong wave for hyperbolic conservation laws in one space dimension, thesis, Courant Institute (1982).

[2] Courant, C. and Friedrichs, K. O., "Supersonic Flow and Shock Waves", Springer-Verlag, 1948.

[3] Dafermos, C., Application of the invariance principle for compact processes, II. Asymptotic behavior of solutions of a hyperbolic conservation law, J. Differential Equations 11 (1972), 416-424.

[4] DiPerna, R., Decay and asymptotic behavior of solutions to nonlinear hyperbolic systems of conservation laws, Indiana Univ. Math. J. 24 (1975), 1047-1071.

[5] DiPerna, R., Decay of solutions of hyperbolic systems of conservation laws with a convex extension, Arch. Ratl. Mech.'s Anal. 69 (1977), 1-46.

[6] Glimm, J., Solutions in the large for nonlinear hyperbolic systems of equations, Comm. Pure Appl. Math. 18 (1965), 695-715.

[7] Glimm, J. and Lax, P., Decay of solutions of systems of nonlinear hyperbolic conservation laws, Amer. Math. Soc. Memoirs 101 (1970).

[8] Keyfitz, B., Solutions with shocks: an example of an L^1-contractive semigroup, Comm. Pure Appl. Math. 24 (1971), 125-132.

[9] Lax, P., Hyperbolic systems of conservation laws, II, Comm. Pure Appl. Math. 10 (1957), 537-566.

[10] Liu, T. P., Asymptotic behavior of solutions of a general system of nonlinear hyperbolic conservation laws, Indiana Univ. J. 27 (1978), 211-253.

[11] Liu, T. P., Admissible solutions of hyperbolic conservation laws, Memoirs of the AMS 30, No. 240 (1981).

[12] Liu, T. P., Decay to N-waves of solutions of general system of nonlinear hyperbolic conservation laws, Comm. Pure Appl. Math. 30 (1977), 585-610.

[13] Liu, T. P., Large-time behavior of solutions of initial and initial-boundary value problem of general system of hyperbolic conservation laws, Comm. Math. Phys. 55 (1977), 163-177.

[14] Liu, T. P., Linear and nonlinear large-time behaviors of solutions of hyperbolic conservation laws, Comm. Pure Appl. Math. 30 (1977), 767-796.

[15] Liu, T. P., The deterministic version of the Glimm scheme, Comm. Math. Phys. 57 (1977), 135-148.

[16] Luskin, L. and Temple, J. B., "The existence of a global weak solution to the nonlinear waterhammer problem", Comm. Pure Appl. Math. 35 (1982), 697-735.

[17] Smoller, J. A., Shock waves and reaction-diffusion equations, Springer-Verlag (1980).

[18] Temple, J. B., Global solution of the Cauchy problem for a class of 2×2 nonstrictly hyperbolic conservation laws, Adv. Appl. Math. 3 (1982), 335-375.

[19] Temple, J. B., Solutions in the large for the nonlinear hyperbolic conservation laws of gas dynamics", J. Diff. Equs. 41, No 1 (1981), 96-161.

[20] Temple, J. B., Systems of conservation laws with coinciding shock and rarefaction curves, Contemporary Mathematics 17 (1983), 143-151.

[21] Temple, J. B., Systems of conservation laws with invariant submanifolds, Trans. Am. Math. Soc. 280, No. 2 (1982), 781-795.

[22] Temple, J. B., No L^1-contractive metrics for systems of conservation laws, Trans. Am. Math. Soc. 288, No. 2, April 1985.

[23] Temple, J. B., Supnorm bounds in Glimm's method, Mathematics Research Center Technical Summary Reports, University of Wisconsin-Madison (to appear).

[24] Temple, J. B., Decay with a rate for noncompactly supported solutions of conservation laws, (to appear) Trans. Am. Math. Soc.

DIFFERENT APPROACH FOR THE RELATION BETWEEN THE KINETIC AND THE MACROSCOPIC EQUATIONS

CLAUDE BARDOS

Centre de Mathématiques Appliquées
Ecole Normale Supérieure
45 Rue d'Ulm Paris 75005
et
C.S.P. Université de Paris Nord
Av. J.B. Clément 93430 Villetaneuse.

I.- INTRODUCTION

This contribution contains both very classical and recent results on the relation between the kinetic and macroscopic equations. The emphasis is made on the variety of results which can be obtained according to the scaling and to the properties of the collision kernel. The non linearity when it appears in the problem introduces a more rich structure and of course leads to the hardest open problems. The basic unknown of the problem is a scalar function $f(x,v,t)$, $(x,v) \in X \times V$, $X \subset \mathbb{R}^d$ $V = \{v \in \mathbb{R}^d / a \leqslant |v| \leqslant b\}$ $X \times V$ is the phase space and in many cases one will have $V = \mathbb{R}^d$. The function $f(x,v,t)$ describes the density of particles which at the point x and at the time t have the velocity v . The macroscopic quantities are the moments of this function :

$$\rho(x,t) = \int f(x,v,t)dv, \quad \rho u = \int vf(x,v,t)dv \ , \quad \rho E = \int |v|^2 f(x,v,t)dv \text{ etc...}$$

and the purpose of the program is the following :

1) one wants to deduce approximate macroscopic equations for the moments,

2) Estimate the error between the real moments and the one given by the macroscopic equations.

II - The Boltzmann Equation and the H theorem

Of course the most classical example is the Boltzmann equation (Boltzmann [10] 1896), which involves the binary collisions in a neutral gaz. The equations are :

(1) $\partial_t f_\varepsilon + v.\nabla_x f_\varepsilon = \varepsilon^{-1} Q(f_\varepsilon, f_\varepsilon), \quad f_\varepsilon(x,v,0) = f_0(x,v)$

where ε denotes the mean free path and where $Q(f,f)$ is given by the following formulas :

(2) $v' = v + (v-v_1,\omega)\omega, \quad v'_1 = v_1 - (v-v_1,\omega)\omega$.

(3) $Q(f,f) = \iint_{\mathbb{R}^3 \times S^2} (f'f'_1 - ff_1)\, q(|v-v_1|,\omega) dv_1 d\omega$.

The relations (2) are consequences of the kinetic conservation of mass, momentum and energy, q denotes the cross section, f_1, f', f_1 denote the value of the function f at the points v_1, v', v'_1 . The macroscopic version of the relations (2) are the relations :

(4) $0 = \int Q(f_\varepsilon, f_\varepsilon) dv = \int |v|^2 Q(f_\varepsilon, f_\varepsilon) dv = \int v_i\, Q(f_\varepsilon, f_\varepsilon) dv \qquad (1 \leqslant i \leqslant 3)$.

Finally the cross section q is a positive function which depends only on the modulus of $v-v_1$ $(=v'-v'_1)$ and of the angle ω . With the Fubini theorem one deduces the so called H theorem :

(5) $\int (1+\text{Log}\, f)\, Q(f,f)\, dv_1 =$

$$1/4 \iiint (f'f'_1 - ff_1)(\text{Log}\, ff'_1 - \text{Log}\, ff_1) q(|v-v_1|) d\omega dv_1 dv \leqslant 0$$

and the relation (5) implies that the only non zero functions which satisfy the relation $Q(f,f) = 0$ are the Maxwellian distributions i.e. the function given by the formula :

(6) $f_{(\rho,u,T)} = \rho(2\pi T)^{-3/2} \exp(-|u-v|^2/2T)$.

Of course any Maxwellian distribution satisfies the relations :

(7) $\int f_{(\rho,u,T)} dv = \rho, \int v\, f_{(\varphi,u,T)} dv = \rho\, , \int (|v|^2/2) f_{(\varphi,u,T)} dv = \rho(|u|^2/2 + 3/2\, T)$

and

$$\int f_{(\rho,u,T)} \mathrm{Log}\, f_{(\rho,u,T)} dv = \rho \mathrm{Log}(\rho^{2/3}/T).$$

The relation between the solution of the compressible Navier Stokes equation (with viscosity and heat dissipation proportional to $\varepsilon > 0$) and the solution of the compressible Euler Equation is still an open problem. However in 1971, P. Lax [20] made the following remark : assume that the solution of the compressible Navier Stokes equation is bounded (in L^{∞}) and converges almost everywhere then its limit is a weak solution of the compressible Navier Stokes Equation which furthermore satisfies the entropy condition. It turns out that a similar remark also holds for the relation between the Boltzmann equation and the compressible Euler Equation. Roughly one could state the following (cf. Bardos and Golse [3] for some details).

THEOREM 1.- *Assume that, for any* $\varepsilon > 0$, *the solution* f_{ε} *of the Boltzmann equation is conveniently bounded and converges almos. everywhere then*

1) *Its limit* f *is a maxwellian distribution* $f_{(\rho,u,T)}$,

2) *The parameters* (ρ,u,T) *of this maxwellian distribution* (ρ,u,T) *satisfy the compressible Euler equation and the entropy condition.*

Proof. (1+ Log f) is the derivative, with respect to f of the expression f Log f therefore we deduce, from (5), the relation :

(8) $\iiint (f'_{\varepsilon} f'_{\varepsilon 1} - f_{\varepsilon} f_{\varepsilon 1}) (\mathrm{Log}\, f'_{\varepsilon} f'_{\varepsilon 1} - \mathrm{Log}\, f_{\varepsilon} f_{\varepsilon 1})\, q(|v-v_1|).d\omega\, dv_1 dv$

$= -4(\varepsilon\, \partial_t \int f_{\varepsilon} \mathrm{Log}\, f_{\varepsilon}\, dv + \varepsilon \nabla . \int v f_{\varepsilon} \mathrm{Log}\, f_{\varepsilon}\, dv).$

and if we can take the limit in the integral of the right hand side (the hypothesis made in Bardos-Golse [3] imply that it is possible) we obtain for $f = \lim_{\varepsilon \to 0} f_{\varepsilon}$:

(9) $\iiint (f'f'_1 - ff_1)(\mathrm{Log}\, f'f'_1 - \mathrm{Log}\, ff_1)\, q(|v-v_1|) dv_1 dv d\omega = 0$.

This relation shows that f is a Maxwellian distribution.

$f = f_{(\rho,u,T)} = \rho/(2\pi T)^{3/2} \exp(-|v-u|^2/2T)$. Multiplication by the moments 1, v and $|v|^2$ yields, with the relations (4) the formulas :

(10) $\partial_t \int f_\varepsilon \, dv + \nabla_x \int v f_\varepsilon dv = 0$

(11) $\partial_t \int v f_\varepsilon \, dv + \nabla_x \int v \otimes v f_\varepsilon \, dv = 0$

(12) $\partial_t \int |v|^2/2 f_\varepsilon dv + \nabla_x \int |v|^2 f_\varepsilon \, dv = 0$.

Once again we assume that we can take the limit in the integrals which appear in the left hand side of (10), (11), (12). For a maxwellian distribution these integrals can be explicitly computed and this gives the relations :

(13) $\partial_t \rho + \nabla.(\rho u) = 0$

(14) $\partial_t \rho u + \nabla(\rho u \otimes u) + \nabla \rho T = 0$

(15) $\partial_t \rho(|u|^2/2 + 3/2 \rho T) + \nabla.(\rho u |u|^2/2 + 3/2 \rho T) + \nabla \rho T u = 0$

which are the compressible Euler equations. Furthermore, from the relation :

$$(f'_\varepsilon f'_{\varepsilon 1} - f_\varepsilon f_{1\varepsilon})(\mathrm{Log}\, f'_\varepsilon f'_{\varepsilon 1} - \mathrm{Log}\, f_\varepsilon f_{\varepsilon 1}) \geq 0 \ ,$$

we deduce that we have :

(16) $\partial_t \int f_\varepsilon \,\mathrm{Log}\, f_\varepsilon \, dv + \nabla. \int v f_\varepsilon \,\mathrm{Log}\, f_\varepsilon \, dv \leq 0$

with the asumptions on the convergeance and the fact that f is a Maxwellian distribution, this relation implies the formula :

(17) $\partial_t \int f \,\mathrm{Log}\, f \, dv + \nabla. \int f \, vf \,\mathrm{Log}\, f \, dv \leq 0$

which, for a Maxwellian distribution gives the classical entropy relation

(18) $\partial_t \, \rho \,\mathrm{Log}(\rho^{2/3}/T) + \nabla_x \, \rho u \,\mathrm{Log}(\rho^{2/3}/T) \leq 0$.

This heuristic derivation shows the difficulties of a real proof for the convergence of f_ε . Such a proof should give an existence result for the solution of the compressible Euler equation. At present such results are known only before the appearance of singularities (which do generally appear cf. Sideris [25]). Therefore the only complete results for $\varepsilon \to 0$ suffer of the same restriction (cf. Nishida [22] and Caflisch [11]).

On the other hand, one could view the Boltzman equation as a regular perturbation of the transport equation

(19) $\partial_t f + v \nabla_x f = 0$

(In this case $(1/\varepsilon)$ will have to be chosen fixed and not too large). This type of idea have already been used by many authors (cf. Klainerman and Ponce [19] and Shatah [24]) to construct a global (in time) smooth solution for the non linear wave equation provided the following conditions are true.

(i) The solution is defined in the whole space (or in the exterior of a bounded domain) and the dimension of the space is large enough.

(ii) The initial data is small enough and localised.

The proof relies on a priori estimates related to the dispersion rate property of the linearised equation. It turns out that similar idea can be used for the Boltzmann equation (and also for other problems; cf. Bardos-Degond [2] for the Vlasov Poisson equation, Horst [15] for the Vlasov Maxwell equation etc...) like in the case of the wave equation one obtains as a by product that for t going to infinity the solution of the non linear equation behaves likes the solution of transport equation. For the transport equation the dispersion property is the following. We assume that the initial condition satisfies the following estimate :

(20) $0 \leqslant f(x,v,0) \leqslant h(x)$

with $h(x) \in L^1(\mathbb{R}^3_x)$. Then an explicit calculation and a change of variable gives, for the solution f of the transport equation : $\partial_t f + v.\nabla_x f = 0$, the following result :

(21) $0 \leqslant \int f(x,v,t)dv = \int f(x-vt,v,0)dv \leqslant \int h(x-vt)dv = t^{-3} \int h(X)dX$.

Therefore the total charge (or density) disperses like $1/t^3$.

<u>THEOREM 2</u>.- *We assume that the cross section* $q(|v-v_1|,\omega)$ *is uniformly bounded, then, for any smooth initial data which satisfies the estimate*:

(22) $0 \leq f_0(x,v) \leq A\, e^{-|x|^2}$

with A *small enough, the Boltzmann equation has a unique smooth solution defined in* $\mathbb{R}_t \times \mathbb{R}^3_x \times \mathbb{R}^3$. *Furthermore there exist two functions* $g^{\pm}(x,v,t)$ *solutions of the transport equation* :

(23) $\partial_t g^{\pm} + v.\nabla_x g^{\pm} = 0$

such that one has :

(24) $\lim_{t\to\pm\infty} \sup_{(x,v)} |f(x,v,t) - e^{|x-vt|^2} g_{\pm}(x,v,t)| = 0$.

<u>Proof</u>. The basic ingredient is the a priori estimate and it will be our unique concern. We write $f(x,v,t) = g(x,v,t)\exp(-|x-vt|^2)$, and we notice that we have :

(25) $(\partial_t + v.\nabla_x)(\exp(-|x-vt|)^2) = 0$.

Therefore we obtain the relation :

(26) $e^{-|x-vt|^2}(\partial_t g + v.\nabla_x g) =$

$$\iint e^{|x-v'_1 t|^2 - |x-v't|^2} g'g'_1 - e^{-|x-vt|^2 - |x-v_1 t|^2} g g_1\, q(|v-v_1|,\omega) dv_1 d\omega.$$

By construction of the collision operator we have :

(27) $|x-vt|^2 - |x-v'_1 t|^2 - |x-v't|^2 = |x-v_1 t|^2$.

Therefore we deduce from (26) and (27) the relation :

(28) $\partial_t g + v.\nabla g = \iint e^{|x-v_1 t|^2}(g'g'_1 - g g_1) q\, dv_1 d\omega$.

Finally with $X(t) = \sup_{(x,v)} g(x,v,t)$ we deduce from (28) the relation :

(29) $$d_{t+}X(t) \leq 2c \iint e^{|x-v_1|^2}(X/t))^2 dv_1 d\omega .$$

The change of variable $X = x - v_1 t$ and a result of local existence (for t small) gives :

(30) $X(t) \leqslant X(0)\,[1 - C(\int_0^t (1+s)^{-3} ds)\, X(0)]^{-1}$.

Therefore, for $X(0) \int_0^{\infty} (1+s)^{-3} ds < 1$ we have a uniform (in time) estimates which leads to global existence, and also to the proof of the asymptotic behaviour.

Remark 1.- The above proof is very simple this is due to the very stringent hypothesis made both on the cross section q and on the initial data. In fact this hypothesis can be relaxed and more realistic care can be handled (cf. Ilner and Shinbrot [17], Hamdache [14], Ilner and Pulvirenti [16], Bellomo Toscani [8] and also the book of Bellomo [7]).

Remark 2.- Taking in account the recent result of Beal [6] for the Broadwell model the following conjecture seems reasonable (surely very hard to prove) :

(i) For any $\varepsilon > 0$ the Boltzmann equation

$\partial_t f_\varepsilon + v \nabla_x f_\varepsilon = (1/\varepsilon)\, \varphi(f_\varepsilon, f_\varepsilon)$ has a global (for $t > 0$) solution ; for t very large this solution behaves like the solution of the equation $\partial_t f + v.\nabla_x f = 0$; for t of the order of ε an initial layer appears and in between the solution behave like a maxwellian $f_{\rho,u,T}$ where (ρ,u,T) are solution of the compressible Euler equation with the entropy condition.

III.- The Radiative Transfert equation and the Porous media equation

The fact that the compressible Euler equation is a hyperbolic system of five equations and five unknowns is closely related to the dimension of the kernel of the collision operator. In fact, the solutions of $Q(f,f) = 0$ describes a five dimensional manifold : $f = \rho(2\pi T)^{-3/2} \exp(-|v-u|^2/2T)$. A change in the structure of the collision kernel leads to an other type of equation obtained with some modifications in the scaling. The most classical example is the neutron transport equation which leads to a diffusion approximation (cf. for a rigourous mathematical treatment, Keller and Larsen [18] or Bensoussan, Lions et Papanicolaou [9]). An other interesting example is the Radiative Transfert Equation which is non linear but which has also a collision type operator with a one dimensionnal kernel. This equation describes the evolution of the density of photons in a stellar atmosphere. It involves the intensity of photons $I(x,v,t,\Omega)$ and the temperature T and it has the following form.

(31) $\partial_t I_\varepsilon + \varepsilon^{-1}\Omega.\nabla_x I_\varepsilon + \varepsilon^{-2}\sigma_\nu(T_\varepsilon)(I_\varepsilon - B_\nu(T_\varepsilon)) = 0$.

(32) $\partial_t T_\varepsilon + \varepsilon^{-2} << \sigma_\nu(T_\varepsilon)(B_\nu(T_\varepsilon) - I_\varepsilon) >> = 0$.

$B_\nu(T)$ is a Planckian function, given (with non physical constants) by the formula

$$B_\nu(T) = \nu^3(\exp -\nu/T - 1)$$

and $\sigma_\nu(T)$ is a positive ν and T depending function which describes the opacity of the media $<< , >>$ denotes the integral

$$<<f>> = 1/4\pi \int_{S^2} \int_0^\infty f(\nu,\Omega)d\nu d\Omega \ .$$

The limit for ε going to zero is "formaly" obtained with an expansion :

$$(I_\varepsilon = I_0 + \varepsilon I_1 + \varepsilon^2 I_{2,} \ T)$$

where I_ε and T are chosen in such a way that (i) the equation (32) is exactly satisfied and (ii) the equation (31) is satisfied up to the order 1 in ε . Since the kernel of the operator $u \rightarrow \sigma_\nu(u)(u - B_\nu(T)) = 0$ are the Planckian functions, one has :

(33) $I_0 = B_\nu(T)$, $I_1 = -(\sigma_\nu(T))^{-1}\ \Omega.\nabla_x B_\nu(T)$,

$$I_2 = -(\sigma_\nu(T))^{-1}(\partial_t B_\nu(T) - \Omega\nabla_x(\sigma_\nu(T))^{-1}\Omega\nabla_x B_\nu(T)) \ .$$

In the equation (32) we use the relation $I_0 = B_\nu(T)$ and the oddness (with respect to Ω) of the function I_1 . This leads to the equation

(34) $\partial_t(T + \langle\langle B_\nu(T) \rangle\rangle) - (1/3)\,\Delta F(T) = 0$

where $F(T)$ is given by the relation.

(35) $F'(X) = \langle\langle (\partial_X B_\nu(X))/\sigma_\nu(X) \rangle\rangle$.

The main difficulty in the mathematical treatment of this formal derivation comes from the fact that the function $\sigma_\nu(T)$ which describes the opacity of the media goes to infinity when T goes to zero. In particular, this will imply that the equation (34) is a degenerate parabolic equation which exhibit most of the properties of this type of equation (existence of a front, existence of a waiting time for the evolution of this front). Numerical computations show no propagation of front for the solution of (31), (32) therefore near the front of the solution of (34) a large error between T_ε and T should exist, leading to the appearance of an internal layer. The relation between (31), (32) and (34) have been derived (with no proof), for the first time, by Larsen, Pomranin and Badahan [20] . For a systematic approach of the above problem one can use either a monotonicity method (cf. Bardos, Golse and Perthame [4] or a compacity method (cf. Bardos, Golse, Perthame and Sentis [5]). For the monotonicity method one assume that the functions

$$T \to \sigma_\nu(T) \text{ and } T \to \sigma_\nu(T)\,B_\nu(T)$$

are for any $\nu \in\,]0,\infty[$ non decreasing and that $\sigma_\nu(T)\,B_\nu(T)$ goes to zero when T goes to zero; then one can show that the operator :

$$\begin{bmatrix} I \\ T \end{bmatrix} \longmapsto \begin{bmatrix} \varepsilon^{-1}\Omega.\nabla_x I + \varepsilon^{-2}\sigma_\nu(T)(I - B_\nu(T)) \\ \langle\langle \sigma_\nu(T)(B_\nu(T) - I) \rangle\rangle \end{bmatrix}$$

is (for any $\varepsilon > 0$) maximal accretive in the space

$L^1(\mathbb{R}^3 \times S^2 \times \mathbb{R}^+) \times L^1(\mathbb{R}^3{}_x)$.

This leads to a proof of the existence and uniqueness of the solution of (32) (33) (use the method of Crandall and Liggett [12] concerning the non linear semi groups). This leads also to an estimate of the error between $(I_\varepsilon, T_\varepsilon)$ and $(B_\nu(T),T)$, given with appropriate initial condition by (34).

However in many classical examples the monotonicity asumptions are note satisfied. In particular, they are not satisfied for the Cramer's opacity which is given by the formula :

(36) $\sigma_\nu(T) = T^{-1/2}\,\nu^{-3}(1-e^{-\nu/T})$.

To overcome this difficulty we propose an alternate method based on a compactness theorem of Golse Perthame and Sentis [13] . This method gives an existence proof for the solution of the R.T.E equation with almost no hypothesis and a proof of the validity of the Rosseland approximation when the opacity is ν is frequency independant. In this case, so called the "grey" problem, with

(37) $\tilde{u} = (4\pi)^{-1} \int_\Omega u(x,\Omega)d\Omega$

we have the :

<u>THEOREM 3</u>.- *In a smooth bounded open domain* X *of* $\mathbb{R}^3$ *we consider the following equation :*

(38) $\partial_t u_\varepsilon + \varepsilon^{-1}\Omega.\nabla u_\varepsilon + \varepsilon^{-2}\sigma(\tilde{u}_\varepsilon)(u_\varepsilon - \tilde{u}_\varepsilon) = 0$

with initial and boundary date.

(39) $u_{\varepsilon|t=0} = u_0(x| \geqslant 0,\ u_{\varepsilon|(\partial X x S^2)_-} = k(x,t) \geqslant 0$.

(In (39) we denote by $(\partial X \times S^2)_-$ the set of points : $(x,\Omega) \in \partial X \times S^2$ such that $\Omega.n(x) < 0$ with $n(x)$ the outward unitary normal to ∂X) . *We assume that* u_0 *and* k *are uniformly bounded in* $L^\infty(X) \times L^\infty(\partial X \times \mathbb{R}^+_t)$ *and* Ω *independant. We assume that the opacity is strictly positive* $(0 < \sigma_m \leqslant \sigma(T), \forall T)$ *and satisfies the following a priori estimate :*

(40) $\sigma(T) \leqslant C\,T^{-1/2}$ $(0 \leqslant T \leqslant 1)$

Then any family of weak solutions of (38°,(39) (cf. Bardos, Golse and

Perthame [5] *converges in* $L^2(\mathbb{R}^+_t \times X)$ *to the solution of the degenerate parabolic equation :*

(41) $\partial_t u - 1/3\, \nabla.(\sigma(u))^{-1} \nabla u = 0$

with initial and boundary condition :

(42) $u(x,0) = u_0(x)$, $u|_{\partial X} = k(x,t)$.

<u>Proof.</u> Multiplication by u_ε and integration over S^2 gives the estimate :

(43) $\partial_t \int u_\varepsilon^2 d\,x\, d\Omega + \varepsilon^{-2} \iint \sigma(\widetilde{u}_\varepsilon)|u_\varepsilon - \widetilde{u}_\varepsilon|^2 d\,x\, d\Omega \leqslant 0$

which implies the relation :

(44) $\int_0^\infty \iint |u_\varepsilon - \widetilde{u}_\varepsilon|^2\, dx\, d\Omega dt \leqslant C\, \varepsilon^2$.

This shows that the sequence u_ε has a weak L^2 limit which is independant of Ω , and that we have :

(45) $\|\sigma^{-1/2}(\widetilde{u}_\varepsilon)\, 1_{u_\varepsilon > 0}(\varepsilon\, \partial_t\, u_\varepsilon + \Omega.\, \nabla_x u_2)\|_{L^2(X \times S^2 \times \mathbb{R}^+)} \leqslant C$

with the hypothesis (40) we deduce that we have :

(46) $\|\varepsilon\, \partial_t\, u_\varepsilon^2\|_{L^2(X \times S^2 \times \mathbb{R}^+)} \leqslant C$.

The relation (46) leads to a regularity result (cf. G.P.S [13] and B.G.P.S [5])

namely that $(u_\varepsilon)^2$ is uniformly bounded in $L^2_{loc}(\mathbb{R}^+; H^{1/2}(X))$. Now with the relation :

(47) $u_\varepsilon^2 - \widetilde{u}_\varepsilon^2 = (u_\varepsilon + \widetilde{u}_\varepsilon)(u_\varepsilon - \widetilde{u}_\varepsilon)$

we deduce that $u_\varepsilon^2 - u_\varepsilon^2$ converges to zero in $L_2(\mathbb{R}_t^+ \times X \times \Omega)$. The integration of (38) with respect to the 0 and 1 order moments give the relation :

(48) $\partial_t \widetilde{u}_\varepsilon + \nabla.\varepsilon^{-1} \widetilde{u_\varepsilon \Omega} = 0$, $\varepsilon^{-1} \sigma(\widetilde{u}_\varepsilon)\, \widetilde{u_\varepsilon \Omega} = -\varepsilon\, \partial_t \widetilde{\Omega u_\varepsilon} - \nabla\, \widetilde{\Omega \otimes \Omega\, u_\varepsilon}$.

Therefore from (48) we deduce that we have :

(49) $\partial_t \tilde{u}_\varepsilon - \nabla\, \sigma(\tilde{u}_\varepsilon)^{-1} \nabla\, \widetilde{\Omega\otimes\Omega\, u_\varepsilon} = O(\varepsilon)$.

Finally one has to show the following relation :

(50) $\lim_{\varepsilon\to 0} \nabla.\sigma(\tilde{u}_\varepsilon)^{-1} \nabla\, \widetilde{\Omega\otimes\Omega\, u_\varepsilon} = \Delta\, F(\lim u_\varepsilon)$

with $F'(u) = 1/(3\sigma(u))$ .From (47) we deduce that u_ε and $\tilde{u}_\varepsilon$ converge in any $L^p\ (X \times S^2 \times [0,T]\ )$ ($T>0$ finite) to a function u . This gives in particular pointwise convergence and the fact that $\sigma^{-1/2}(\tilde{u}_\varepsilon)$ converge to $\sigma^{-1/2}(u)$, in L^2 . On the other hand $\sigma^{-1/2}(\tilde{u}_\varepsilon)\ \nabla\ \Omega\otimes\Omega\ u_\varepsilon$ converge in L^2 weak to a limit q and one has :

(51) $\lim_{\varepsilon\to 0} \sigma^{-1}(\tilde{u}_\varepsilon)\, \nabla\, \widetilde{\Omega\otimes\Omega\, u_\varepsilon} = \sigma^{-1/2}(u)\tilde{q}$.

To compute $\tilde{q}$ one uses the relation :

(52) $\tilde{q}\ \sigma^{1/2}(u)u = \lim_{\varepsilon\to 0} \nabla_x \Omega \otimes \Omega\, u_\varepsilon^{\,2}/2 = \nabla_x\, \Omega\otimes\Omega\, u^2/2$

and with the formula :

(53) $\tilde{q}\ \sigma^{1/2}(u)u = (1/6)\ \nabla\, u^2$

the result follows.

IV.- <u>The diffusion approximation for a collision less gaz in a tube with a purely diffusive boundary.</u>

This final example shows how the diffusion may appear without any collision or scattering operator in the domain but with a diffusion coming from the boundary. To increase the influence of the boundary we consider the following domain : $X_\varepsilon = \{(x,z) \in \mathbb{R} \times]-\varepsilon,\varepsilon[\}$ and we assume that the gaz may have any velocity but that the effect of the collision between the particles can be neglected. Changing t into t/ε and z into εz leads to the equation :

(54) $\partial_t f_\varepsilon + \varepsilon^{-1} v\, \partial_x f_\varepsilon + \varepsilon^{-2} w \partial_z f_\varepsilon = 0$ for $(x,z) \in \mathbb{R} \times]-1,1[$.

As a boundary condition we consider the relations :

(55) $f_\varepsilon(x, \pm 1, v, w, t)_{|\pm w<0} = M \iint_{\pm w'>0} |w'| f(x, \pm 1, v', w', t) dw'.dv'$.

In (55) M is a maxwellian type function, which is given by the relation :

(56) $M = C|w| \exp(-(v^2+w^2)/2)$

where C is a normalising constant given by :

(57) $\iint_{w>0} |w| M.dvdw = 1 \;\; ; \; C = 1/2\pi$.

The solution of the problem (53), (54) is given by a ε dependant linear semi group on contraction (in the space $L^1(\mathbb{R}_x \times]-1,1[\times \mathbb{R}^2_{v,w})$. The boundary condition is an absolute diffusion, the amount of particle emitted on the boundary depends only of the angle with the outward normal and of the total amount of incoming particles. Since the problem is linear the formal expansion

(58) $f_\varepsilon = f_0 + \varepsilon f_1 + \varepsilon^2 f_2 + v_\varepsilon$

leads to an estimate for the reminder r_ε . We have

(59) $w\, \partial_z f_0 = 0 \, , \, v\, \partial_x f_0 + w\, \partial_z f_1 = 0 \, , \, \partial_t f_0 + v\, \partial_x f_1 + w\, \partial_z f_2 = 0$.

The first equation of (59) gives, taking in account, the boundary condition :

(60)
$$f_0(x,t,v,w) = \begin{cases} M \iint_{w'>0} |w'| f_0(x,-1,v',w',t) dw' & w>0 \\ M \iint_{w'<0} |w'| f_0(x,1,v',w',t) dw' & w<0 . \end{cases}$$

Therefore one has :

(61) $f_0(x,t,v,w) = M\, h^{\pm}_0(x,t)$ for $\pm\, w>0$

finally with the boundary condition we have $h^+ = h^-$.

More generally, an integration over $]-1,1[\times R_v x R_w$ shows that the relation :

(62) $\iiint h(z,v,w) dvdwdz = 0$

is a necessary condition for the solvability of the problem $w\,\partial_z f = h$ with the boundary condition (55). It is also a sufficient condition. Indeed the unique solution satisfying (62) is then given by :

$$(63)\qquad f(z,v,w) = \begin{cases} \int_{-1}^{z} h(s,v,w)\,ds/w & z > 0 \\ -\int_{z}^{1} h(s,v,w)ds/w + M\int_{-1}^{1}\int_{w>0} h(s,v,w)dsdvdw. \end{cases}$$

This remark gives for f_1 for the formula :

$$(64)\qquad f_1(z,v,w) = \begin{cases} -v\,e^{-(v^2+w^2)/2}\,\partial_x h_0(z+1) \\ -v\,e^{-(v^2+w^2)/2}\,\partial_x h_0(1-z) + 2M\,\partial_x h_0 \end{cases}$$

Finally, to obtain a solution f_2 for the last equation of (59) h_0 has to satisfy a solvability condition which reads.

$$(65)\qquad \partial_t h_0 - \alpha\,\partial_x^{\,2} h_0 = 0$$

where α is a positive constant (Details will be available in Babowsky, Bardos and Platkowsky [1].

V.- Conclusion.

The link between the different examples is mostly based on the method used (introduction of a small parameter and a priori estimates), other approach using in particular probabilistic tools do exist in the litterature. On the other hand these examples show the variety of results which may be obtained and how they depend on the structure of the collision operator, the boundary condition etc... Similar problem do appear in many other branches of mathematical physic (cf. for instance F. Poupault [22] for semi-conductors).

REFERENCES

[1] H. BABOVSKI, C. BARDOS and T. PLATOWSKI, *The diffusion limit for a Knudsen gaz in a tube with diffusive boundary condition. To appear.*

[2] C. BARDOS and P. DEGOND, *Global existence for the Vlasov Poisson equation in three spaces variables with small initial data,* Ann. Inst. Henri Poincaré 2, (2), 101-118 (1985).

[3] C. BARDOS and F. GOLSE, *Differents aspects de la notion d'entropie au niveau de l'équation de Boltzmann et de Navier-Stokes.* Note CRAS t.229 série A (7), 225-228 (1984).

[4] C. BARDOS, F. GOLSE et B. PERTHAME, *The Rosseland Approximation for Radiative Transfert equation,* submitted to Comm. In Pure and Appl. Math.

[5] C. BARDOS, F. GOLSE, B. PERTHAME et R. SENTIS, *The non accretive Radiative transfert equation existence of solutions of Rosseland Approximation.* To appear.

[6] T. BEALE, *Large time behavior of discrete velocity Boltzmann equations* To appear.

[7] N. BELLOMO, *Lecture Note on Boltzmann equation.* (to appear).

[8] N. BELLOMO and G. TOSCANI, *On the Cauchy Problem for the non linear Boltzmann Equation,* J. Math. phys. 1 (1985), 334-338.

[9] A. BENSOUSSAN, J.L. LIONS and G. PAPANICOLAOU, *Boundary layers and homogenization of transport processes,* J. Publ. R.I.M.S, Kyoto Univ. 15 (1979), 53-157.

[10] L. Boltzmann Vorlesungen uber Gas theorie Liepzig, 1896.

[11] R. CAFLISCH, *The fluid dynamic limit of the non linear Boltzmann equation,* Comm. Pure Appl. Math. 33 (1980), 651-666.

[12] M.G. CRANDALL and T.S LIGGETT, *Generation of semi-groups of non linear transformation on general Banach spaces* , Am. J. of Math. 93, (1971), 265-297.

[13] F. GOLSE, B. PERTHAME et R. SENTIS, *Un résultat de compacité pour le: équations de transport et application au calcul de la limite de la valeu, propre principale d'un opérateur de Transport.*

[14] K. HAMDACHE, *Existence in the large and asympytotic behaviour for th Boltzmann equation.* Jap. Jour. of Applied Math. 2 (1985), 1-15.

[15] E. HORST, to appear.

[16] R. ILNER and P. PULVIRENTI, *Global validity of the Boltzmann Equatio for a two dimensional rare gas in vacuum.* Prép. 92 Fach. Mathematik, Kaiserlautern.

[17] R. ILNER and M. SHINBROT, *The Boltzmann equation. Global existenc for a rare gas in an infinite vacuum.* Comm. Math. Phys. 95, (1984), 117-126.

[18] J.B KELLER and E. LARSEN, *Asymptotic solutions of neutron transpor problems.* J. Math. PHys. 15 (1974), 75-81.

[19] S. KLAINERMAN and G. PONCE, *Global small amplitude solutions to no linear evolution equations,* Comm. Pure Appl. Math. 36, 1 (1983), 133-141.

[20] E.W. LARSEN, G;C POMRANING and V.C BADHAN, *Asymptotic analysis o. radiative transfer problems,* J. Quant, Spec. Radia. Trans. 29, 4, 285-310, 1983.

[21] P. LAX, *Hyperbolic systems of Conservation Laws and th mathematical Theory of Shock waver.* S.I.A.M Regional Conference serie in Math. 11 (1973).

[22] T. NISHIDA, *Fluid dynamical limit of the non linear Boltzman equation to the level of the compressible Euler equation.* Comm. Math. Phys. 61, 119-148 (1978).

[23] F. POUPAULT, To appear.

[24] J. SHATAH, *Global existence to non linear evolution equation.* J.Diff. Eq. 46-3 (1982), 409-425

[25] T.SIDERIS, *Formation of singularities in three-Dimensiona Compressible Fluids.* To appear.

INTEGRABLE TRANSPORT PROCESSES

J. M. Greenberg
Department of Mathematics
University of Maryland Baltimore County
Catonsville, Maryland 21228

I. INTRODUCTION

In this note we shall discuss a class of nonlinear transport processes with quadratic interaction terms. The remarkable fact about these processes is that through an elementary change of variables, basically the Hopf-Cole transformation, they are equivalent to their linearized part and are integrable; that is the one particle distribution function for these processes may be obtained by quadrature. These models differ from certain familiar models of the Boltzman equation like the Broadwell equations in that they are endowed with only one conservation law.

For any integer $N \geq 1$, we have models with flow velocities $v_k = \frac{k}{N}$, $k = 0, \pm 1, \pm 2 \ldots, \pm N$, and one with velocities $v_k = \frac{k}{N}$, $k = \pm 1, \pm 2, \ldots, \pm N$. The former evolve as

$$\varepsilon\left(\partial_t + \frac{k}{N}\partial_x\right)f^k + \alpha\left(f^k - \frac{1}{2N+1}\sum_{j=-N}^{N} f^j\right) + \frac{f^k}{2N+1}\left(\sum_{j=-N}^{N}\frac{j}{N}f^j - \frac{k}{N}\sum_{j=-N}^{N} f^j\right) = 0,$$

$$k = 0, \pm 1, \ldots, \pm N, \tag{1}$$

where $f^k = f^k(x,t)$, $\alpha = 0$ or 1, and $0 < \varepsilon$, while the latter satisfy

$$\varepsilon\left(\partial_t + \frac{k}{N}\partial_x\right)f^k + \alpha\left(f^k - \frac{1}{2N}\sum_{\substack{j=-N\\ j\neq 0}}^{N} f^j\right) + \frac{f^k}{2N}\left(\sum_{\substack{j=-N\\ j\neq 0}}^{N}\frac{j}{N}f^j - \frac{k}{N}\sum_{\substack{j=-N\\ j\neq 0}}^{N} f^j\right) = 0,$$

$$k = \pm 1, \pm 2, \ldots \pm N, \tag{2}$$

where $f^k = f^k(x,t)$, $\alpha = 0$ or 1 and $0 < \varepsilon$. The continuous version of (1) and (2) is also a model of the desired type. It satisfies

$$\varepsilon\left(\partial_t + v\partial_x\right)f + \alpha\left(f - \tfrac{1}{2}\int_{-1}^{1} f(x,t,\mu)d\mu\right) + f\left(\int_{-1}^{1}\mu f(x,t,\mu)d\mu - v\int_{-1}^{1} f(x,t,\mu)d\mu\right) = 0,$$

$$-1 \leq v \leq 1 \tag{3}$$

where $f = f(x,t,v)$, $\alpha = 0$ or 1, and $0 < \varepsilon$. Solutions of (1)-(3) satisfy the conservation law

$$\frac{\partial q_0}{\partial t} + \frac{\partial q_1}{\partial x} = 0. \tag{4}$$

---

*This research was partially funded by the National Science Foundation under MCS 8219806.

When (1) holds

$$q_o := \sum_{j=-N}^{N} f^j \quad \text{and} \quad q_1 := \sum_{j=-N}^{N} \frac{j}{N} f^j ; \tag{5}$$

when (2) holds q_o and q_1 are given by (5) with the term f^o deleted; and finally when (3) holds

$$q_o := \int_{-1}^{1} f(x,t,\mu)d\mu \quad \text{and} \quad q_1 := \int_{-1}^{1} \mu f(x,t,\mu)d\mu. \tag{6}$$

For all three systems these are the only conserved moments. The existence of these conservation laws means that if we introduce the potential ϕ defined by

$$\frac{\varepsilon\phi_x}{\phi} := -q_o \quad \text{and} \quad \frac{\varepsilon\phi_t}{\phi} := q_1 \tag{7}$$

and the distribution functions

$$F^k(x,t) = f^k(x,t,v)\phi \quad \text{and} \quad F(x,t,v) = f(x,t,v)\phi(x,t), \tag{8}$$

then if f^k satisfies (1) or (2) or $f(x,t,v)$ satisfies (3) the functions F^k and $F(x,t,v)$ satisfy the linearized versions of (1)-(3) respectively, that is the equations

$$\varepsilon\left(\partial_t + \frac{k}{N}\partial_x\right)F^k + \alpha\left(F^k - \frac{1}{2N+1}\sum_{j=-N}^{N} F^j\right) = 0, \ k = 0, \pm 1, \ldots, \pm N, \tag{I}$$

or

$$\varepsilon\left(\partial_t + \frac{k}{N}\partial_x\right)F^k + \alpha\left(F^k - \frac{1}{2N}\sum_{\substack{j=-N \\ j\neq o}}^{N} F^j\right) = 0, \ k = \pm 1, \pm 2, \ldots, N, \tag{II}$$

and

$$\varepsilon\left(\partial_t + v\partial_x\right)F + \alpha\left(F - \frac{1}{2}\int_{-1}^{1} F(x,t,\mu)d\mu\right) = 0, \ -1 \leq v \leq 1. \tag{III}$$

The converse also holds for solutions of (I)-(III). One observes that solutions of (I)-(III) satisfy

$$\frac{\partial Q_o}{\partial t} + \frac{\partial Q_1}{\partial x} = 0. \tag{9}$$

When F^k satisfies (I),

$$Q_o := \sum_{j=-N}^{N} F^j \quad \text{and} \quad Q_1 := \sum_{j=-N}^{N} \frac{j}{N} F^j; \tag{10}$$

when F^k satisfies (II), Q_o and Q_1 are given by (10) with the term F^o deleted; and finally when (III) holds,

$$Q_o := \int_{-1}^{1} F(x,t,\mu)d\mu \quad \text{and} \quad Q_1 := \int_{-1}^{1} \mu F(x,t,\mu)d\mu. \tag{11}$$

In all cases we introduce ϕ by

$$\varepsilon\phi_x = -Q_o \quad \text{and} \quad \varepsilon\phi_t = Q_1 \tag{12}$$

and define f^k and f by (8) as appropriate. The fact that (I), (II), or (III) hold implies that f^k or f satisfy (1), (2) or (3). Of course this construction is predicated on the assumption that the potential ϕ introduced in (13) is positive.

In the next section we shall exploit the equivalence of (1)-(3) with their linearized parts (I)-(III). We shall examine special solutions which correspond to (i) simple equilibria and (ii) solutions with shock waves. We shall also show that if one is willing to neglect exponentially decaying transients, then there is a diffusion equation which governs the asymptotic evolution of the mean $q_o(x,t)$. For definiteness we shall restrict our attention exclusively to the equations (3) and (III). Similar results obtain for the other systems.

II. SPECIAL SOLUTIONS AND DIFFUSIVE APPROXIMATES

In this section we shall focus on the two systems

$$\varepsilon\left(\partial_t+v\partial_x\right)f+\alpha\left(f-\frac{1}{2}\int_{-1}^{1}f(x,t,\mu)d\mu\right)+f\left(\int_{-1}^{1}\mu f(x,t,\mu)d\mu - v\int_{-1}^{1} f(x,t,\mu)d\mu\right)= 0 \tag{NL}$$

and

$$\varepsilon\left(\partial_t + v\partial_x\right)F + \alpha\left(F - \frac{1}{2}\int_{-1}^{1}F(x,t,\mu)d\mu\right) = 0 \tag{L}$$

when $0 < \varepsilon$, $-1 \leq v \leq 1$, and $\alpha = 0$ or 1. Once again solutions of one generate solutions of the other through the transformations (6), (7), (8), (11), and (12) of the preceding section. The first thing to note is that when $\alpha = 0$

solutions to the linear problem (L) are given by

$$F(x,t,v) = F_0(x-vt,v) \tag{1}$$

where $F_0(\cdot,\cdot)$ is the initial value for F. The mean Q_0 is given by

$$Q_0(x,t) = \int_{-1}^{1} F_0(x-vt,v)dv \tag{2}$$

and if $F_0(\cdot,\cdot)$ is a nonnegative function satisfying

$$\sup_{-1\leq v\leq 1} F_0(\xi,v) \leq \bar{F}_0(\xi) \tag{3}$$

where

$$\int_{-\infty}^{\infty} \bar{F}_0(\xi)d\xi = \bar{F} < \infty, \tag{4}$$

then

$$Q_0(x,t) = \int_{-1}^{1} F_0(x-vt,v)dv \leq \frac{\bar{F}}{t}. \tag{5}$$

If $F(\cdot,\cdot) \geq 0$ and $\int_{-\infty}^{\infty}\int_{-1}^{1} F_0(x,v)dvdx < \infty$, then (1) generates a solution to (NL) with $\alpha = 0$ as follows:

$$\phi(x,t) := \delta + \frac{1}{\varepsilon}\int_{x}^{\infty}\int_{-1}^{1} F_0(\xi-vt,v)dvd\xi, \quad \delta > 0, \tag{6}$$

and

$$f(x,t,v) := F(x,t,v)/\phi(x,t). \tag{7}$$

With this construction

$$q_0(x,t) := \int_{-1}^{1} f(x,t,v)dv \quad \text{and} \quad q_1(x,t) := \int_{-1}^{1} vf(x,t,v)dv \tag{8}$$

satisfy

$$-q_0(x,t) = \frac{\varepsilon\phi_x}{\phi} \quad \text{and} \quad q_1(x,t) = \frac{\varepsilon\phi_t}{\phi} \tag{9}$$

and therefore f satisfies (NL) with $\alpha = 0$.

We shall now turn our attention to the more interesting case where $\alpha = 1$. The first thing we note is that the linear equation (L) has separable solutions

$$F_{(q_0,q_1)}(x,t,v) := \frac{q_0 \exp((q_1 t - q_0 x)/\varepsilon)}{2(1 + q_1 - q_0 v)} \tag{10}$$

provided q_1 and q_0 obey the dispersion relation

$$q_1 = q_0 \coth q_0 - 1. \tag{11}$$

These solutions are of importance because they correspond to spatially homogeneous equilibria of the nonlinear system (NL); that is they represent the Maxwellian's for (NL). To see this we observe that

$$Q_0(x,t) := \int_{-1}^{1} F_{(q_0,q_1)}(x,t,v)dv$$

is, by virtue of (11), given by

$$Q_0(x,t) = q_0 \exp((q_1 t - q_0 x)/\varepsilon). \tag{12}$$

Equation (12) of the preceding section implies that

$$\phi = \exp((q_1 t - q_0 x)/\varepsilon) , \tag{13}$$

and the last two relations imply that

$$f(x,t,v) = \frac{q_0}{2(1+q_1-q_0 v)}, \quad q_0(x,t) := \int_{-1}^{1} f(x,t,v)dv \equiv q_0, \text{ and}$$

$$q_1(x,t) := \int_{-1}^{1} v f(x,t,v)dv \equiv q_1. \tag{14}$$

Our interest in these solutions when $q_0 \searrow 0$.

Linear combinations of solutions of the type given by (10) are also of interest. A sum of two such solutions generates a travelling wave solution to the nonlinear equation (NL). To see this let $q_0^\ell > q_0^r > 0$ be arbitrary constants,

$$q_1^\ell = q_0^\ell \coth q_0^\ell - 1 \quad \text{and } q_1^r = q_0^r \coth q_0^r - 1, \tag{15}$$

and

$$F = \frac{A q_0^\ell \exp((q_1^\ell t - q_0^\ell x)/\varepsilon)}{2(1 + q_1^\ell - q_0^\ell v)} + \frac{B q_0^r \exp((q_1^r t - q_0^r x)/\varepsilon)}{2(1 + q_1^r - q_0^r v)} \tag{16}$$

where A and B are positive constants. The functions $Q_0 := \int_{-1}^{1} F(x,t,v)dv$, ϕ, and f are readily obtained and the results are

$$Q_0(x,t) = Aq_0^{\ell}\exp((q_1^{\ell}t - q_0^{\ell}x)/\varepsilon) + Bq_0^{r}\exp((q_1^{r}t - q_0^{r}x)/\varepsilon), \qquad (17)$$

$$\phi = A\exp((q_1^{\ell}t - q_0^{\ell}x)/\varepsilon) + B\exp((q_1^{r}t - q_0^{r}x)/\varepsilon)\ , \qquad (18)$$

and

$$f := F/\phi. \qquad (19)$$

If we now let

$$c = \frac{q_1^{\ell} - q_1^{r}}{q_0^{\ell} - q_0^{r}} = \frac{q_0^{\ell}\coth q_0^{\ell} - q_0^{r}\coth q_0^{r}}{q_0^{\ell} - q_0^{r}}\ , \qquad (20)$$

we find that ϕ and f may be rewritten as

$$\phi = \exp\delta t\left[A\exp(-q_0^{\ell}\xi/\varepsilon) + B\exp(-q_0^{r}\xi/\varepsilon)\right] \qquad (21)$$

and

$$f = \frac{\dfrac{Aq_0^{\ell}\exp(-q_0^{\ell}\xi/\varepsilon)}{2(1 + q_1^{\ell} - q_0^{\ell}v)} + \dfrac{Bq_0^{r}\exp(-q_0^{r}\xi/\varepsilon)}{2(1 + q_1^{r} - q_0^{r}v)}}{A\exp(-q_0^{\ell}\xi/\varepsilon) + B\exp(-q_0^{r}\xi/\varepsilon)} \qquad (22)$$

where

$$\xi := x - ct \text{ and } \delta = q_1^{r} - cq_0^{r} = q_1^{\ell} - cq_0^{\ell}. \qquad (23)$$

If we now exploit the definitions

$$q_0(x,t) := \int_{-1}^{1} f(x,t,v)dv \quad \text{and } q_1(x,t) := \int_{-1}^{1} vf(x,t,v)dv, \qquad (24)$$

we obtain

$$q_o(x,t) = \frac{Aq_o^{\ell}\exp(-q_o^{\ell}\xi/\varepsilon) + Bq_o^{r}\exp(-q_o^{r}\xi/\varepsilon)}{A\exp(-q_o^{\ell}\xi/\varepsilon) + B\exp(-q_o^{r}\xi/\varepsilon)}, \tag{25}$$

and

$$q_1(x,t) = \frac{Aq_1^{\ell}\exp(-q_o^{\ell}\xi/\varepsilon) + Bq_1^{r}\exp(-q_o^{r}\xi/\varepsilon)}{A\exp(-q_o^{\ell}\xi/\varepsilon) + B\exp(-q_o^{r}\xi/\varepsilon)} \tag{26}$$

where again $\xi = x - ct$. These solutions represent travelling waves solutions to (NL) and converge to shock's as $\varepsilon \to 0^+$. The fact that $q_o^{\ell} > q_o^{r} > 0$ implies that

$$\lim_{\varepsilon\to o^+} f(x-ct,v) = \begin{cases} \dfrac{q_o^{\ell}}{2(1 + q_1^{\ell} - q_o^{\ell}v)}, & (x-ct) < 0 \\ \dfrac{q_o^{r}}{2(1 + q_1^{r} - q_o^{r}v)}, & (x-ct) > 0 \end{cases} \tag{27}$$

and

$$\lim_{\varepsilon\to o^+} (q_o(x-ct),\ q_1(x-ct)) = \begin{cases} (q_o^{\ell}, q_1^{\ell}), & (x-ct) < 0 \\ (q_o^{r}, q_1^{r}), & (x-ct) > 0 \end{cases}. \tag{28}$$

Once again, the wave speed c is given by (20).

We now turn our attention to the initial value problem for (L) with $\alpha = 1$, namely the problem

$$\varepsilon(\partial_t + v\partial_x)F + (F - \frac{1}{2}\int_{-1}^{1} F(x,t,\mu)d\mu) = 0,\ -\infty < x < \infty,\ -1 < v < 1,\ t > 0, \tag{L}$$

and

$$F(x,0^+,v) := F_o(x,v),\ -\infty < x < \infty \text{ and } -1 < v < 1. \tag{IC}$$

Solutions of (L) and (IC) are representable as

$$F(x,t,v) = e^{-t/\varepsilon} F_o(x-vt,v) + \frac{1}{2\varepsilon}\int_o^t e^{-(t-s)/\varepsilon} Q_o(x-v(t-s),s)ds \tag{29}$$

where again $Q_o(x,t) := \int_{-1}^{1} F(x,t,\mu)d\mu$. Moreover, (29) implies that Q_o satisfies the integral equation

$$Q_o(x,t) = e^{-t/\varepsilon}\int_{-1}^{1} F_o(x-vt,v)dv + \frac{1}{2\varepsilon}\int_o^t e^{-(t-s)/\varepsilon}\int_{-1}^{1} Q_o(x-v(t-s),s)dvds. \tag{30}$$

A simple analysis of the iterates

$$\left.\begin{aligned} &Q_o^{(o)}(x,t) := e^{-t/\varepsilon}\int_{-1}^{1} F_o(x-vt,v)dv \\ &\text{and for } n \geq 0 \\ &Q_o^{(n+1)}(x,t) := e^{-t/\varepsilon}\int_{-1}^{1} F_o(x-vt,v)dv + \frac{1}{2\varepsilon}\int_o^t e^{-(t-s)/\varepsilon}\int_{-1}^{1} Q_o^{(n)}(x-v(t-s),s)dvds \end{aligned}\right\} \tag{31}$$

guarantees that for nonnegative data, F_o, (30) has a unique nonnegative solution, Q_o, and the representation (29) guarantees that the density F is nonnegative on its domain. We also have the elementary upper bound

$$Q_o(x,t) \leq \int_{-1}^{1} \max_x F_o(x,v)dv, \tag{32}$$

and if the data $F_o \geq 0$ satisfies

$$\int_{-\infty}^{\infty}\int_{-1}^{1} F_o(x,v)dvdx = \bar{Q}_o < \infty, \tag{33}$$

then the solution of (30) satisfies

$$\int_{-\infty}^{\infty} Q_o(x,t)dx = \bar{Q}_o, \quad t \geq 0. \tag{34}$$

For such data we are guaranteed the existence of a positive potential ϕ satisfying

$$-\varepsilon\phi_x = Q_o(x,t) \text{ and } \varepsilon\phi_t = Q_1, \tag{35}$$

infact for any $\delta > 0$

$$\phi = \delta + \frac{1}{\varepsilon}\int_x^{\infty} Q_o(\xi,t)d\xi, \tag{36}$$

does the job. Such solutions clearly generate solutions of the nonlinear problem (NL) via $f(x,t,v) = F(x,t,v)/\phi(x,t)$.

These results, though fundamental, give little insight into the detailed structure of F and its determining moment $Q_o(x,t) := \int_{-1}^{1} F(x,t,\mu)d\mu$. To obtain these results we study the Fourier–Laplace transforms of F and Q_o. In the sequel we shall let

$$\hat{F}(w,p,v) = \int_{-\infty}^{\infty} e^{\frac{-iwx}{\varepsilon}} \left[\int_{0}^{\infty} e^{\frac{-pt}{\varepsilon}} F(x,t,v)dt \right] dx \tag{37}$$

and

$$\hat{Q}_o(w,p) = \int_{-\infty}^{\infty} e^{\frac{-iwx}{\varepsilon}} \left[\int_{0}^{\infty} e^{\frac{-pt}{\varepsilon}} Q_o(x,t)dt \right] dx = \int_{-1}^{1} \hat{F}(w,p,v)dv. \tag{38}$$

With these definitions, F and Q_o are recaptured via the inversion formulas

$$F(x,t,v) = \frac{-i}{4\pi^2\varepsilon^2} \int_{-\infty}^{\infty} e^{\frac{iwx}{\varepsilon}} \left[\int_{C=\{p=a+is,-\infty<s<\infty\}} e^{\frac{pt}{\varepsilon}} \hat{F}(w,p,v)dp \right] dw \tag{39}$$

and

$$Q_o(x,t) = \frac{-i}{4\pi^2\varepsilon^2} \int_{-\infty}^{\infty} e^{\frac{iwx}{\varepsilon}} \left[\int_{C=\{p=a+is,-\infty<s<\infty\}} e^{\frac{pt}{\varepsilon}} \hat{Q}_o(w,p,)dp \right] dw \tag{40}$$

where the contour C lies to the right of all singularities of F and Q_o.

The fact that F satisfies (L) and (IC) implies that

$$\hat{F}(w,p,v) = \frac{\hat{Q}_o(w,p)}{2(1+p+iwv)} + \frac{\varepsilon \hat{F}_o(w,v)}{(1+p+iwv)} \tag{41}$$

and that

$$\left[1 - \frac{1}{2} \int_{-1}^{1} \frac{dv}{(1+p+iwv)} \right] \hat{Q}_o(w,p) = \varepsilon \int_{-1}^{1} \frac{\hat{F}_o(w,v)dv}{(1+p+iwv)} \tag{42}$$

where

$$\hat{F}_o(w,v) := \int_{-\infty}^{\infty} e^{\frac{-iwx}{\varepsilon}} F_o(x,v)dx. \tag{43}$$

To proceed a few words are in order the function

$$H(p,w) := \frac{1}{2} \int_{-1}^{1} \frac{dv}{(1+p+iwv)}. \tag{44}$$

H is analytic in $p = r - 1 +$ is and can be written as

$$H = U(r,s,w) + i\, V(r,s,w) \tag{45}$$

where for $r > 0$, $s > 0$, and $w > 0$

$$U(r,s,w) = \frac{1}{2w}\left[\arctan\left(\frac{w-s}{r}\right) + \arctan\left(\frac{w+s}{r}\right)\right] \tag{46}$$

and

$$V(r,s,w) = \frac{1}{2w}\,\mathrm{Log}\left[\frac{r^2 + (w-s)^2}{r^2 + (w+s)^2}\right]^{1/2}. \tag{47}$$

Moreover, U and V have the following symmetries:

$$U(-r,s,w) = -U(r,s,w) \quad \text{and} \quad V(-r,s,w) = V(r,s,w), \tag{48}$$

$$U(r,-s,w) = U(r,s,w) \quad \text{and} \quad V(r,-s,w) = -V(r,-s,w), \tag{49}$$

$$U(r,s,-w) = U(r,s,w) \quad \text{and} \quad V(r,s,-w) = V(r,s,w), \tag{50}$$

and for any $w \neq 0$, H has a branch cut given by (51) below

$$B_w = \{p = -1 + is,\ -|w| < s < |w|\}. \tag{51}$$

We also note that for $|w| < \pi$ the equation $H(p,w) = 1$ has exactly one solution

$$p = w \cot w - 1 + io\ , \tag{52}$$

while for $|w| > \pi$ it has none.

Equipped with this information we find that

$$\hat{Q}_o(w,p) := \varepsilon \int_{-1}^{1} \frac{\hat{F}_o(w,\mu)\,d\mu}{(1+p+iwv)} \Big/ (1 - H(p,w)) \tag{53}$$

and that for any $a \geq 0$

$$\frac{1}{2\pi i\varepsilon}\int_{C=\{p=a+is,-\infty<s<\infty\}} e^{\frac{pt}{\varepsilon}}\hat{Q}_o(w,p)\,dp = \begin{cases} \dfrac{w^2}{\sin^2 w}\, e^{\frac{(w\cot w-1)t}{\varepsilon}} \displaystyle\int_{-1}^{1} \frac{\hat{F}_o(w,\mu)\,d\mu}{(w\cot w+iw\mu)} + \frac{1}{2\pi i\varepsilon}\int_{C_*=\{p=a_*+is,-\infty<s<\infty\}} e^{\frac{pt}{\varepsilon}}\hat{Q}_o(w,p)\,dp, \\ \quad \text{for } 0 \leq |w| < \frac{\pi}{2} \text{ and } -1 < a_* < w\cot w - 1, \\ \dfrac{1}{2\pi i\varepsilon}\displaystyle\int_{C_*=\{p=a_*+is,-\infty<s<\infty\}} e^{\frac{pt}{\varepsilon}}\hat{Q}_o(w,p)\,dp, \quad \text{for } \frac{\pi}{2} < |w| \text{ and any } -1 < a_* < 0. \end{cases} \tag{54}$$

The last result implies that to within terms decaying as $e^{\frac{-\lambda t}{\varepsilon}}$, $0 < \lambda < 1$, the moment Q_0 and density F may be written as

$$\left.\begin{aligned} Q_0 &= \frac{1}{2\pi\varepsilon}\int_{-\pi/2}^{\pi/2}\frac{(w^2\exp(((w\cot w-1)t+iwx)/\varepsilon)}{\sin^2 w}\int_{-1}^{1}\frac{\hat{F}_0(w,\mu)d\mu\, dw}{(w\cot w + iw\mu)}, \\ &\text{and} \\ F &= \frac{1}{4\pi\varepsilon}\int_{-\pi/2}^{\pi/2}\frac{w^2\exp(((w\cot w-1)t+iwx)/\varepsilon)}{\sin^2 w(w\cot w + iwv)}\int_{-1}^{1}\frac{\hat{F}_0(w,\mu)d\mu\, dw}{(w\cot w + iw\mu)}. \end{aligned}\right\} \quad (55)$$

To obtain the small ε behavior of Q_0 and F we recall that

$$\hat{F}_0(w,\mu) := \int_{-\infty}^{\infty} e^{\frac{-iwx}{\varepsilon}} F_0(x,\mu)dx$$

and therefore that

$$\hat{F}_0(\varepsilon k,\mu) = \int_{-\infty}^{\infty} e^{-ikx} F_0(x,\mu)dx$$

and

$$F_0(x,\mu) = \frac{1}{2\pi}\int_{-\infty}^{\infty} e^{ikx}\hat{F}_0(\varepsilon k,\mu)dk.$$

This identities imply that

$$\frac{1}{2\pi\varepsilon}\int_{-\pi/2}^{\pi/2}\frac{w^2 e^{\frac{iwx}{\varepsilon}}}{\sin^2 w}\int_{-1}\frac{\hat{F}_0(w,\mu)d\mu dw}{w\cot w + iw\mu}$$

$$= \frac{1}{2\pi}\int_{-\pi/2\varepsilon}^{\pi/2\varepsilon}\frac{\varepsilon^2 k^2 e^{ikx}}{\sin^2(\varepsilon k)}\int_{-1}^{1}\frac{\hat{F}_0(\varepsilon k,\mu)d\mu dk}{\varepsilon k\cot(\varepsilon k) + i\varepsilon k\mu}$$

$$\sim \int_{-1}^{1} F_0(x,\mu)d\mu + O(\varepsilon)$$

and

$$\frac{1}{2\pi\varepsilon}\int_{-\pi/2}^{\pi/2}\frac{w^2 e^{\frac{iwx}{\varepsilon}}}{\sin^2 w(w\cot w + iwv)}\int_{-1}^{1}\frac{\hat{F}_0(w,\mu)d\mu dw}{w\cot w + i\mu w}$$

$$= \frac{1}{2\pi}\int_{-\pi/2\varepsilon}^{\pi/2\varepsilon}\frac{\varepsilon^2 k^2 e^{ikx}}{\sin^2(\varepsilon k)(\varepsilon k\cot\varepsilon k + i\varepsilon kv)}\int_{-1}^{1}\frac{\hat{F}_0(\varepsilon k,\mu)d\mu dk}{(\varepsilon k\cot\varepsilon k + i\mu\varepsilon k)}$$

$$\sim \int_{-1}^{1} F_0(x,\mu)d\mu + O(\varepsilon),$$

and therefore the convolution theorem implies that

$$Q_0(x,t) \sim \left[\int_{x-t}^{x+t} \Gamma(x-\xi,t,\varepsilon)\int_{-1}^{1}F_0(\xi,\mu)d\mu d\xi\right](1 + O(\varepsilon)) \tag{56}$$

and

$$F(x,t,v) \sim Q_0(x,t)/2$$

where

$$\Gamma(x,t,\varepsilon) := \frac{1}{2\pi\varepsilon}\int_{-\pi/2}^{\pi/2} e^{((w\cot w-1)t + iwx)/\varepsilon}dw \tag{57}$$

$$\sim \frac{1}{2\sqrt{\pi\varepsilon t}\,K_*}\, e^{(q_*\coth q_*-1)t-q_*x)/\varepsilon}$$

where, for $-1 < \frac{x-\xi}{t} < 1$, q_* is the unique solution of

$$\coth q_* - \frac{q_*}{\sinh^2 q_*} = \frac{x-\xi}{t} \tag{58}$$

and

$$K_* = \sqrt{\frac{q_*\coth q_* - \sinh q_*}{\sinh^3 q_*}} > 0. \tag{59}$$

Equation (55) - (57) imply that to within exponentially decaying transients and algebraically small terms in ε Q_0 evolves as the solution of

$$\frac{\partial Q_0}{\partial t} - \frac{1}{\varepsilon}(\varepsilon\partial_x\coth(\varepsilon\partial_x) - 1)Q_0 = 0$$

and

$$Q_o(x,0^+) = \int_{-1}^{1} F_o(x,v)dv \;.$$

This result implies that the moment

$$q_o := \frac{\varepsilon\, Q_0(x,t)}{\varepsilon\delta + \int_x^\infty Q_o(\xi,t)d\xi}, \quad \delta > 0$$

asymptotically evolves as

$$e^{-\frac{1}{\varepsilon}\int_x^\infty q_o(\xi,t)d\xi}\,[\varepsilon\partial_t - \varepsilon\partial_x \coth(\varepsilon\partial_x) + 1]\,e^{\frac{1}{\varepsilon}\int_x^\infty q_o(\xi,t)d\xi} = 0$$

and takes on the data

$$q_o(x,0+) = \frac{\varepsilon\, Q_0(x,0_+)}{\varepsilon\delta + \int_x^\infty Q_0(\xi,0^+)d\xi} \;.$$

The integral representations (55) may also be used to show that the solution to the Riemann Problem for (NL) converges to the one of the travelling waves discussed previously in this section. The Riemann problem for (NL) requires solving

$$\varepsilon(\partial_t+v\partial_x)f+(f - \frac{1}{2}\int_{-1}^{1} f(x,t,\mu)d\mu) + f(\int_{-1}^{1}\mu f(x,t,\mu)d\mu - v\int_{-1}^{1} f(x,t,\mu)d\mu)=0, \quad \text{(NL)}$$

$$-\infty < x < \infty,\ -1 < v < 1,\ t > 0,$$

$$f(x,0^+,v) = \begin{cases} \dfrac{q_o^\ell}{2(1+q_1^\ell - q_o^\ell v)}, & x < 0 \\[2ex] \dfrac{q_o^r}{2(1 + q_1^r - q_o^r v)}, & x > 0 \end{cases} \qquad \text{(IC)}$$

where again

$$0 < q_o^r < q_o^\ell,\quad q_1^\ell = q_o^\ell \coth q_o^\ell - 1, \text{ and } q_1^r = q_o^r \coth q_o^r - 1,$$

and by virtue of (6), (7), (8), (11), and (12) of the previous section solving this problem is equivalent to solving the following problem for the linear equation (L):

$$\varepsilon(\partial_t + v\partial_x)F + (F - \frac{1}{2}\int_{-1}^{1}F(x,t,\mu)d\mu) = 0, \quad -\infty < x < \infty, \; -1 < v < 1, \; t > 0, \tag{L}$$

and

$$F(x,0^+,v) = \begin{cases} \dfrac{q_0^\ell e^{-q_0^\ell x/\varepsilon}}{2(1 + q_1^\ell - q_0^\ell v)}, & x < 0 \\ \dfrac{q_0^r e^{-q^r x/\varepsilon}}{2(1 + q_1^r - q_0^r v)}, & x > 0 . \end{cases} \tag{IC}$$

The solution to (L) and (IC) may be written as

$$F(x,v,t) = \frac{q_0 e^{(q_1^\ell t - q_0^\ell x)/\varepsilon}}{2(1 + q_1^\ell - q_0^\ell v)} + G \tag{60}$$

where G satisfies

$$\varepsilon(\partial_t + v\partial_x)G + (G - \frac{1}{2}\int_{-1}^{1}G(x,t,\mu)d\mu)=0, \quad -\infty < x < \infty, \; -1 < v < 1, \; t > 0 \tag{61}$$

and

$$G(x,0^+,v) = G_0(x,v) := \begin{cases} 0, & x < 0 \\ \dfrac{q_0^r e^{-q_0^r x/\varepsilon}}{2(1 + q_1^r - q_0^r v)} - \dfrac{q_0^\ell e^{-q_0^\ell x/\varepsilon}}{2(1 + q_1^\ell - q_0^\ell v)}, & x > 0. \end{cases} \tag{62}$$

Noting that

$$\hat{G}_0(w,\mu) = \frac{\varepsilon q_0^r}{2(1 + q_1^r - q_0^r v)} \frac{1}{(q_0^r + iw)} - \frac{\varepsilon q_0^\ell}{2(1 + q_1^\ell - q_0^\ell v)} \frac{1}{(q_0^\ell + iw)} \tag{63}$$

and that the identities (20), (53), and (63) imply that

$$\int_{-1}^{1} \frac{\hat{G}_0(w,\mu)d\mu}{(w\cot w + iw\mu)} = \frac{\varepsilon}{(w\cot w + iw\coth q_0^r)} - \frac{\varepsilon}{(w\cot w + iw\coth q_0^\ell)}, \tag{64}$$

we find that the dominant terms of G and $Q_o^G := \int_{-1}^{1} G(x,t,v)dv$ are given by

$$G = \frac{1}{4\pi\varepsilon}\int_{-\pi/2}^{\pi/2} \frac{w^2 e^{((w\cot w-1)t+iwx)/\varepsilon}}{(w\cot w + iwv)\sin^2 w}\left[\frac{1}{(w\cot w+iw\coth q^r)_o} - \frac{1}{(w\cot w+iw\coth q_o^\ell)}\right]dw \quad (65)$$

and

$$Q_o^G = \frac{1}{2\pi\varepsilon}\int_{-\pi/2}^{\pi/2} \frac{w^2 e^{((w\cot w-1)t+iwx)/\varepsilon}}{\sin^2 w}\left[\frac{1}{(w\cot w+iw\coth q_o^r)} \quad \frac{1}{(w\cot w+iw\coth q_o^\ell)}\right]dw. \quad (66)$$

We now turn to the asymptotic evaluation of Q_o^G as t/ε tends to infinity. We let $x = ct + \varepsilon\xi$ where $c = \dfrac{q_o^\ell \coth q_o^\ell - q_o^r \coth q_o^r}{q_o^\ell - q_o^r}$ is the speed of the travelling wave described in (15)-(28) of this section and observe that $Q_o^G(ct+\varepsilon\xi,t)$ may be written as

$$Q_o^G = \frac{1}{2\pi}\int_{-\pi/2}^{\pi/2} \frac{w^2 \exp((w\cot w-1\ iwc)t/\varepsilon)\exp(iw\xi)}{\sin^2 w}\left[\frac{1}{w\cot w+iw\coth q_o^r} - \frac{1}{w\cot w+iw\coth q_o^\ell}\right]dw.$$

(67)

Our strategy in evaluating this integral is to deform the horizontal contour from $[-\pi/2, \pi/2]$ into one running through the stationary point of the phase

$$\psi := w\cot w - 1 + icw. \quad (68)$$

The stationary point is of the form $w = iq_*, q_o^r < q_* < q_o^\ell$, where q_* satisfies

$$\frac{d}{dq_*}(q_*\coth q_*) = \coth q_* - \frac{q_*}{(\sinh q_*)^2} := \hat{c}(q_*) = c := \frac{1}{q_o^\ell - q_o^r}\int_{q_o^r}^{q_o^\ell} \hat{c}(q)dq \quad (69)$$

Moreover $\dfrac{d^2\psi}{dw^2}(w = iq_*)$ may be written as

$$\frac{d^2\psi}{dw^2}(w = iq_*) = \frac{2}{\sinh q_*^2}(1 - q_*\coth q_*) := -2K^2(q_*). \quad (70)$$

The fact that for all $q \in (-\infty,\infty)$

$$-1 < \coth q - \frac{q}{\sinh^2 q} < 1 \text{ and } \frac{2}{\sinh^2 q}(1 - q \coth q) < 0 \tag{71}$$

guarantees that for any $q_o^r < q_* < q_o^\ell$

$$0 < c < 1 \text{ and } K^2(q_*) > 0. \tag{72}$$

This in turn implies that Q_o^G may be rewritten as

$$Q_o^G = q_o^r e^{(q_o^r \coth q_o^r - 1 - q_o^r c)t/\varepsilon} e^{-q_o^r \xi} + R \tag{73}$$

where

$$R = \frac{1}{2\pi} \int_{c_{[-\pi,\pi]}} \frac{w^2 \exp((w\cot w - 1 + iwc)t/\varepsilon)\exp(iw\xi)}{\sin^2 w} \left[\frac{1}{(w\cot w + iw\coth q_o^r)} - \frac{1}{(w\cot w + iw\coth q_o^\ell)} \right] dw \tag{74}$$

and $c_{[-\pi,\pi]}$ is any contour of the type shown below:

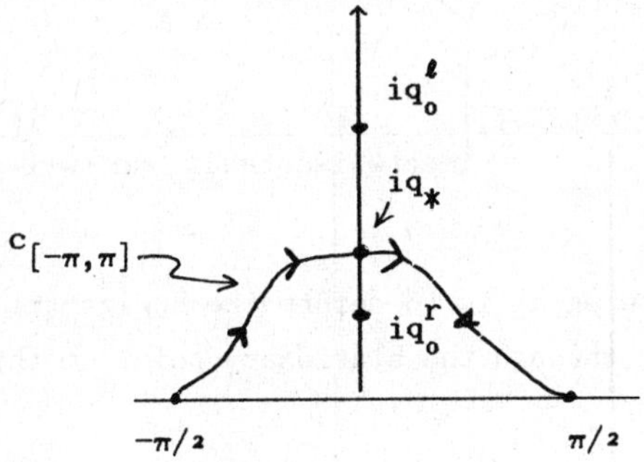

Moreover, the asymptotic value of the integral R as $t/\varepsilon \to \infty$ is given by

$$R = \frac{\varepsilon^{1/2} q_* e^{((q_* \coth q_* - 1 - q_* c)t/\varepsilon)} e^{-q_* \xi}}{2\, K(q_*) \sqrt{\pi t}\, \sinh q_*} \left[\frac{1}{(\coth q_* - \coth q_o^r)} - \frac{1}{(\coth q_* - \coth q_o^\ell)} \right] \tag{75}$$

and thus (20), (23), (60), and $x = ct + \varepsilon\xi$ imply that $Q_o^F := \int_{-1}^{1} f(x,t,v)dv$ satisfies

$$Q_o^F(ct + \varepsilon\xi, t) \sim e^{\delta t}\left[q_o^{\ell} e^{-q_o^{\ell}\xi} + q_o^r e^{-q_o^r \xi}\right]$$

$$+ \frac{\varepsilon^{1/2} q_* \, e^{\delta_* t/\varepsilon} e^{-q_*\xi}}{2\, K(q_*)\sqrt{\pi t}\, \sinh q_*}\left[\frac{1}{(\coth q_* - \coth q_o^r)} - \frac{1}{(\coth q_* - \coth q_o^{\ell})}\right] \tag{76}$$

where

$$\delta = q_o^{\ell}\coth q_o^{\ell} - 1 - c q_o^{\ell} = q_o^r \coth q_o^r - 1 - c q_o^r, \tag{77}$$

$$\delta_* = q_* \coth q_* - 1 - c q_*, \tag{78}$$

and

$$\delta > \delta_*. \tag{79}$$

The last inequality implies that at t/ε tends to infinity the first term of (76) dominates, that is that

$$Q_o^F(ct + \varepsilon\xi, t) \sim e^{\delta t/\varepsilon}\left[q_o^{\ell} e^{-q_o^{\ell}\xi} + q_o^r e^{-q_o^r\xi}\right]. \tag{80}$$

Analogous computations also yield

$$F(ct + \xi, t, v) \sim e^{\delta t/\varepsilon}\left[\frac{q_o^{\ell} e^{-q_o^{\ell}\xi}}{2(q_o^{\ell}\coth q_o^{\ell} - q_o^{\ell} v)} + \frac{q_o^r e^{-q_o^r \xi}}{2(q_o^r \coth q_o^r - q_o^r v)}\right] \tag{81}$$

and these when combined with (6), (7), (8), (11), and (12) of the first section yield the desired asymptotics for solutions of the Riemann Problem.

REFERENCES

[1] I. Kolodner, On the Carleman's Model for the Boltzman Equation and its Generalizations, Annali di Mathematica pura ed applicata, IV, Vol. LXIII (1963), 11-32.

[2] K. Inoue and T. Nishida, On the Broadwell Model of the Boltzman equation for a simple discrete velocity gas, Appl. Math. Optim., 3 (1976), 27-49.

[3] J. Thomas Beale, Large-Time Behavior of the Broadwell Model for a Discrete Velocity Gas (preprint).

Vol. 1117: D.J. Aldous, J.A. Ibragimov, J. Jacod, Ecole d'Été de Probabilités de Saint-Flour XIII - 1983. Édité par P.L. Hennequin. IX, 409 pages. 1985.

Vol. 1118: Grossissements de filtrations: exemples et applications. Seminaire, 1982/83. Edité par Th. Jeulin et M. Yor. V, 315 pages. 1985.

Vol. 1119: Recent Mathematical Methods in Dynamic Programming. Proceedings, 1984. Edited by I. Capuzzo Dolcetta, W.H. Fleming and T. Zolezzi. VI, 202 pages. 1985.

Vol. 1120: K. Jarosz, Perturbations of Banach Algebras. V, 118 pages. 1985.

Vol. 1121: Singularities and Constructive Methods for Their Treatment. Proceedings, 1983. Edited by P. Grisvard, W. Wendland and J.R. Whiteman. IX, 346 pages. 1985.

Vol. 1122: Number Theory. Proceedings, 1984. Edited by K. Alladi. VII, 217 pages. 1985.

Vol. 1123: Séminaire de Probabilités XIX 1983/84. Proceedings. Edité par J. Azéma et M. Yor. IV, 504 pages. 1985.

Vol. 1124: Algebraic Geometry, Sitges (Barcelona) 1983. Proceedings. Edited by E. Casas-Alvero, G.E. Welters and S. Xambó-Descamps. XI, 416 pages. 1985.

Vol. 1125: Dynamical Systems and Bifurcations. Proceedings, 1984. Edited by B.L.J. Braaksma, H.W. Broer and F. Takens. V, 129 pages. 1985.

Vol. 1126: Algebraic and Geometric Topology. Proceedings, 1983. Edited by A. Ranicki, N. Levitt and F. Quinn. V, 423 pages. 1985.

Vol. 1127: Numerical Methods in Fluid Dynamics. Seminar. Edited by F. Brezzi, VII, 333 pages. 1985.

Vol. 1128: J. Elschner, Singular Ordinary Differential Operators and Pseudodifferential Equations. 200 pages. 1985.

Vol. 1129: Numerical Analysis, Lancaster 1984. Proceedings. Edited by P.R. Turner. XIV, 179 pages. 1985.

Vol. 1130: Methods in Mathematical Logic. Proceedings, 1983. Edited by C.A. Di Prisco. VII, 407 pages. 1985.

Vol. 1131: K. Sundaresan, S. Swaminathan, Geometry and Nonlinear Analysis in Banach Spaces. III, 116 pages. 1985.

Vol. 1132: Operator Algebras and their Connections with Topology and Ergodic Theory. Proceedings, 1983. Edited by H. Araki, C.C. Moore, Ş. Strătilă and C. Voiculescu. VI, 594 pages. 1985.

Vol. 1133: K.C. Kiwiel, Methods of Descent for Nondifferentiable Optimization. VI, 362 pages. 1985.

Vol. 1134: G.P. Galdi, S. Rionero, Weighted Energy Methods in Fluid Dynamics and Elasticity. VII, 126 pages. 1985.

Vol. 1135: Number Theory, New York 1983–84. Seminar. Edited by D.V. Chudnovsky, G.V. Chudnovsky, H. Cohn and M.B. Nathanson. V, 283 pages. 1985.

Vol. 1136: Quantum Probability and Applications II. Proceedings, 1984. Edited by L. Accardi and W. von Waldenfels. VI, 534 pages. 1985.

Vol. 1137: Xiao G., Surfaces fibrées en courbes de genre deux. IX, 103 pages. 1985.

Vol. 1138: A. Ocneanu, Actions of Discrete Amenable Groups on von Neumann Algebras. V, 115 pages. 1985.

Vol. 1139: Differential Geometric Methods in Mathematical Physics. Proceedings, 1983. Edited by H. D. Doebner and J. D. Hennig. VI, 337 pages. 1985.

Vol. 1140: S. Donkin, Rational Representations of Algebraic Groups. VII, 254 pages. 1985.

Vol. 1141: Recursion Theory Week. Proceedings, 1984. Edited by H.-D. Ebbinghaus, G.H. Müller and G.E. Sacks. IX, 418 pages. 1985.

Vol. 1142: Orders and their Applications. Proceedings, 1984. Edited by I. Reiner and K. W. Roggenkamp. X, 306 pages. 1985.

Vol. 1143: A. Krieg, Modular Forms on Half-Spaces of Quaternions. XIII, 203 pages. 1985.

Vol. 1144: Knot Theory and Manifolds. Proceedings, 1983. Edited by D. Rolfsen. V, 163 pages. 1985.

Vol. 1145: G. Winkler, Choquet Order and Simplices. VI, 143 pages. 1985.

Vol. 1146: Séminaire d'Algèbre Paul Dubreil et Marie-Paule Malliavin. Proceedings, 1983–1984. Edité par M.-P. Malliavin. IV, 420 pages. 1985.

Vol. 1147: M. Wschebor, Surfaces Aléatoires. VII, 111 pages. 1985.

Vol. 1148: Mark A. Kon, Probability Distributions in Quantum Statistical Mechanics. V, 121 pages. 1985.

Vol. 1149: Universal Algebra and Lattice Theory. Proceedings, 1984. Edited by S. D. Comer. VI, 282 pages. 1985.

Vol. 1150: B. Kawohl, Rearrangements and Convexity of Level Sets in PDE. V, 136 pages. 1985.

Vol 1151: Ordinary and Partial Differential Equations. Proceedings, 1984. Edited by B.D. Sleeman and R.J. Jarvis. XIV, 357 pages. 1985.

Vol. 1152: H. Widom, Asymptotic Expansions for Pseudodifferential Operators on Bounded Domains. V, 150 pages. 1985.

Vol. 1153: Probability in Banach Spaces V. Proceedings, 1984. Edited by A. Beck, R. Dudley, M. Hahn, J. Kuelbs and M. Marcus. VI, 457 pages. 1985.

Vol. 1154: D.S. Naidu, A.K. Rao, Singular Pertubation Analysis of Discrete Control Systems. IX, 195 pages. 1985.

Vol. 1155: Stability Problems for Stochastic Models. Proceedings, 1984. Edited by V.V. Kalashnikov and V.M. Zolotarev. VI, 447 pages. 1985.

Vol. 1156: Global Differential Geometry and Global Analysis 1984. Proceedings, 1984. Edited by D. Ferus, R.B. Gardner, S. Helgason and U. Simon. V, 339 pages. 1985.

Vol. 1157: H. Levine, Classifying Immersions into $\mathbb{R}^4$ over Stable Maps of 3-Manifolds into $\mathbb{R}^2$. V, 163 pages. 1985.

Vol. 1158: Stochastic Processes – Mathematics and Physics. Proceedings, 1984. Edited by S. Albeverio, Ph. Blanchard and L. Streit. VI, 230 pages. 1986.

Vol. 1159: Schrödinger Operators, Como 1984. Seminar. Edited by S. Graffi. VIII, 272 pages. 1986.

Vol. 1160: J.-C. van der Meer, The Hamiltonian Hopf Bifurcation. VI, 115 pages. 1985.

Vol. 1161: Harmonic Mappings and Minimal Immersions, Montecatini 1984. Seminar. Edited by E. Giusti. VII, 285 pages. 1985.

Vol. 1162: S.J.L. van Eijndhoven, J. de Graaf, Trajectory Spaces, Generalized Functions and Unbounded Operators. IV, 272 pages. 1985.

Vol. 1163: Iteration Theory and its Functional Equations. Proceedings, 1984. Edited by R. Liedl, L. Reich and Gy. Targonski. VIII, 231 pages. 1985.

Vol. 1164: M. Meschiari, J.H. Rawnsley, S. Salamon, Geometry Seminar "Luigi Bianchi" II – 1984. Edited by E. Vesentini. VI, 224 pages. 1985.

Vol. 1165: Seminar on Deformations. Proceedings, 1982/84. Edited by J. Ławrynowicz. IX, 331 pages. 1985.

Vol. 1166: Banach Spaces. Proceedings, 1984. Edited by N. Kalton and E. Saab. VI, 199 pages. 1985.

Vol. 1167: Geometry and Topology. Proceedings, 1983–84. Edited by J. Alexander and J. Harer. VI, 292 pages. 1985.

Vol. 1168: S.S. Agaian, Hadamard Matrices and their Applications. III, 227 pages. 1985.

Vol. 1169: W.A. Light, E.W. Cheney, Approximation Theory in Tensor Product Spaces. VII, 157 pages. 1985.

Vol. 1170: B.S. Thomson, Real Functions. VII, 229 pages. 1985.

Vol. 1171: Polynômes Orthogonaux et Applications. Proceedings, 1984. Edité par C. Brezinski, A. Draux, A.P. Magnus, P. Maroni et A. Ronveaux. XXXVII, 584 pages. 1985.

Vol. 1172: Algebraic Topology, Göttingen 1984. Proceedings. Edited by L. Smith. VI, 209 pages. 1985.

Vol. 1173: H. Delfs, M. Knebusch, Locally Semialgebraic Spaces. XVI, 329 pages. 1985.

Vol. 1174: Categories in Continuum Physics, Buffalo 1982. Seminar. Edited by F.W. Lawvere and S.H. Schanuel. V, 126 pages. 1986.

Vol. 1175: K. Mathiak, Valuations of Skew Fields and Projective Hjelmslev Spaces. VII, 116 pages. 1986.

Vol. 1176: R.R. Bruner, J.P. May, J.E. McClure, M. Steinberger, H_∞ Ring Spectra and their Applications. VII, 388 pages. 1986.

Vol. 1177: Representation Theory I. Finite Dimensional Algebras. Proceedings, 1984. Edited by V. Dlab, P. Gabriel and G. Michler. XV, 340 pages. 1986.

Vol. 1178: Representation Theory II. Groups and Orders. Proceedings, 1984. Edited by V. Dlab, P. Gabriel and G. Michler. XV, 370 pages. 1986.

Vol. 1179: Shi J.-Y. The Kazhdan-Lusztig Cells in Certain Affine Weyl Groups. X, 307 pages. 1986.

Vol. 1180: R. Carmona, H. Kesten, J.B. Walsh, École d'Été de Probabilités de Saint-Flour XIV – 1984. Édité par P.L. Hennequin. X, 438 pages. 1986.

Vol. 1181: Buildings and the Geometry of Diagrams, Como 1984. Seminar. Edited by L. Rosati. VII, 277 pages. 1986.

Vol. 1182: S. Shelah, Around Classification Theory of Models. VII, 279 pages. 1986.

Vol. 1183: Algebra, Algebraic Topology and their Interactions. Proceedings, 1983. Edited by J.-E. Roos. XI, 396 pages. 1986.

Vol. 1184: W. Arendt, A. Grabosch, G. Greiner, U. Groh, H.P. Lotz, U. Moustakas, R. Nagel, F. Neubrander, U. Schlotterbeck, One-parameter Semigroups of Positive Operators. Edited by R. Nagel. X, 460 pages. 1986.

Vol. 1185: Group Theory, Beijing 1984. Proceedings. Edited by Tuan H.F. V, 403 pages. 1986.

Vol. 1186: Lyapunov Exponents. Proceedings, 1984. Edited by L. Arnold and V. Wihstutz. VI, 374 pages. 1986.

Vol. 1187: Y. Diers, Categories of Boolean Sheaves of Simple Algebras. VI, 168 pages. 1986.

Vol. 1188: Fonctions de Plusieurs Variables Complexes V. Séminaire, 1979–85. Edité par François Norguet. VI, 306 pages. 1986.

Vol. 1189: J. Lukeš, J. Malý, L. Zajíček, Fine Topology Methods in Real Analysis and Potential Theory. X, 472 pages. 1986.

Vol. 1190: Optimization and Related Fields. Proceedings, 1984. Edited by R. Conti, E. De Giorgi and F. Giannessi. VIII, 419 pages. 1986.

Vol. 1191: A.R. Its, V.Yu. Novokshenov, The Isomonodromic Deformation Method in the Theory of Painlevé Equations. IV, 313 pages. 1986.

Vol. 1192: Equadiff 6. Proceedings, 1985. Edited by J. Vosmansky and M. Zlámal. XXIII, 404 pages. 1986.

Vol. 1193: Geometrical and Statistical Aspects of Probability in Banach Spaces. Proceedings, 1985. Edited by X. Femique, B. Heinkel, M.B. Marcus and P.A. Meyer. IV, 128 pages. 1986.

Vol. 1194: Complex Analysis and Algebraic Geometry. Proceedings, 1985. Edited by H. Grauert. VI, 235 pages. 1986.

Vol.1195: J.M. Barbosa, A.G. Colares, Minimal Surfaces in $\mathbb{R}^3$. X, 124 pages. 1986.

Vol. 1196: E. Casas-Alvero, S. Xambó-Descamps, The Enumerative Theory of Conics after Halphen. IX, 130 pages. 1986.

Vol. 1197: Ring Theory. Proceedings, 1985. Edited by F.M.J. van Oystaeyen. V, 231 pages. 1986.

Vol. 1198: Séminaire d'Analyse, P. Lelong – P. Dolbeault – H. Skoda. Seminar 1983/84. X, 260 pages. 1986.

Vol. 1199: Analytic Theory of Continued Fractions II. Proceedings, 1985. Edited by W.J. Thron. VI, 299 pages. 1986.

Vol. 1200: V.D. Milman, G. Schechtman, Asymptotic Theory of Finite Dimensional Normed Spaces. With an Appendix by M. Gromov. VIII, 156 pages. 1986.

Vol. 1201: Curvature and Topology of Riemannian Manifolds. Proceedings, 1985. Edited by K. Shiohama, T. Sakai and T. Sunada. VII, 336 pages. 1986.

Vol. 1202: A. Dür, Möbius Functions, Incidence Algebras and Power Series Representations. XI, 134 pages. 1986.

Vol. 1203: Stochastic Processes and Their Applications. Proceedings, 1985. Edited by K. Itô and T. Hida. VI, 222 pages. 1986.

Vol. 1204: Séminaire de Probabilités XX, 1984/85. Proceedings. Edité par J. Azéma et M. Yor. V, 639 pages. 1986.

Vol. 1205: B.Z. Moroz, Analytic Arithmetic in Algebraic Number Fields. VII, 177 pages. 1986.

Vol. 1206: Probability and Analysis, Varenna (Como) 1985. Seminar. Edited by G. Letta and M. Pratelli. VIII, 280 pages. 1986.

Vol. 1207: P.H. Bérard, Spectral Geometry: Direct and Inverse Problems. With an Appendix by G. Besson. XIII, 272 pages. 1986.

Vol. 1208: S. Kaijser, J.W. Pelletier, Interpolation Functors and Duality. IV, 167 pages. 1986.

Vol. 1209: Differential Geometry, Peñíscola 1985. Proceedings. Edited by A.M. Naveira, A. Ferrández and F. Mascaró. VIII, 306 pages. 1986.

Vol. 1210: Probability Measures on Groups VIII. Proceedings, 1985. Edited by H. Heyer. X, 386 pages. 1986.

Vol. 1211: M.B. Sevryuk, Reversible Systems. V, 319 pages. 1986.

Vol. 1212: Stochastic Spatial Processes. Proceedings, 1984. Edited by P. Tautu. VIII, 311 pages. 1986.

Vol. 1213: L.G. Lewis, Jr., J.P. May, M. Steinberger, Equivariant Stable Homotopy Theory. IX, 538 pages. 1986.

Vol. 1214: Global Analysis – Studies and Applications II. Edited by Yu.G. Borisovich and Yu.E. Gliklikh. V, 275 pages. 1986.

Vol. 1215: Lectures in Probability and Statistics. Edited by G. del Pino and R. Rebolledo. V, 491 pages. 1986.

Vol. 1216: J. Kogan, Bifurcation of Extremals in Optimal Control. VIII, 106 pages. 1986.

Vol. 1217: Transformation Groups. Proceedings, 1985. Edited by S. Jackowski and K. Pawalowski. X, 396 pages. 1986.

Vol. 1218: Schrödinger Operators, Aarhus 1985. Seminar. Edited by E. Balslev. V, 222 pages. 1986.

Vol. 1219: R. Weissauer, Stabile Modulformen und Eisensteinreihen. III, 147 Seiten. 1986.

Vol. 1220: Séminaire d'Algèbre Paul Dubreil et Marie-Paule Malliavin. Proceedings, 1985. Edité par M.-P. Malliavin. IV, 200 pages. 1986.

Vol. 1221: Probability and Banach Spaces. Proceedings, 1985. Edited by J. Bastero and M. San Miguel. XI, 222 pages. 1986.

Vol. 1222: A. Katok, J.-M. Strelcyn, with the collaboration of F. Ledrappier and F. Przytycki, Invariant Manifolds, Entropy and Billiards; Smooth Maps with Singularities. VIII, 283 pages. 1986.

Vol. 1223: Differential Equations in Banach Spaces. Proceedings, 1985. Edited by A. Favini and E. Obrecht. VIII, 299 pages. 1986.

Vol. 1224: Nonlinear Diffusion Problems, Montecatini Terme 1985. Seminar. Edited by A. Fasano and M. Primicerio. VIII, 188 pages. 1986.

Vol. 1225: Inverse Problems, Montecatini Terme 1986. Seminar. Edited by G. Talenti. VIII, 204 pages. 1986.

Vol. 1226: A. Buium, Differential Function Fields and Moduli of Algebraic Varieties. IX, 146 pages. 1986.

Vol. 1227: H. Helson, The Spectral Theorem. VI, 104 pages. 1986.

Vol. 1228: Multigrid Methods II. Proceedings, 1985. Edited by W. Hackbusch and U. Trottenberg. VI, 336 pages. 1986.

Vol. 1229: O. Bratteli, Derivations, Dissipations and Group Actions on C*-algebras. IV, 277 pages. 1986.

Vol. 1230: Numerical Analysis. Proceedings, 1984. Edited by J.-P. Hennart. X, 234 pages. 1986.

Vol. 1231: E.-U. Gekeler, Drinfeld Modular Curves. XIV, 107 pages. 1986.